Subtlety in Relativity

Subtlety in Relativity

Sanjay Moreshwar Wagh

Pan Stanford Publishing

Published by

Pan Stanford Publishing Pte. Ltd.
Penthouse Level, Suntec Tower 3
8 Temasek Boulevard
Singapore 038988

Email: editorial@panstanford.com
Web: www.panstanford.com

British Library Cataloguing-in-Publication Data
A catalogue record for this book is available from the British Library.

Subtlety in Relativity

ISBN 978-981-4774-44-4 (Hardcover)
ISBN 978-1-351-23832-8 (eBook)

Contents

Preface

Why This Book?

Einstein's ideas rocked the world during the twentieth century. The twin paradoxes of special relativity, the warped space-time of his theory of gravity, the bending of light by the sun, black holes, the expanding universe, etc., have formed headlines across media in the past and do so even in the present.

At the same time, Planck's idea of the quantum of light had challenged the classical Newtonian ideas. Schrödinger, Heisenberg, Bohr, Born, Dirac, and others developed the initial ideas (of Planck, Einstein, and de Broglie) into a mathematical theory of the quantum world. This probabilistic quantum theory predicts only the chance of an event. But as Einstein would say, this theory does not describe the constituents of that event.

Then, we have to ask, is it that the physical world is based on only chance happenings? Is our world governed only by the laws of chance? Or, as Einstein, Schrödinger, and some other physicists had believed, is our world governed by deterministic laws?

Schrödinger's cat paradox and the Einstein–Podolsky–Rosen paradox are famous interesting examples of some perplexing issues underlying probabilistic ideas of quantum theory.

These issues have surfaced periodically One reads about our world being deterministic or indeterministic. The imaginary world of physicists meets the real world of us all! How is this so? This has been the subject of many discussions of the past.

A reader may then begin to wonder, why is another book being written on these old issues? Don't we already know enough? Haven't we heard enough already? What has changed since then?

The justification of this book rests with the following:

This book is the only one that has been written after the author's discovery of a new way in which the wave phenomena happen. In the history of science, this is the first time we have realized this new way, the emission origin of the waves, of producing wave phenomena. This drastically changes most issues of the old debate about the world being deterministic or probabilistic.

The features of the phenomena of nature dictate, indeed, our theoretical constructions. Still, the concepts of theory remain free constructions of the human mind. However, this is seldom mentioned in the literature on issues relating to quantum theory.

(Young) Einstein had, emphatically and very aptly, said that "concepts that have proved to be useful in ordering things easily acquire such an authority over us that we forget their human origin and accept them as invariable. Then they become "necessities of thought," "given a priori," etc. The path of scientific progress is then, by such errors, barred for a long time."

(Old) Einstein too had continued with this line of thought to say in later years that "the prejudice . . . consists in the faith that facts by themselves can and should yield scientific knowledge without free conceptual construction. Such a misconception is possible only because one does not easily become aware of the free choice of such concepts, which, through verification and long usage, appear to be immediately connected with the empirical material."

This applies also to the debate about the issues of the quantum ideas. Adherents of determinism could be accused of prejudice, and so could be those favoring indeterminism.

One may also quote Pauli, a physicist: "If new features of the phenomena of nature are discovered that are incompatible with the system of theories assumed at that time, the question arises, which of the known principles used in the description of nature are general enough to comprehend the new situation and which have to be modified or abandoned."

The newly discovered emission origin for the wave phenomena is not incompatible with the ideas of quantum theory; rather, this new and novel way (in which waves can get generated) justifies the use of the mathematical and probabilistic methods of quantum theory. We are not required to modify quantum theory then. However, the emission origin for the waves shows that quantum theory is statistically incomplete, and that too in precisely Einstein's sense!

Underlying the ideas of quantum theory, there exists then a certain, previously unexplored, conceptual framework. Is this the theory that Einstein and others were looking for then? How is this way of thinking related to the ideas of relativity? Is this a relativistic theory in the usual sense of this word? This is the journey we proceed on in this book.

Concepts Are Free Creations of the Human Mind

What do we mean by physical understanding of Nature? How do we relate it to experiments we perform, to observations we make of the Universe? How is it that the "concepts of physics" are the free creations of the human mind?

The purpose of a physical theory is to explain observations of nature. Concepts of a physical theory are the tools to formulate relations between them; and such relations are, ultimately, the predictions of that theory. Ultimate or testable relations are to be expressed so as to be applicable or usable in the context of an observation and/or experiment.

An experiment is an intentional arrangement of physical bodies that then aims to test a relation of observable quantities as implied by the theory under considerations. Then, an experimental device is an arrangement of bodies that responds according to a certain relationship of observable quantities, under an appropriate control of change to other observable quantities. An experiment must necessarily therefore possess the underlying theory it is attempting to verify or negate.

An experiment uses many experimental devices at the same time. Results of any experiment are dependent on the implicit assumption that all devices respond according to the correspondingly verified relationships of all of the involved observable quantities.

On the basis of the responses of the devices used within an experiment, we then arrive at the observed relationship of some observable quantities. This observed relationship is then the basis of the phenomenological theory underlying the corresponding experimental observation. A phenomenological theory then means the conceptual medley of the workings of experimental devices used within an experiment. Any experiment has such a theory underlying it, always. Such a theory underlying an experiment is, however, not the theory that we aim at in theoretical physics.

A phenomenological theory is not a complete theoretical understanding of the underlying behavior of physical bodies.

Such a theory only provides us guidelines for formulating a correspondingly complete theory, which transcends limits of that specific experiment and helps us predict results of new experiments.

Notice that the phenomenological theory of the new experiments can be quite different than that of the experiment we began with.

The same phenomenological theory of one experiment can be consistent with more than one underlying, more complete, theoretical frameworks of concepts. Only the results of a new experiment may then decide in favor of one of many competing theories by providing for them a new phenomenological theory to be consistent with.

Out of many competing theories, we choose the one agreeing with the new phenomenological theory, for it has a proven wider applicability. Then concepts of the theory of wider applicability are to be accepted as more appropriate for the description of nature. This is, incidentally, the sole purpose behind the act of performing an experiment.

An observation, in the sense of astronomy, is a naturally occurring arrangement of bodies leading to a result as if an experiment has been performed in the above sense. The formulation of a phenomenological theory underlying an observation is then based, first, on imagining an arrangement of bodies to correspond to that naturally occurring one. This is a mandatory step we have to take in such situations first.

On having imagined satisfactory arrangement of bodies to correspond to naturally occurring arrangement of bodies, we then follow the same steps as those leading to a phenomenological theory of observation. This is modeling a physical system; and we may neglect some bodies as being irrelevant to the situation to simplify the model.

We check predictions of a model against observations. If any kind of discrepancy is seen between the results of the model and those of the observation, then we have a choice of changing the model or call into question the theory assumed in the construction of the model. This is involved and laborious, no doubt. But in astronomy as well as in the atomic and subatomic world, we have no options than to resort to it.

When results of an experiment agree with the prediction of the theory, a relation of observable quantities is supported, but not all the concepts of that theory! Concepts remain free creations of the human mind.

An "ugly" experimental fact can destroy a "beautiful" theory. We will encounter many examples of this in the sequel. We will see how Descartes and many others compared light with sound waves. Concepts underlying waves of pressure in sound did not, however, agree with the observed properties of light, specifically, its polarization properties.

We will see that Newton's corpuscular picture of light also did not lead to explanations of the wave properties of light. We will see how Robert Brown's concepts of atoms of living matter had to be abandoned. Concepts of cold and hot radiation got abandoned, also.

Originally, Young, and Fresnel next, had realized that the vibrations of light can be taking place in a direction transverse to that of its propagation; in contrast to the longitudinal vibrations of particles in a sound wave. Thus, we needed to not abandon but modify the concept of a wave when applying to light. The concept of wave was not then abandoned but modified suitably so as to be consistent with the polarization properties of light. In other words, we had discovered a new type of wave, a transverse wave.

With Young's idea of transverse wave for light, Newton's corpuscular picture of light went out of favor. For over a century, the corpuscular concept of light was forgotten and efforts were directed at detection of the medium of the propagation of light as a transverse wave. The picture of light as a wave was as per Maxwell's theory of electromagnetism.

Since efforts to detect the medium of propagation of light were inconclusive and mutually contradictory, Lorentz proposed that an electromagnetic wave is not a wave propagating in any medium but is rather to be looked upon as wavy or oscillatory changes of the electric and magnetic vector fields existing in space, which can be free.

But Hertz's discovery of photoelectric effect led to Lenard's subsequent experimental investigations. Einstein's explanation of photoelectric effect, based on Planck's hypothesis of the quantum of light, and Millikan's subsequent experimental investigations forced the return of the corpuscular picture of light in the form of a quantum of light.

Light thence acquired a "schizophrenic" existence: depending on the experimental setup involving light, it was imagined by Heisenberg to exhibit the wave nature or the quantum nature.

Louis de Broglie's daring hypothesis that matter, customarily or usually considered to be corpuscular in nature, must also exhibit the wave nature received great experimental confirmation by diffraction of electrons and neutrons by crystalline matter. Not only light but also matter acquired the aforementioned schizophrenic existence as a consequence.

In total conformity with de Broglie's relation, $\lambda = h/\mathbf{p}$, where h is Planck's constant and λ is the wavelength associated with the physical body in question having a momentum $\mathbf{p}$, Schrödinger put forward a suitable equation for the waves associated with a body of mass m.

Independently, Heisenberg put forward a matrix formulation of the same phenomenon, and Schrödinger then showed its equivalence with his own formulation of an equation for the waves associated with a physical body, or the famous Schrödinger's equation.

Max Born then showed that the solutions of the Schrödinger equation possess interpretation as probability; the amplitude of the solution is the probability density of finding a physical body under considerations at a spatial location and at an instant of time.

These developments led to quantum theory, which had provided only a probabilistic description of the nature. As far as the mathematical framework of this quantum theory is considered and its probabilistic character is concerned, it is unexceptional. It led to many theoretical as well as experimental advances. Many of its predictions have been experimentally verified, and much of the modern technology is a consequence of the understanding of nature gained on the basis of this theory.

However, importantly, notice that quantum theory does not explain the schizophrenic existence of physical bodies, for it is based on de Broglie's revolutionary hypothesis. This theory does not therefore explain de Broglie's relation; rather it assumes this relation.

Demonstration of Heisenberg's uncertainty relation using quantum theory is reminiscent of the fact that it is based on de Broglie's relation $\lambda = h/\mathbf{p}$. The position of the body will be indeterminate within the wavelength $\delta x = \lambda$, and momentum will be indeterminate within $\delta\mathbf{p} = h/\lambda$, and we have $\delta\mathbf{p} \times \delta x \approx h$, which is Heisenberg's indeterminacy relation.

It does not constitute an explanation of this relation or that of de Broglie's relation. Rather, any such demonstration of the uncertainty relations only shows that the mathematical framework of quantum theory has successfully incorporated de Broglie's relation.

The issue of the explanation of de Broglie's relation or, equivalently, that of Planck's relation, $\in = h\nu$, then remains open, and out of the reach of quantum theory, importantly.

Then, how can any physical body know in advance what kind of experimental setup it is going to encounter? How can it be a wave and a quantum at the same time? Such questions led Einstein to say, in 1927, that "what nature demands from us is not a quantum theory or a wave theory; rather, nature demands from us a synthesis of these two views . . ." Quantum theory is not this synthesis, Einstein believed.

In this context, we will see that the emission wave mechanism provides such a synthesis then . . . even when each quantum moves along straight line path; it explains how the quanta, and not a single quantum, can be producing a wavy pattern of their numbers.

This then brings us to the following discussion of what we mean by observable physical quantity. Science is, importantly, based on measurable quantities related to the natural bodies.

From observations of bodies in nature, we formulate common concepts, applicable to them all. Concepts, which are our "free" creations related to a natural body, are not that natural body. If a concept does not agree with observations, we need to abandon or change it as required.

Measurements involve a specific arrangement of bodies. The measurement of a quantity for a body, in general, involves a specifically created arrangement of natural bodies in which we compare the value of that quantity for that body with its value for a reference body. We also assume that the creation of the arrangement of bodies for measurement does not uncontrollably affect the bodies and the value of a quantity is a real number.

One basic principle of physics is that no quantity should be introduced that cannot, at least in principle, be measured. It distinguishes science, and therefore physics, from other (nonscientific) thought systems. Such principles of science establish its practical utility.

Measurable are the physical quantities, and bodies to which these concepts apply are the physical bodies. Then, physical bodies are hypothetical and are defined to always obey our conceptions.

Quantities that can be directly measured in an arrangement of physical bodies are to be called directly observable quantities. In other words, a physical quantity is directly measurable when its value can be ascertained within a single attempt of its measurement. In contrast to directly observable quantities, a quantity is indirectly observable if its value has to be necessarily inferred from those of the directly observable quantities.

Now, the question arises as to which physical quantities are directly and which ones are indirectly observable quantities. The issue of some physical quantity being directly observable or not is determined by the nature of the corresponding concept and its interrelationships with other concepts.

As an example of a quantity that is indirectly observable, consider probability. We cannot measure it in a single arrangement of bodies. This is so because we may repeat the same arrangement of natural bodies many times, note the measured value for a specific physical quantity for every instance of the arrangement of bodies, and then determine the probability for the specific value of its measurement.

We tacitly assume that the arrangement of bodies is repeatable, in other words, the experiment of measurement of a physical quantity is repeatable as many times as we wish. Repeatability of experiments is quite an important principle of physics, and science, in general.

Furthermore, we may also assume that the physical quantity with which the probability is being associated is directly observable. We are free to conceptually associate probability with both directly and indirectly observable physical quantities.

The structure of concepts and their interrelationships, in totality, are the theoretical construction. In contrast to the above case with probability being an indirectly observable quantity posited after the introduction of some directly observable physical quantities, like position, we may begin with the concept of probability for the theoretical construction.

In this case, we associate a priori probability for the value of the physical quantity like position prior to or with disregard to its measurement.

Probability provided by the distribution of errors of measurement of the position of a body can itself, for example, be taken to be the a priori probability for the value of the position of that body without measurement. Einstein had stressed this issue on many occasions.

With the a priori probability associated with the value of any physical quantity, it is then necessarily indirectly observable within this conceptual framework, for we have only the likelihood of its value within such a theoretical construction that disregards or completely ignores the procedure for its measurement. Probability continues to be indirectly observable, for we need to repeat the measurement to verify its a priori distribution. That is to say, probability is not any directly observable physical quantity and is to be inferred from the ensemble properties, always.

For quantum theory, the above nature of probability underlies its incomplete character as a theory, Einstein had argued.

Einstein describes this situation in succinct words: "It is the theory which decides what we can observe." ("Observe" means "directly observe" in our sense.) In quantum theory, no direct measurement, but only indirect measurement, of the value of physical quantity is permitted.

Now, as we will see in the sequel, universal relativity is the most general theory about the physical world, for its mathematical framework is independent of how we may represent a physical body.

Then its explanations can be expected to be based on minimally formulated assumptions about characters of physical bodies and their interactions.

Universal relativity begins by recognizing that the natural or the inertial state of motion of a physical body is as prescribed by Galileo. It recognizes that a body of nonzero inertia has nonzero momentum.

Then, it recognizes that any body of vanishing inertia is only a momentumless energy quantum.

With the above mutually consistent and minimally formulated assumptions about the characteristics of physical bodies of nature, universal relativity aims to explain all physical phenomena on the basis of their possible interactions. It is a theory of "the reality" as it exists independent of the act of observation.

Organization of This Book

This book is divided into two parts. The first part is less mathematical and more conceptual in its orientation. The second part focuses on mathematical ideas needed to implement physical concepts.

Part I deals with how certain physical concepts developed historically. It is more accessible to a general reader. This discussion is neither chronologically ordered nor complete in all the historical details. It is only kaleidoscopic in character. Nevertheless, it aims to provide an overview of how some physical concepts got developed, with some getting modified and some others getting abandoned in view of results of experiments. A reader is advised to read Part I carefully.

In Chapter 1, we begin with the history of the wave theory of light. Then, we discuss how our ideas of electricity and magnetism evolved to form our concept of an electromagnetic wave. Next, we explore as to how the atomic nature of matter was discovered. Interaction of atomic matter and radiation is the subject of our further explorations into the history of related ideas. In particular, we elaborate on the role of thermodynamics in the formulation of statistical methods in physics also.

Next, we discuss the idea of what a quantum is. The notion of a quantum jump in energy is distinguished from that of a quantum jump in space. Beginning with paradoxes related to the ideas of a quantum, we then discuss the emission origin of the wave of quanta as a newly discovered manner in which waves of particles get produced in nature.

In Chapter 2, ideas of the theory of relativity and what the word "relativity" actually means are our focus.

Our account of this interesting history must, necessarily, begin with Galileo's ideas of the inertia of a body and its inertial state of motion. We then explore the structure of Newton's idea of a force vis-à-vis Galileo's notion of inertia. Likewise, to Einstein's special relativity, Newton's theory is also 4D, three of space and one of time, except that time is absolute in Newton's considerations. We stress that massless bodies do not obey Newton's laws, in particular, the Newtonian law of addition of velocities. This observation escaped notice in the past.

In the sequel, we discuss how Einstein's idea of relative time is untenable and that time is absolute, that is to say, it runs at the same rate for all observers irrespective of their state of motion.

A genuine theory of relativity is, necessarily, a theory of everything, that is, its formalism must encompass all physical bodies. Einstein's general principle of relativity is then a statement of point of view that needs to be adopted for formulating such a theory.

However, this principle is silent about how we may implement it mathematically.

We then discover the universal principle of relativity that overcomes the aforementioned lacuna of the general principle of relativity by providing us the nature of mathematics needed to implement it. This mathematical framework is that of category theory.

Before we explore the ideas of category theory, we explore the way a massless quantum of light needs to be treated in universal relativity. In this discussion, we emphasize that Einstein's ensemble interpretation of the probabilistic quantum theory is the right point of view. In other words, we emphasize that the (usual) probabilistic quantum theory is incomplete in precisely Einstein's sense.

In Chapter 3, we discuss the Doppler effect. We point out that a historical mistake in the derivation of the standard Doppler shift formula eventually leads to contradiction of special relativity with experiments. Then Einstein's concept of relative time gets experimentally rejected.

In this chapter, we further discuss how observed Doppler shifts must be interpreted to arrive at physically proper results. This discussion is aimed at astronomers and astrophysicists.

Part II deals with technical matters related to ideas of the universal theory of relativity. It begins with ideas of category theory that is essential to ideas of universal relativity.

In Chapter 4, we develop the notion of what we mean by a category. We focus on the most general definition of a category.

In Chapter 5, we discuss properties of arrows and objects that form a category. Some of these properties are crucial to defining measures within the categorical context.

In Chapter 6, functors as arrows connecting categories are our focus. We also discuss the equivalence of categories.

In Chapter 7, we explore the concept of universal association by functors, and in Chapter 8, the concept of adjunction or adjoint situation of categories is developed.

A categorical notion of "measure" is developed in Chapter 9. The notion of a measure is crucial to the development of further ideas. After developing relevant ideas of category theory, we discuss their applications to ideas of universal relativity in Chapter 10.

Readers interested in further technical details of physical concepts and their development may refer to references 1 to 24 for general physics, atomic theory, and astronomy and astrophysics; references 25 to 42 for quantum theory; references 43 to 48 for the Doppler effect, Lorentz transformations, and applications; references 49 to 52 for Einstein's ideas and the theory of relativity; references 53 to 58 for category theory; references 58 to 64 for measure theory and dynamical systems; and references 65 to 83 for ideas and stages of development of universal relativity.

Note: All figures in Chapter 1 and Fig. 2.2 (Chapter 2) have been taken from the following references:

Wagh SM and Deshpande DA (2013). *Essentials of Physics*, Vol. I, PHI Learning, New Delhi, Copyright © 2013 by PHI Learning. Reprinted with permission of PHI Learning.

Wagh SM and Deshpande DA (2013). *Essentials of Physics*, Vol. II, PHI Learning, New Delhi, Copyright © 2013 by PHI Learning. Reprinted with permission of PHI Learning.

Sanjay M. Wagh

Central India Research Institute,
Nagpur, India

Part I

Development of Physical Concepts

Chapter 1

The Light Quantum

In this chapter, we consider the development of ideas about the kind of physical body light is and lead up to the concept of a light quantum. We begin with the wave theory of light.

1.1 Wave Theory of Light

During some era, sound got established as a propagating sequence of *compression* and *rarefaction* of matter, a wavy disturbance of static pressure, a pressure wave propagating in a material, such as air. The motion of material particle is in the direction of the propagation of the (sound) wave, and such waves are called *longitudinal waves* now.

René Descartes (1596–1664) had assumed light to be like a sound wave, that light is a (longitudinal) pressure wave propagating in a subtle medium that he called "ether."

Originally, this Greek word, *ether*, had meant *the blue sky* or *the upper air* as compared to air at the sea level, and later, it only meant *that which occupied the great celestial sphere.* Descartes brought ether into science by postulating that it has mechanical properties.

He had assumed that particles of ether are in motion continually. For him, the motion of a single particle of ether involved motion of them all, as they come into contact with each other in a chain, forming a vortex. His theory of vortices had formed his picture of the cosmos.

Subtlety in Relativity
Sanjay M. Wagh

ISBN 978-981-4774-44-4 (Hardcover), 978-1-351-23832-8 (eBook)
www.panstanford.com

Descartes had been the originator of the mechanistic view of the world, which imagined that the (inanimate) world works as an automatic mechanism, that is, every physical phenomenon is understandable by a suitable mechanical model of the involved bodies.

However, Descartes's assumption that light is a longitudinal wave of static pressure propagating within ether ran contrary to some properties of light, discovered subsequently by others.

*

A law of refraction of light was experimentally discovered by Willebrord Snell (1591–1676), professor of mathematics at Leyden, Holland. Descartes used Snell's law but did not acknowledge Snell for it.

*

Robert Hooke (1635–1703), an assistant to Robert Boyle (1627–1691) to begin with, had been the secretary of the Royal Society of London at a later stage. He performed experiments in optics, analyzed the propagation of waves in detail, and visualized the concept of a wavefront. These considerations were theoretical, marching to the fully developed theory of waves. He had postulated light to be very rapid vibrations, of a small amplitude, of ether particles. For him, light consisted of a traveling wavefront with particles of ether vibrating along the direction of motion of the wavefront of light. He considered colors of light to be a property of these vibrations. Nevertheless, Hooke's ideas of light relied on Descartes's ideas, for he also imagined particles of ether to be vibrating along the direction of motion of the propagating wavefront of light.

*

Isaac Newton (1642–1727) was beginning a scientific career at the Trinity College, Cambridge University, England, then. During 1666–1672, he performed many experiments in optics and had formulated his law of gravitation as well. He was the first to formulate the laws of mechanics of bodies; laws based on inertia and (implicit notion) of force were formulated by Galileo Galilee (1564–1642).

Newton joined the Trinity College in the summer of 1661. Then the presently world famous Cambridge University was in very bad shape.

Still, Cambridge had an excellent library; and Newton was a voracious reader of Euclid's book *Elements*, Descartes's book

La Geometrie, etc. He later recalled how he read Descartes's book—he would read a few pages and get utterly stumped. Then he would start all over again from page 1 and begin anew, penetrating a little deeper than before until he got stumped again. Then he would start all over again from page 1! Remarkable that he mastered geometry without the assistance of any tutor or professor.

Only one professor, Isaac Barrow (1630–1677), could be said to be qualified to assist Newton. Barrow was the Lucasian chair of mathematics, a position to which Newton himself was appointed later.

Newton was left to himself to study and had begun conducting experiments on the nature of light, color, and vision in the then indifferent educational atmosphere at Cambridge University.

In one experiment, Newton stared at the sun for an excessively long time to dutifully record the spots and flashes that affected his vision for days afterward! In fact, he had to confine himself to darkness for some time to let the images gradually fade. Newton was very lucky indeed that he did not damage the retinas of his eyes permanently!

In another experiment, this curious student of science wanted to investigate the effect of the eyeball's shape on distorting and altering vision. For this purpose, he pushed a bodkin between his eye and the bone so as to reach the backside of the eyeball. This courageous student, Newton, could then control the shape of his eyeball by moving the end of the bodkin jutting outside his eye. With the bodkin in place in his eye, he then illustrated the effects of the eyeball's shape on distorting and altering vision by drawing how the bodkin slid under and behind the eyeball, with the positions of the bodkin nicely marked with letters from a to g. Newton was simply lucky not to have permanently damaged his eye, either by physical injury to the muscles of the eye or by getting an incurable infection in the eye!

Early in 1665, Newton discovered what we know as the generalized binomial theorem, or a formula dealing with binomial expansion. Discovering this method had great influence on Newton's later works related to differential calculus and integral calculus.

As a Lucasian professor, Newton's main job was to deliver regular lectures on mathematical topics. We may imagine his class to be full of enthusiastic and eager-to-listen-to-the-great-man type of students. But we are in for a surprise here! Newton's contemporary said that so few went to hear him (Newton) and fewer yet understood him that oft times he did, in a manner, for want of listeners, read to the walls. That was, however, before Newton's reputation had grown, mainly through the circulation of his extraordinary unpublished treatises.

Newton did not consider light to be vibrations of ether but considered light to be consisting of corpuscles or particles existing independently of ether. He imagined luminous bodies, those that shine, to be emitting these corpuscles. He further imagined the color of light to be a definite property of its corpuscles. Newton did subscribe to the notion of ether but for quite different reasons. He imagined vibrations of tether to be affecting the light corpuscles in their motions, if at all. In defending their respective points of view, Newton and Hooke entered into an unpleasant controversy.

(To some historians of science, Newton appears to be noncommittal to the actual nature of light, whether light is corpuscular or wavy. It is, however, a general impression of the later physicists that Newton held the corpuscular view of light, which is also consistent with his ideas regarding the nature and motions of other bodies.)

In the beginning of 1666, Newton obtained a triangular prism "to try therewith the celebrated Phenomena of Colors." On the basis of Snell's law of refraction, there should not have been any difference between the length and breadth of the spectrum obtained using a prism. The observed difference of length and breadth surprised Newton.

In a memoir of 1671, Newton wrote:

> Having darkened my chamber, and made a small hole in my window-shuts, to let in a convenient quantity of the Sun's light, I placed my Prism at his entrance, that it might be thereby refracted to the opposite wall. It was at first a very pleasing divertisement, to view the vivid and intense colours produced thereby; but after a while applying myself to consider them more circumspectly, I became surprised to see them in an oblong form, which, according to the received laws of Refraction, I expected should have been circular.

As the records reveal, the length of the colored spectrum he had obtained was five times more than its breadth. After more experiments, he discovered the true explanation that ordinary white light consists of a mixture of rays of every color and that the elongation of the spectrum (as he had seen) is due to the differences in the refractive power of the glass for different colors.

Newton inferred:

> Colours are not qualifications of light, derived from refractions, or reflections off natural bodies, but original and connate properties. . . . Some rays are disposed to exhibit a red colour and no other, some a yellow and no other, some a green and no other, and so of the rest.

He further went on to state:

> The species of colour, and degree of refrangibility proper to any particular sort of rays, is not mutable by refraction, or by reflection from natural bodies, or by any other cause, that I could yet observe. When any one sort of rays hath been well parted from those of other kinds, it hath afterwards obstinately retained its colour, notwithstanding my utmost endeavours to change it.

In conformity with his other ideas, Newton considered color to be an intrinsic characteristic of light and inferred that it must be associated with some definite quality of the luminous corpuscles (of light).

When light from the sun falls on a thin film of oil, beautiful colors are displayed. These colors had fascinated many, including Newton. They all had undertaken to explain the formation of such colors theoretically but had failed to explain this phenomenon on the basis of Newton's corpuscular picture of light that light consists of fast-moving corpuscular bodies.

Newton's deep interest in the colors of thin films led him to ingenious experiments related to light. In one experiment, Newton observed concentric dark and bright rings, Newton's rings (Fig. 1.1). However, he had no theoretical explanation for the formation of these (Newton's) rings.

*

Christian Huygens (1629–1695) greatly improved, as well as extended, the wave theory of light. He studied how waves propagate in space. He had formulated a well-defined principle, now called the Huygens principle (Fig. 1.2). On its basis, he formulated the wave theory of the reflection and refraction of waves. In those times, this was a great achievement for the wave theory of light.

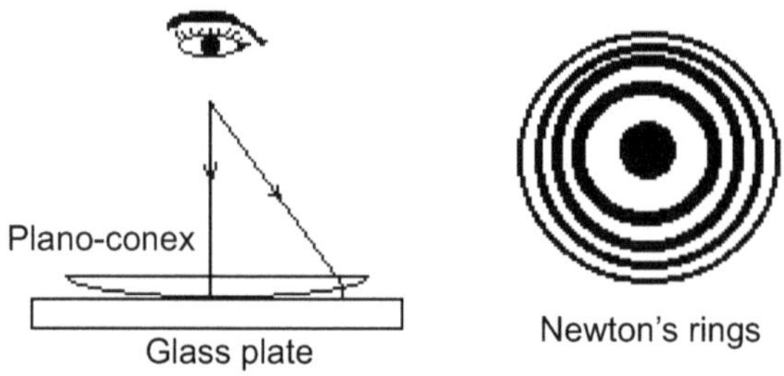

Figure 1.1 Newton's rings. Reprinted with permission from Ref. [23]. Copyright © 2013 by PHI Learning.

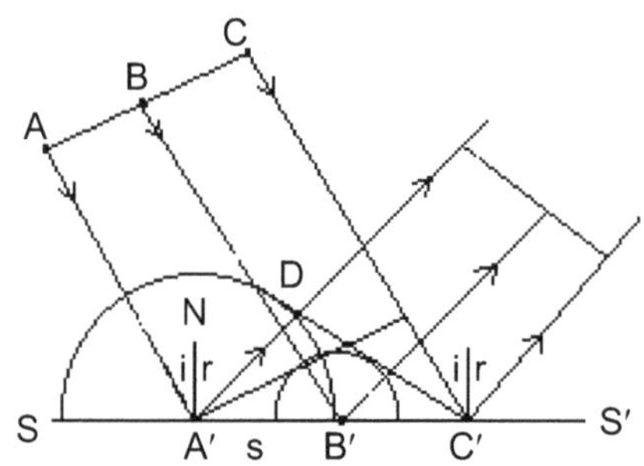

Figure 1.2 Huygens's explanation of reflection. Reprinted with permission from Ref. [23]. Copyright © 2013 by PHI Learning.

Historically, Huygens's research on light came later than Newton's discoveries of 1666–1672. While experimenting with Iceland crystals, Huygens discovered a curious property of light, its double refraction (Fig. 1.3). A ray of light passing through a crystal of tourmaline along the axis of the crystal produced two images of the source, when only one image was expected by Huygens. Light passing through the crystal breaks into two parts, which move along two different paths, the two refracted rays. Then one object when viewed through the crystal forms two separate images. This was called the phenomenon of double refraction.

Not Huygens but Newton, in 1717, attempted the first explanation of double refraction. Newton, using Huygens's observations, showed that vibrations of particles of ether along the direction of a light wave are contrary to Huygens's results. Newton's explanation of Huygens's observations involved supposing that a corpuscle of light has sides, similar to a cube.

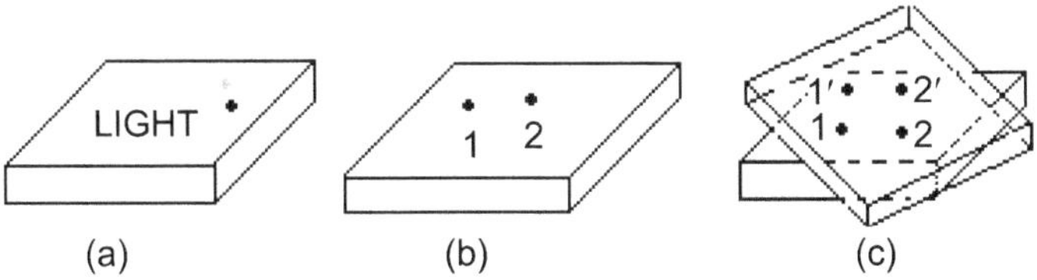

Figure 1.3 Double refraction. Reprinted with permission from Ref. [23]. Copyright © 2013 by PHI Learning.

*

Historically, Descartes assumed that light moves faster in a denser medium than in a rarer medium. He obtained the laws of reflection and refraction of light, which agreed with experiments.

However, Pierre de Fermat (1601–1665) had also derived the laws of reflection and refraction of light. With this purpose, he proposed that nature always acts along the shortest course, Fermat's principle of least time. He thus assumed light to be moving more slowly in a denser medium than in a rarer medium, contrary to what Descartes had assumed. Having assumed this, Fermat derived the laws of reflection and refraction of light and obtained the same laws as those of Descartes! Descartes's assumption was in error.

The controversy that ensued from these contradictory results continued for about a century. Pierre Louis Moreau de Maupertuis (1698–1759) revived it. Although he had not resolved the conflict, mathematical works of Maupertuis were used by Leonhard Euler (1707–1783) and Joseph Louis Lagrange to develop a principle that nature is a dynamical system that obeys an important principle—the principle of least action.

*

Etienne-Louis Malus (1775–1812) analyzed the light of the setting sun with a rhombic crystal of Iceland spar. He noticed that the two images of the sun created by the crystal were of very different intensities (Fig. 1.4).

In January 1808, the French Academy of Sciences proposed the topic "To Furnish a Mathematical Theory of Double Refraction, and to Confirm It by Experiment" for its prize in physics for 1810. Malus was, then, a young colonel of engineers in service with Napoleon's expedition in Egypt. While conducting experiments toward the end of 1808 in Paris, Malus noticed that light reflected from glass also shows a property similar to that noticed by Huygens in rays having experienced double refraction in Iceland spar. Malus also discovered that light refracted at the surface of any transparent material has, to some degree, the same property noted by Huygens, and he called this property *polarization*. To Malus, this discovery was very important. So, without waiting for the prize competition, he communicated it to the academy in December 1808 and published it the following month. It is then history that in 1811, Malus won the prize of the French Academy of Sciences for 1810. The successful discoverer of polarization was elected to the academy, but his health broke down due to the Egyptian campaign, and he died the following year.

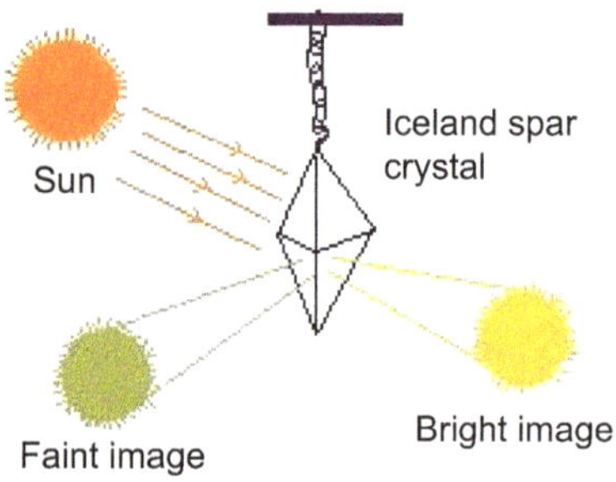

Figure 1.4 Observations of Malus. Reprinted with permission from Ref. [23]. Copyright © 2013 by PHI Learning.

The following was also realized as an example of polarization of light. Consider a source emitting visible light and two plates of crystalline tourmaline, as shown in Fig. 1.5. Let us view the source through the two plates arranged to be in between the source and the eye.

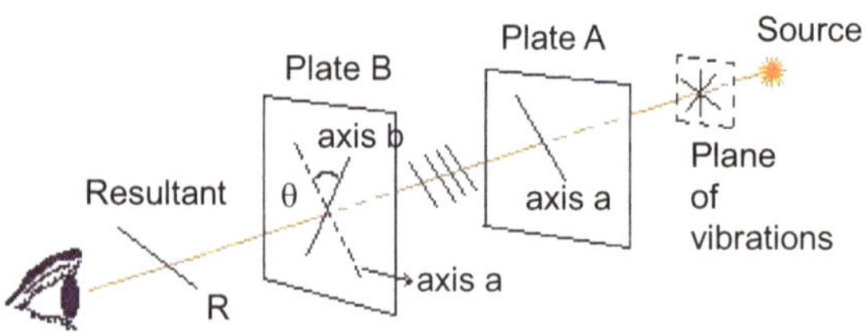

Figure 1.5 Polarization of light. Reprinted with permission from Ref. [23]. Copyright © 2013 by PHI Learning.

Let us keep one plate fixed and rotate the other plate about the source and the eye axis. Then there are two positions of the plate for which the light transmitted to the eye has a large intensity and two other positions of the plate for which the light transmitted to the eye has almost zero intensity.

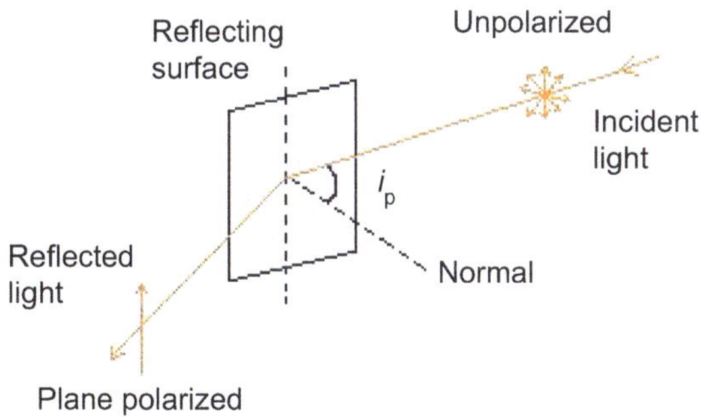

Figure 1.6 Polarization of light on reflection. Reprinted with permission from Ref. [23]. Copyright © 2013 by PHI Learning.

Malus's memoir submitted to the French Academy of Sciences is rich with experimental and theoretical results. He had noted that polarization of reflected light is, in general, incomplete, that is, the reflected rays display only imperfectly the properties of light that has undergone double refraction. With considerable accuracy, he had measured the polarizing angles for glass and water (Fig. 1.6).

*

David Brewster (1781–1868) showed in 1818 that light is completely polarized by reflection when the reflected and refracted rays are at right angles to each other. Brewster also made another discovery of fundamental importance to the theory of double refraction. Most had believed that double refraction is always of the kind occurring in the crystal of Iceland spar. But Brewster found that in a large class of crystals there are two axes along which there is no double refraction. Such crystals are called biaxial crystals, and crystals of Iceland spar type are called uniaxial crystals.

*

About a century after Newton, Thomas Young (1773–1829) and Augustine Jean Fresnel (1788–1827) championed the wave theory of light. Young explained the formation of Newton's rings on the basis of the wave theory of light. After their pioneering works, essentially no adherents were left for Newton's corpuscular point of view of light, which could not explain their observations.

It was also realized very soon that luminous or hot bodies emit heat radiation, and the properties of heat radiation were also found to be the same as those of (visible) light. Physicists began calling even (visible) light *radiation*. By the beginning of the nineteenth century, the wave picture of light firmly established itself and the view that radiation is a wave in a medium of ether was accepted by most. Newton's corpuscular point of view of light went into oblivion, and no references to it appear later.

Newton's corpuscular picture of light could not explain diffraction of light (Fig. 1.7), while the wave picture could, as shown by Fresnel. (However, the reader may wish to note here that these are not the failures of the corpuscular picture, as we will see later, in the emission mechanism of waves.)

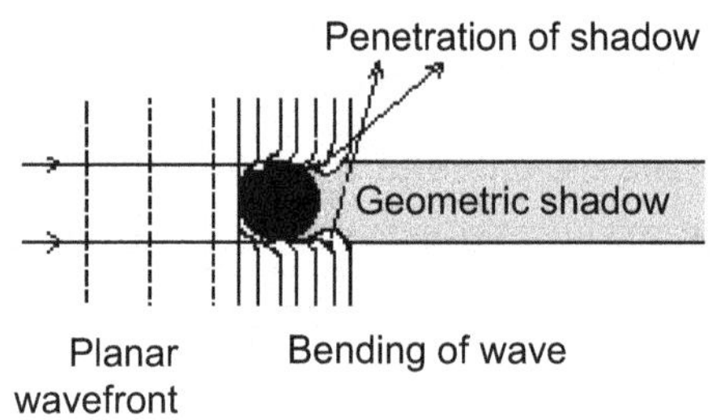

Figure 1.7 Diffraction of light. Reprinted with permission from Ref. [23]. Copyright © 2013 by PHI Learning.

*

In 1815, Fresnel submitted a memoir to the French Academy of Sciences about his theory of diffraction. The commission constituted by the academy consisted of P. S. Marquis de Laplace, Simon Denis Poisson (1781–1840), Jean-Baptiste Biot (1774–1862), Dominique Francois Jean Arago (1786–1853), and others, who were zealous supporters of Newton's corpuscular theory. But Fresnel's memoir was accepted by them. Impressed by Fresnel's arguments, Arago sought his friendship, and they worked together thereafter.

Later, Arago and Fresnel showed that two pencil beams of light that are polarized in planes at right angles do not interfere under the circumstances in which ordinary light shows interference, but they always give on their reunion the same intensity of light, whatever be the difference of the paths the two beams of light have.

Pondering these results, Young realized how to reconcile the polarization and the wave picture of light—the vibrations of light were executed at a right angle to the direction of its propagation. In an article titled "Chromatics," for the supplement to *The Encyclopaedia Britannica*, Young had written that "if we assume as a mathematical postulate, on the undulating theory, that a transverse motion may be propagated in a direct line, we may derive from this assumption a tolerable illustration of the subdivision of polarized light by reflection in an oblique plane by supposing the polar motion to be resolved into two constituents."

In a letter to Arago, written on April 29, 1818, Young had compared light to the undulations of a cord agitated by one of its extremities. In fact, this analogy had been given by Hooke in a communication to the Royal Society on February 15, 1671. But Hooke had not, during his times, appreciated the point advocated by Young in 1818.

Interestingly, almost 80 years before Young, John Bernoulli (Jr.) (1710–1790) also rejected this very idea while following the general ideas of his father, John Bernoulli (Sr.) (1667–1748), who considered light to be like vibrations of a stretched string. Vibrations of a stretched string are seen to be transverse to a string, while Bernoulli (Jr.) considered light as longitudinal waves, following Descartes's ideas.

Arago showed Young's letters to Fresnel, who at once recognized the merit of Young's idea as a genuine explanation of their experiment, the Arago–Fresnel experiment.

The theory of the propagation of waves in elastic media was then unknown. With wonderful insights, Fresnel headed for the precise direction in which this theory needed to be developed. From this point of view, he addressed the propagation of light in crystals. Aware of Brewster's discovery that many crystals do not conform to the type of Iceland spar and his laws, the Brewster laws, Fresnel developed the theory for biaxial crystals. We owe to Fresnel the theory of polarization of light.

It is greatly to the most brilliant of Fresnel's efforts that we owe the theory underlying polarization. At the end of his memorable publication, Fresnel said:

> The theory which I have adopted and the simple constructions which I have deduced from it, have this remarkable character, that all the unknown quantities are determined together by the solution of the problem. We find at the same time the velocities of the ordinary ray

> and of the extraordinary ray, and their planes of polarization. Physicists who have studied attentively the laws of Nature will feel that such simplicity and such close relations between the different elements of the phenomenon are conclusive in favour of the hypothesis on which they are based.

In 1827, the Royal Society awarded the Rumford medal to Fresnel. Arago conveyed it to Fresnel, who died eight days after receiving it.

*

William Rowan Hamilton (1805–1865), a royal astronomer of Ireland, pointed out in 1832 the striking consequence of Fresnel's theory: that a single ray from a point within the crystal propagating in the direction of one of Fresnel's four (conical) points must be divided into an infinite number of rays along a conical surface on emergence. Then a corresponding ray incident externally on the crystal should be divided within the crystal into an infinite number of refracted rays propagating along a conical surface. In 1833, Humphrey Lloyd verified such unexpected consequences of Fresnel's theory.

*

Now, waves propagating on water are not longitudinal but what we call transverse waves. Any cork floating on water moves up and down as a wave propagates on water. The motion of the cork in water is not in the direction of the propagation of the wave but perpendicular to it, quite unlike the motion of particles of air when sound propagates.

Of importance now is the fact that a wave propagates in all the available directions within the medium of its propagation. If we drop a stone at a location in a pond, we see waves move away from that location in all the available directions. Where a stone lands on water, it displaces particles of water in all the directions available there. These displaced particles of water, in their turn, displace more particles of water in a region around that location. Then this process, a train of the displacement of water particles, is seen to move on the surface as a wave. Naturally, this wave propagates within water in all directions away from the location of the landing of the stone on water. The motion of particles of water is seen to be such as to result into only the vertical displacement of a floating body, like a cork.

This wave is a disturbance of water, a complicated motion of particles of water. Particles (of water) are set into motion in some manner, and it is the pattern of the motion of these particles that is seen as a propagating wave. Wherever no water exists, no such wave will, evidently, be seen. We say that such a wave requires a medium for its propagation. If the medium is a thick and viscous liquid, like some syrup, then the wave does not propagate long distances in it. Waves get attenuated quickly. This is understood as being due to the dissipation of the energy of the motion of the constituents of the medium of the wave. Properties of the medium determine various characteristics of the wave, such as its speed of propagation and how the waves disperse within it. This is the modern perspective of how such a wave forms and propagates.

*

Great mathematicians, like Augustine-Louis Cauchy (1789–1857), Hamilton, and George Green (1793–1841), provided mathematical foundations to the equations of wave propagation. Green had developed the theory of reflection and the polarization of light on its reflection.

*

James MacCullagh (1809–1847), from Dublin, and Franz Ernst Neumann (1798–1895), from Konigsberg, concluded on performing mathematical calculations of the properties of ether that it is impossible to explain optical phenomena by comparing ether with an elastic solid of the ordinary type, ordinarily resisting compression and distortion. The properties of ether are not to be those of the ordinary matter. Specifically, MacCullagh showed through detailed calculations that ether is needed to be rotationally elastic. No such rotationally elastic material was known at the time.

*

William Thomson (Lord Kelvin) (1824–1907) designed two mechanical models possessing rotational elasticity. In 1889, these were great achievements, indeed. These models dispelled any of the remaining doubts about the rotationally elastic property of ether.

Joseph Boussinesq (1842–1929) asserted that all space, whether within bodies or outside them, is occupied by identical ether, the same everywhere in inertia and elasticity. In 1867, he showed that mathematically, all ethereal processes are to be represented by two kinds of equations—the first kind expressing the invariable equation of motion of ether and the second kind expressing the interaction of matter with ether.

*

With this short historical review, let us see what the wave theory of light had effectively explained. If we place an obstacle in the path of a wave form traveling in a pond, we observe that the wave form bends around that object. In this context, Francesco Grimaldi (1613–1663), a Jesuit father, and Hooke had also observed that there exists some illumination within the geometrical shadow of a completely opaque body. This is quite like the bending of a wave around any obstacle in its path. Grimaldi called this phenomenon diffraction. Since its discovery, the diffraction of light has played an important role in our concepts about the nature of light. Any theory of light must explain results on the basis of its polarization and diffraction.

So we simply say that light bends around opaque objects and diffracts.

If we observe mixing of two (almost similar) wave forms in water coming from two different directions, quite a characteristic pattern of large and small amplitudes is noticeable. This pattern of the mixing of the two waves is called their interference pattern, and we say that waves interfere.

Monochromatic light from two different sources is also seen to produce an interference. This is also seen with oil spilled over a road and when light reflecting off the oil produces a display of beautiful bands of colors.

In those times, Young's double-slit experiments producing the interference pattern of light could be explained only by assuming that light is a wave and not using the corpuscular picture (Fig. 1.8).

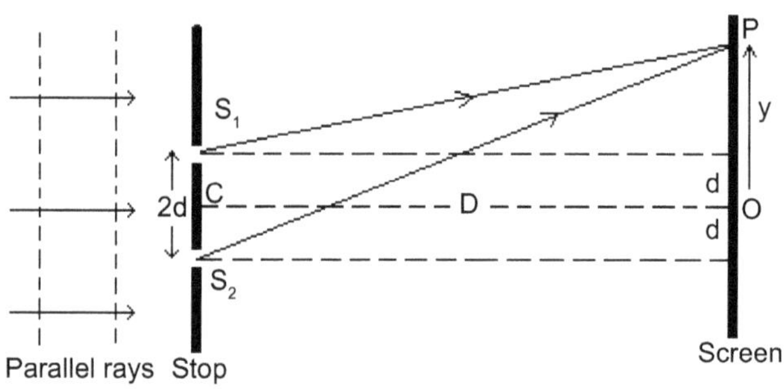

Figure 1.8 Young's experiment. Reprinted with permission from Ref. [23]. Copyright © 2013 by PHI Learning.

In 1801, Young had written about the phenomenon of interference:

> Suppose a number of equal waves of water to move upon the surface of a stagnant lake, with a certain constant velocity, and to enter a narrow channel leading out of the lake; suppose then another similar cause to

have excited another equal series of waves, which arrive at the same channel, with the same velocity, and at the same time with the first. Neither series of waves will destroy the other, but their effects will be combined; if they enter the channel in such a manner that the elevations of one series coincide with those of the other, they must together produce a series of greater joint elevations; but if the elevations of one series are so situated as to correspond to the depressions of the other, they must exactly fill up those depressions, and the surface of the water must remain smooth. Now, I maintain that similar effects take place whenever two portions of light are thus mixed; and this I call the general law of the interference of light.

*

If we view the sky through a tourmaline crystal, the view can get quite darkened when we rotate that crystal. Tourmaline crystal polarizes light, and we say that the tourmaline crystal is a Polaroid. We can explain this observation by assuming that light has a state of polarization.

*

Claude-Louis Marie Henry Navier (1785–1836), from Paris, had the correct equations of the vibratory motion of crystal. George Gabriel Stokes (1819–1903) developed the theory of crystal optics. We know these equations as the Navier–Stokes equations.

Huygens's observations related to the double refraction of light with an Iceland spar crystal can be explained as being due to the transverse wave of light passing through an optically anisotropic material. These Polaroid materials were also engineered. (Thin films of suitable Polaroid material are employed for windowpanes to block the view of the inside from the outside, but the outside can be viewed from the inside. Polaroid sunglasses are also available these days.)

These properties of radiation could be explained assuming only that light is a wave, like that propagating on water.

*

Of course, it took quite a while to firmly establish that light is a transverse wave. Still, there remained many questions regarding ether—its properties, its rigidity, etc. These questions led to further experiments establishing properties of ether. A variety of ingenious experiments were then designed and performed. That these were experiments performed in laboratories is proof enough that technology had advanced, too.

What then is the inferred nature of ether these experiments had indicated to us?

Simultaneous to these developments of the wave theory of light, the theories of electricity and magnetism also developed. It is the history of these theories that is now our concern. This history also relates to light and our understanding of its nature.

1.2 Electromagnetism and Light

Historically, the phenomena of electricity and magnetism were known much before the advent of modern science. A mineral called amber was known to attract bodies when rubbed. Magnetic iron ore was also known to attract iron. These and related observations were not developed into an understanding, however.

During the thirteenth century, Pierre de Maricourt realized a fact of great significance. He made a globule out of naturally magnetic material, laid it on a needle on a table, and marked the line along which the needle set itself. When he covered the entire surface of the globule with such lines, he noted that lines formed circles over it, just like the meridians of longitude circle the globe. These circles on the globule met at two points, which were like the North and South Poles of the earth (see Fig. 1.9).

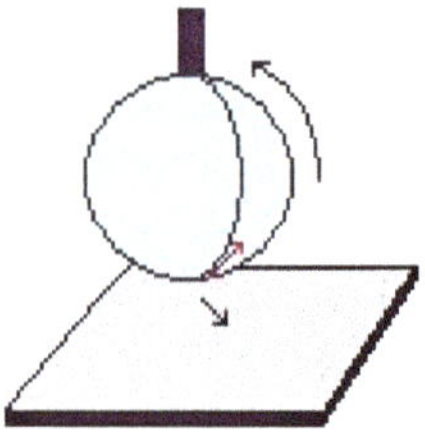

Figure 1.9 De Maricourt's observations. Reprinted with permission from Ref. [23]. Copyright © 2013 by PHI Learning.

Peregrinus noticed that the way in which two magnets set them and the way they attract or repel each other depend on the mutual positions of the poles of the two magnets.

*

William Gilbert (1540–1603) developed these observations further and discovered that the earth is itself a gigantic magnet (Fig. 1.10). He propounded that magnetic forces were such that the north-seeking pole of every magnet attracts the south-seeking pole of every other magnet.

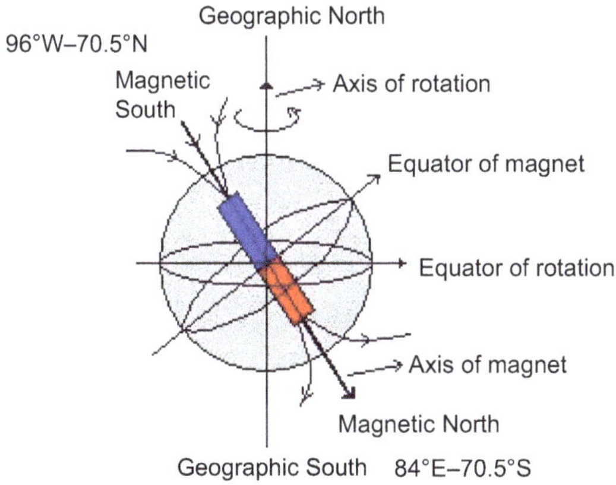

Figure 1.10 Earth as a magnet. Reprinted with permission from Ref. [23]. Copyright © 2013 by PHI Learning.

These were profound discoveries in those times, and these were used for navigational purposes soon.

John Michell (1726–1793), from Cambridge, discovered the law of force between magnetic poles. It was analogous to Newton's law of gravitation.

Gilbert also showed that many bodies, when rubbed, display the phenomenon shown by amber. He conjectured that there is a common origin for the force of attraction in all these bodies. This common force he called by the name of *electric*, a name that is retained even today.

The electric phenomena, Gilbert had conjectured, are due to something—an electric fluid—that under the influence of rubbing is released by amber and similar substances, while that something remains imprisoned in these substances otherwise.

*

Charles-Francois du Fay (1698–1739), a superintendent of the gardens of the King of France, discovered that there exist two kinds of electric fluid. The one to be found in transparent substances like glass and crystal he called vitreous electricity, while the other to be found in resinous substances like amber and sealing wax he called resinous electricity. It was also discovered soon that some substances (metals) transfer electricity between bodies, and these were called conductors.

Benjamin Franklin (1706–1790) called vitreous electricity positive and resinous electricity negative. Thus emerged the terminology of positive charge and negative charge, used in the modern era (Fig. 1.11). Franklin also formulated the law of conservation of electric charge—the total charge remains the same in a process.

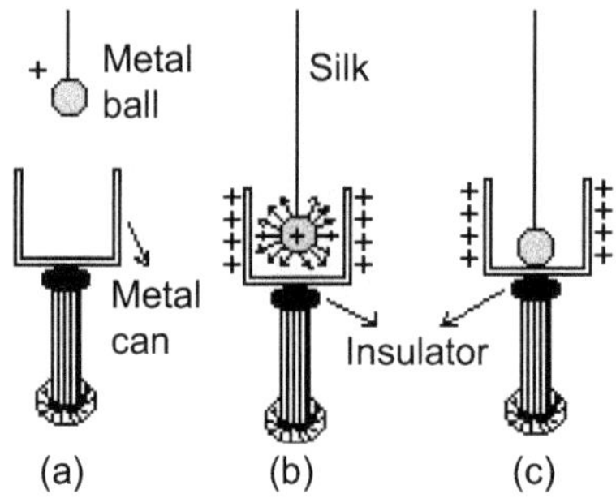

Figure 1.11 Franklin's experiments. Reprinted with permission from Ref. [23]. Copyright © 2013 by PHI Learning.

*

Of prime significance was the collaboration of Franklin and Joseph Priestley (1733–1804), the discoverer of oxygen. At Franklin's suggestions, Priestley performed experiments in 1766 and stated his conclusions in 1767 that "the attraction of electricity is subject to the same laws with that of gravitation, and is therefore according to the squares of the distances . . . "

Charles Augustine Coulomb (1736–1806) established a detailed quantitative law about the attraction of electricity.

Although it was stated first by Priestley, the modern world knows this law simply as Coulomb's law of force between electric charges.

*

Henry Cavendish (1731–1810) performed experiments on conductance of electricity, the electricity-storing capacity of substances, etc.

Poisson developed the clear mathematical theory of static electricity, a theory to which other great mathematicians, such as Lagrange and Laplace, also contributed significantly.

*

Luigi Galvani (1737–1808), from Bologna, Italy, as well as Alessandro Volta (1745–1827), from Pavia, Italy, performed many experiments on electricity using a variety of chemicals.

Humphry Davy (1778–1829), too, contributed to these developments.

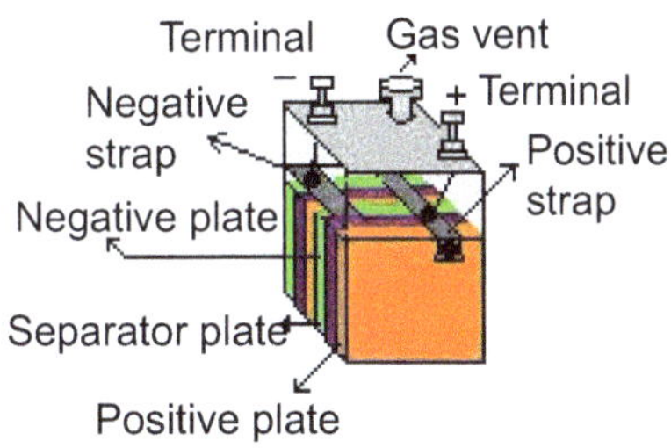

Figure 1.12 Chemical battery. Reprinted with permission from Ref. [23]. Copyright © 2013 by PHI Learning.

These investigations led to chemical batteries (Fig. 1.12), and many laws related to the deposition of chemicals at the electrodes of a battery were then discovered. These developments had ushered physical science into a new era of its own. Many experiments, unthinkable to perform until then, were now feasible, such was the rapid development of the new technology of those times.

*

Hans Christian Oersted (1777–1851) investigated the action of electricity on magnetic materials. Oersted's efforts provided many useful experimental results. Analysis of Oersted's experimental results was, however, provided by Biot and by Felix Savart (1791–1841). The law of force between a wire conducting electricity and a magnet is now known as the Biot–Savart law.

Georg Simon Ohm (1787–1854) provided empirical laws of electric flow, or electric current, through materials. Modern students of physics know these as Ohm's laws. Ohm's experiments also led many others to theories regarding the nature of electrical materials.

*

Andre-Marie Ampere (1775–1836) discovered that two electric current–carrying wires attract each other while carrying current in the same direction, while they repel each other if the current flows in opposite directions.

James Clerk Maxwell (1831–1879), who formulated a proper mathematical theory of electromagnetism, was impressed by Ampere's memoir of 1825 and spoke of it as "one of the most brilliant achievements in science." He further said:

> The whole, theory and experiment, seems as if it had leaped, full grown and full armed, from the brain of the "Newton of electricity". It is perfect in form and unassailable in accuracy; and it is summed up in a formula from which all the phenomena may be deduced, and which must always remain the cardinal formula of electrodynamics.

But Oliver Heaviside (1850–1925) differed in opinion. In 1888, he said:

> It has been stated, on no less authority than that of the great Maxwell, that Ampere's law of force between a pair of current elements is the cardinal formula of electrodynamics. If so, should we not be always using it? . . . Surely there is some mistake. I do not in the least mean to rob Ampere of the credit of being the father of electrodynamics; I would only transfer the name of cardinal formula to another due to him, expressing the mechanical force on an element of a conductor supporting current in any magnetic field - the vector product of current and induction. There is something real about it, . . . it is fundamental; and, as everybody knows, it is in continual use, either actually or virtually (through electromotive force), both by theorists and practicians.

*

At one time, Humphry Davy had written that "the magnetism seems directly as the quantity of electricity which they transmit." Magnetism and electricity are related to each other. This fact was getting recognized with the studies being conducted.

*

Michael Faraday (1787–1867), a young journeyman at a bookbinder's shop, wrote a letter to Davy in 1812 that he wanted to escape from trade and looked for an opportunity in a scientific laboratory. He had also attended Davy's public lectures and provided to Davy with his letter the neat, handwritten notes of those lectures. These historical papers are now respectfully preserved at the Royal Society of London. Faraday educated himself by reading the volumes that came to him for binding. He had mentioned *The Encyclopaedia Britannica* and Mrs. Jane Marcet's *Conversations on Chemistry*. Impressed by the sincerity of the writer, Davy arranged for his interview and appointed Faraday as an assistant. Faraday succeeded Davy as the director of that laboratory after Davy's death in 1829.

A journalist once asked Davy as to what he thought his greatest discovery is. He had to his credit many important scientific discoveries until then, including the miners' lamp (Davy's lamp), which warned them of hazardous gases inside the mines. Davy had replied, "Michael Faraday, undoubtedly."

*

The undisputed father of experimental methods in physics, Galileo Galilei had conceptually grasped all the essentials of the mechanics of bodies prior to Newton; and Newton expressed them mathematically, while also adding new ideas, like force, of his own. Similarly, Faraday also had conceptually grasped all the essentials of the phenomena of electricity, magnetism, and electromagnetism, which were later expressed mathematically by James Clerk Maxwell, who also added important new concepts to electromagnetism.

*

Faraday's illustrious career included many scientific discoveries of great importance to electricity and magnetism.

An electrically charged body was already known to induce or cause an accumulation of an opposite charge on another body. This is electrostatic induction (Fig. 1.13). Michael Faraday (1787–1867) asked if the same holds for the flow of electric charges—electric current. Indeed, this holds, he found, but only transiently. The induced current is perceived only when the original current is either started or stopped, that is, on its variation. Ampere had narrowly missed discovering this.

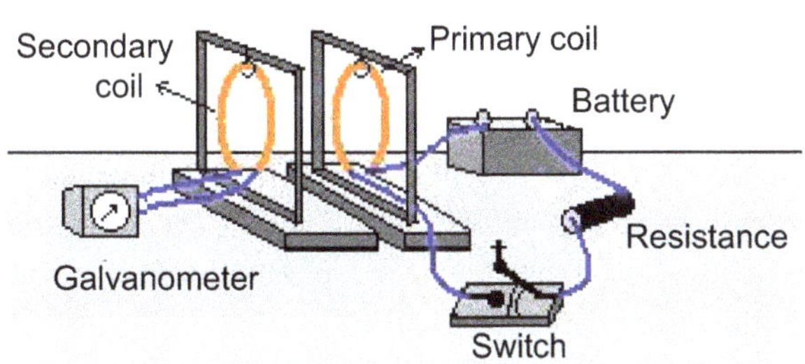

Figure 1.13 Induction of electric current. Reprinted with permission from Ref. [23]. Copyright © 2013 by PHI Learning.

Much before Faraday, it was known that iron filings kept on a piece of paper align themselves along clearly discernible curves when a magnet is brought underneath the paper (Fig. 1.14).

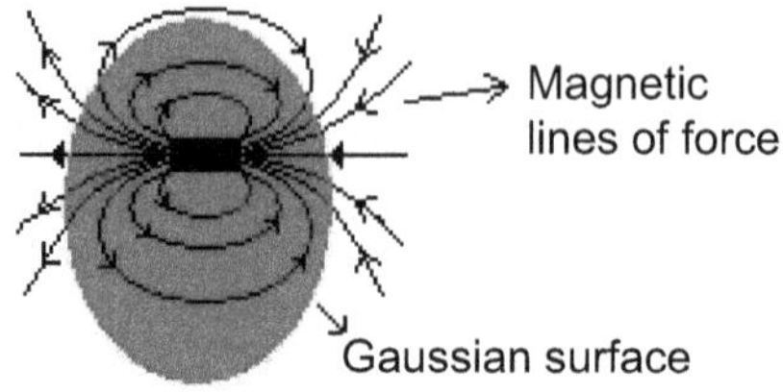

Figure 1.14 Lines of magnetic force. Reprinted with permission from Ref. [23]. Copyright © 2013 by PHI Learning.

This observation gave Faraday the idea of defining the lines of magnetic force, or the curves whose direction at a point coincided with the direction of the magnetic intensity at that point. He conceived the idea that all of the space is filled with lines of magnetic force, and defined the tubes of magnetic force. These ideas led to quantitative studies of induced currents.

The law of the induction of currents is entirely Faraday's discovery. He, not being a mathematician himself, had not contributed to the development of the mathematical theory of his ideas. But his ideas were pivotal to further developments of the theory.

*

Consider an arrangement, originally by Arago, historically six years before Faraday had discovered the phenomenon of induction using a metallic disc. It is usually called Arago's disc, a metallic disc free to rotate about a vertical axis. The lower part of the shaft, acting as the axis of rotation of Arago's disc, is connected to a rotor using a cogwheel. The rotation of the shaft can then introduce the rotations of the metallic disc. A magnetic needle, coaxial with Arago's disc, is placed atop a casing of glass within which Arago's disc is mounted. The magnetic needle is also free to rotate about its axis but has no connection whatsoever with the shaft of the disc, that is, the magnetic needle and Arago's disc are independent to rotate in this experimental arrangement (Fig. 1.15).

With the help of the rotor, Arago rotated the metal disc continually, not expecting the magnetic needle sitting coaxially atop the glass casing of the disc to rotate, as it is not connected with the shaft of the disc. But he found that with time, the magnetic needle begins to rotate in the same direction as that of the disc! The question is, what causes its rotation? Arago could not explain the aforementioned participation of the magnetic needle in the rotation of the metallic disc. Faraday's theory of magnetic induction explained this observation on the basis of the formation of induced currents in the disc. As the disc rotates, the flux of magnetic induction of the needle through any noncoaxial circle on the disc changes, inducing current in it. These are eddy currents.

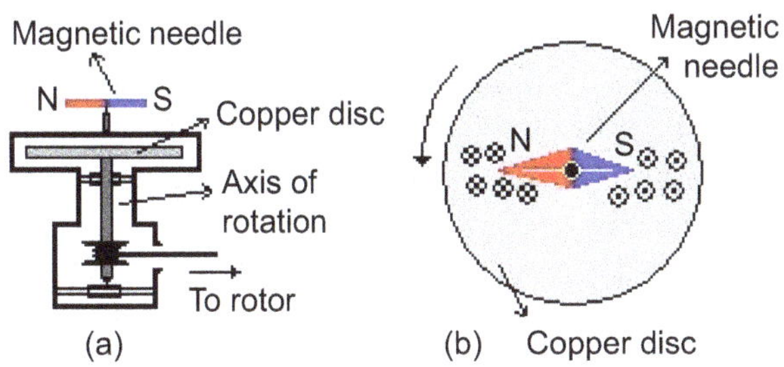

Figure 1.15 Arago's disc. Reprinted with permission from Ref. [23]. Copyright © 2013 by PHI Learning.

Faraday's theory of magnetic induction explained many observations on the basis of induced currents. The existence of eddy currents was subsequently explained by Leon Foucault (1819–1898) in 1895.

*

Oersted, Ampere, and Faraday—this trio very clearly showed the existence of the connection between electricity and magnetism. It is the first example of unification of forces in the history of science. Many would strive in the future for unification of all forces of nature.

*

Faraday's discoveries have been fundamental to our ideas of the electromagnetism. It would be befitting to elaborate further on some of his illustrious discoveries.

Faraday had studied the relation of electricity and chemistry. It is to Faraday that we owe the modern terminology of electrodes. An electrode at which oxygen, chlorine, acids, etc., are evolved in any electrochemical cell he called *anode*, while the one at which metals, alkalis, bases, etc., are evolved he called *cathode*. He called bodies that directly decompose on the passage of electric current *electrolytes*. The parts of an electrolyte that form on the passage of electric current he called *ions*. Ions that travel to the anode he named *anions*, and those ions that travel to the cathode he named *cations*.

Faraday also verified that the rate at which an electrolyte gets deposited depends solely on the intensity of the electric current passing through it. On comparison of different electrolytes, he found that the mass of any ion liberated by a given quantity of electricity is directly proportional to its chemical equivalent, that is, to the amount needed to combine with some standard mass of a standard element. (In modern language, 96,580 coulombs of electric charge must pass through the electrolyte to liberate each ion of a number of grams equal to the quotient of the atomic weight by the valence. An explanation of this fact necessitates atomic matter.)

Christian Friedrich Schonbein (1799–1868), the discoverer of ozone, was Faraday's friend. He and Faraday developed important early ideas about dielectric materials.

As a consequence of these studies perhaps, Faraday came to believe in action propagating step by step through a medium by the influence of adjacent particles. He wrote in 1838 that

> Magnetic action may be communicated to a distance by the action of the intervening particles, in a manner having a relation to the way in which the inductive forces of static electricity are transferred to a distance; the intervening particles assuming for the time more or less of a peculiar condition.

*

In 1820, John Herschel (1792–1871) discovered that the outward form of the quartz crystal is connected with its property of rotating the plane of polarization of light. He also noticed that the rectilinear electric current, deflecting a magnetic needle left and right at all locations around it, has similar (helicoid) dissymmetry. He then speculated that the plane of polarization (of light) would be deflected by magnetoelectricity.

Many had attempted to discover the connection between magnetism and light. But none had succeeded, not even Herschel, but Faraday had established this. His first experiment of 1834 had failed, but he succeeded in 1845 to establish that the plane of polarization of light passing through glass rotates parallel to the lines of force of the magnetic field.

In 1846, E. F. Wartmann, a Swedish physicist, showed that this exact phenomenon also happens with heat radiation, when it is passed through rock salt placed in a strong magnetic field.

George B. Airy (1801–1892) provided the analytical, mathematical way of representing this effect.

*

In 1846, Faraday speculated about the electromagnetic theory of light or radiation. He suggested that light and radiant heat might be transverse vibrations propagated along the lines of force, which permeate the whole space. This marks the beginning of Faraday's concept of a field. For the theory of electromagnetism, Maxwell mathematically perfected this.

In 1846, Faraday also discovered types of magnetizable bodies, which he called *diamagnetic*, a name retained even in the present era.

During Faraday's time, Bernhard Riemann (1826–1866), Karl Friedrich Gauss (1777–1855), Oliver Heaviside (1850–1925), and Wilhelm Weber (1804–1890) developed the mathematical theory of electricity and magnetism.

James Prescott Joule, Hermann von Helmholtz (1821–1890), Nicolas Leonard Sadi Carnot (1796–1832), and others experimented with heat.

Ohm had discovered in 1826 that a conductor carrying current dissipates energy. Joule established the law of heating in a current-carrying wire, resistance, in 1841. These results established that the flow of current causes evolving of heat in a material carrying current.

A natural question to investigate then was whether a metal whose two ends are at different temperatures has an electric current flowing in it only as a consequence of the temperature difference between its two ends. The answer is in the negative when the conducting material is a pure chemical element. There is no current flowing in a rod of iron, copper, gold, silver, or tungsten whose two ends are maintained at different temperatures.

Oersted had discovered the connection between electricity and magnetism in 1820. Not long after this, Thomas Johann Seebeck (1770–1831), from Berlin, discovered in 1822 a connection between electricity and heat. He discovered that an electric current, quite small in strength, flows through a circuit of two different metals by merely disturbing their thermal equilibrium. For example, rods of copper and bismuth soldered at two extremities carry current when the ends are at different temperatures. This was termed *thermoelectricity*, and the effect was called the *Seebeck effect*.

In Faraday's laboratory notebook, we find an interesting remark under the date July 28, 1836: "Surely the converse of thermoelectricity aught to be obtained experimentally. Pass current through a circuit of antimony and bismuth."

Unknown to Faraday, Jean Charles Peltier (1785–1845), from Paris, had already performed such an experiment in 1834. Peltier had discovered that heat evolves at the junctions of dissimilar metals on the passage of current, indeed. But he had also observed that a junction that heats up with the passage of current cools when the direction of current is reversed. This surprising result was called the *Peltier effect*. It is quite distinct from the ordinary Joule heating of a resistance carrying current.

*

Gustav Robert Kirchhoff (1824–1887) showed that electric disturbance is propagated within a current-carrying wire with the speed of light.

*

In 1846, Faraday speculated about the electromagnetic theory of light or radiation. He suggested that light and radiant heat might be transverse vibrations propagated along the lines of force, which permeate the whole space. This marked the beginning of Faraday's concept of a field. For the theory of electromagnetism, Maxwell mathematically perfected this. Maxwell formulated the mathematical theory of electromagnetism in 1857. He showed that light can be, mathematically, looked upon as a wave of electric and magnetic vectors oscillating in magnitude along the direction of its propagation. Both vectors are perpendicular to the direction of propagation of the wave. This picture of an electromagnetic wave, a transverse wave, explained all the previous observations and experiments. It provided predictions of electromagnetic waves in particular.

*

Ludwig Lorenz (1829–1891), from Copenhagen, Denmark, had, independently of Maxwell, proposed a mathematical theory of electromagnetism. The two theories differ in their treatment, and most physicists of today believe that Maxwell's is the proper formalism of electromagnetism.

*

In 1873, Maxwell had published, at Cambridge, his *Treatise on Electricity and Magnetism*. In this treatise, Maxwell had disseminated and elaborated upon many of Faraday's ideas at great length.

Heinrich Rudolph Hertz (1857–1894) undertook experimental verification of Maxwell's prediction of (radio) electromagnetic waves.

*

As a historical fact, David Edward Hughes (1830–1900) had, about seven years before Hertz, produced and detected electromagnetic waves. In 1881, Hughes demonstrated these experiments before members of the Royal Society, who wrongly believed that the results of the experiments could be explained differently. Discouraged, Hughes did not publish those results—not until quite afterward. Thus, the priority of publication belongs to Hertz.

Unaware of the experiments by Hughes, Hertz designed an apparatus to produce an electromagnetic wave in the laboratory in 1888 and also detected it at another location within the laboratory (Fig. 1.16). Hertz's experimental demonstration confirmed Maxwell's theory. Since then, this theory is the basis of our household electrical appliances, the broader means of (radio) communication, and so many other usages.

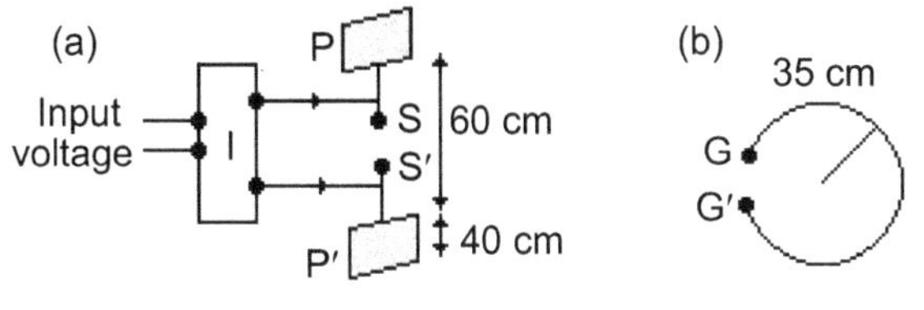

Figure 1.16 Hertz's experiment. Reprinted with permission from Ref. [23]. Copyright © 2013 by PHI Learning.

Hertz's experiments had showed Maxwell's electromagnetic waves to be similar in all respects to visible light. Joseph Larmor (1857–1942) said in 1900 that "the discoveries of Hertz left no further room for doubt that the physical scheme of Maxwell constituted a real formulation of the underlying unity in physical dynamics."

*

In Maxwell's theory, light possesses a finite velocity of motion, its propagation as a wave. Many then measured the velocity of light and found the value of 3×10^{10} cm/s, denoted by the famous symbol *c*.

Astronomical observations of satellites of Jupiter were the first to provide the speed of light. The finite velocity of propagation of light has consequences. One leading consequence was examined by Christian Doppler (1803–1855), known today as Doppler's effect. He showed that the motion of a source of light relative to an observer modifies the period or the frequency of oscillations of light received by the observer. This is like the depression of the pitch of sound when its source is moving away from the receiver of the sound.

Doppler's principle is routinely used in astronomy today to infer stellar velocities, as was suggested and shown first by Armand Hippolyte Louis Fizeau in 1848.

*

Maxwell's theory of electromagnetism did not totally dispel ether, however. Questions arise, such as what is the physical meaning of the electric vector, the magnetic vector? Answers were sought using mechanical models for ether along the lines of Lord Kelvin's models, indicating rotationally elastic properties, as were then inferred by MacCullagh.

Fresnel adopted Young's suggestion that refractive powers of materials depend on the concentration of ether in them, and had assumed that the density of ether is proportional to the square of its refractive index. On assuming further that part of ether within it is carried by the body in motion, one that is in excess of that in vacuum, Fresnel provided an expression for the absolute velocity of light passing through matter relative to the moving matter of that body.

Fizeau confirmed, in 1851, that Fresnel's formula is empirically correct. M. Hoek confirmed in 1868 that the result does not depend on the orientation of the apparatus relative to the direction of the terrestrial motion. This indicated that ether is not carried along by moving bodies.

*

In 1878, Maxwell wrote an article titled "Ether" for *The Encyclopaedia Britannica*. He hoped that astronomical observations of Jupiter's satellites would detect the drag of ether on material bodies. He thought that laboratory experiments would not detect the tiny effects.

Maxwell had underrated the capabilities or powers of laboratory experimentation, Albert Abraham Michelson (1852–1931) had complained in 1881. Michelson devised an interference experiment at Helmholtz's laboratory, Potsdam, during his postdoctoral research (Fig. 1.17). However, Michelson did not detect any drag of ether.

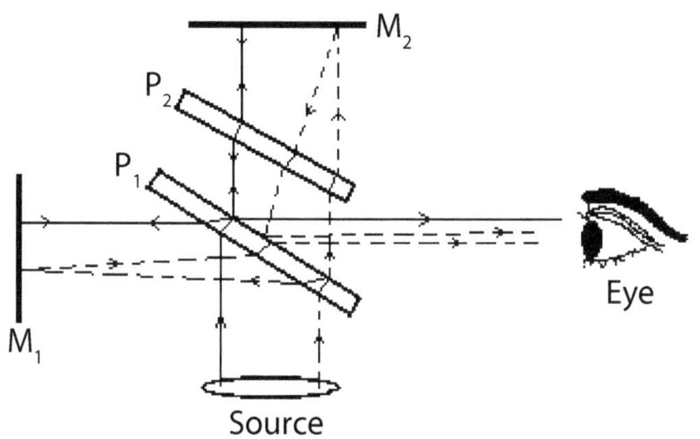

Figure 1.17 The Michelson–Morley experiment. Reprinted with permission from Ref. [23]. Copyright © 2013 by PHI Learning.

Hendrik Antoon Lorentz (1853–1928) pointed out an error in Michelson's theory regarding this experiment. Michelson redevised and reperformed the experiment in 1892 along the lines of the Potsdam one, this time in collaboration with Edward Williams Morley (1838–1923), in Cleveland, USA. This Michelson–Morley experiment, too, did not detect any drag of ether. This indicated that the hypothesis of stationary ether is in error, for the moving bodies carry ether along with them.

Michelson himself, Lorentz, Lord Kelvin, Lord Rayleigh (J. W. Strutt) (1842–1919), and others were disappointed with the null result of the Michelson–Morley experiment, for it defied earlier results.

In 1897, Michelson checked whether the relative motion of the earth and ether varies with the vertical height above the earth's surface. Again, he concluded that the earth's influence on ether extends many thousands of kilometers.

*

Fizeau and Hoek's experiments required a stationary ether, while that of Michelson and Morley required ether to be carried along by the bodies in motion with them. These results began to cast doubts on the theory, but no one questioned the existence of ether. A definite theoretical impasse had therefore been reached.

*

In 1889, George Francis Fitzgerald (1851–1901) pointed out that the experimental results of Fizeau, Hoek, Michelson, and Morley could all be reconciled with the theory if it is assumed that a material body contracts in the direction of its motion.

This was not any absurd suggestion, for everyone believed in the existence of ether. As moving bodies drag ether, it should resist this drag. A body was imagined to possess an excess of the density of ether anyway; it was then thinkable, indeed, that a body contracts due to the drag on it. This was the physical contraction of the body in motion.

Some five months later, Lorentz adopted Fitzgerald's idea and had begun thinking about how bodies might contract when in motion, relative to ether. Thereafter, Fitzgerald's length contraction won great favor in ever-widening circles, and eventually it became the basis of all the theoretical investigations related to motion of bodies through ether.

*

In discussing the influence of the motion of the earth on electrical phenomena, Lorentz was led to transform the equations of ether as well as those of electrical charges to the frame of reference moving with the earth. While doing so, he introduced the transformation of the time coordinate only as a matter of a suitable mathematical procedure. These transformations were confined only to Maxwell's electromagnetism and were not considered to be playing any role in general physics.

Fitzgerald's length contraction and Lorentz's time transformation produced together formulas that agreed with experimental results.

*

It is under these general circumstances that in 1905, Albert Hermann Einstein (1879–1955) stated that the Lorentz–Fitzgerald transformations be applied not just to the equations of the electromagnetism but also to general situations in physics. This is the basis of Einstein's special theory of relativity. It hoped to explain phenomena by the property of their equations being the same under these transformations.

If Maxwell's equations retained their mathematical form under the Lorentz–Fitzgerald transformations, then the equations of motion of material bodies are also needed to retain their form under these same transformations. Else, we would be able to ascertain the existence of ether on observing motions of bodies. Boussinesq, Heaviside, and others had stressed that the equations of ether and those of the motions of bodies are closely related.

The theory of light—Maxwell's theory of electromagnetism—was, thus, reduced to only one kind of a set of equations among the many others invariant under the Lorentz–Fitzgerald transformations. There are many other such equations, of course. (But these transformations are special, linear in coordinates. In 1916, Einstein conceived the idea of a theory whose equations hold under the general transformations. This was the basis of his general theory of relativity. We will consider this later.)

*

Hermann Minkowski (1864–1909) showed that Einstein's equations of special relativity could be succinctly expressed in terms of a mathematical notion of 4D quantities, like four vectors. Thus, with Minkowski was born the legend of the 4D space-time.

Nevertheless, the three dimensions of ordinary space and time had existed all through the history of physics. Even Newton's theory has four dimensions, except that time is not transformed from one observer to another. Contrary to general impression, the idea of 4D space-time is not related to the ideas of special relativity.

*

Many had conveniently forgotten ether. But a few remembered it, nevertheless, and not without reason. In particular, Michelson raised the issue as late as in 1927. In his book *Studies in Optics*, published in 1927, he wrote:

> The existence of the Ether appears to be inconsistent with the theory (special relativity) But, without a medium how can the propagation of light waves be explained? How [to] explain the constancy of propagation, the fundamental assumption (at least of the restricted theory); if there is no medium?

Indeed, special relativity does not invalidate the concept of ether. This theory shows only that the equations of electromagnetism have a curious property of being invariant under the Lorentz–Fitzgerald transformations and that this is consistent with the equations of motion of bodies interacting with ether, which are also invariant under the same transformations of the four coordinates. This alone does not dispel ether.

It was Lorentz who brought about a definite change in the perspective about ether. Lorentz had appealed to the concept of a field, in the manner speculated by Faraday, in 1846, that light and radiant heat might be transverse vibrations propagated along the lines of force, which permeate the whole space.

The Faraday–Maxwell concept of the (electromagnetic) field was to be the center stage of physics hereafter.

Modern physical theories also postulate the field (but different in conception than the field of Faraday–Maxwell's imaginations) and investigate the consequences of field interactions.

*

Developments in another direction also took place as the theory of electromagnetism developed. These concerned the heat and laws governing exchange of heat. Next, we consider this history.

Maxwell had predicted the existence of an electromagnetic wave. Hertz produced and detected an electromagnetic wave. Experiments indicated that heat radiation is like Maxwell's electromagnetic waves. Then the theory of electromagnetic radiation needed to be consistent also with our understanding of heat radiation.

We now turn to these developments. We consider the atomic theory of matter first.

1.3 Atomic Theory of Matter

The Greeks had a philosophical notion of an atom as an extremely tiny particle of matter admitting no division. Kanaad, an Indian sage, too, had proposed the atomic view of matter during the pre-Christ era. He had, however, considered an atom to be divisible into further parts. He considered differences in the properties of matter to be basically superficial. Ultimately, these differences, he had argued, should vanish when we divide the atom into more parts. He believed in the oneness of everything of the universe.

But these philosophical notions of what an atom should be like were never confronted with their reality, that is, with the physical existence of an atom by way of any experiments. This situation about the physical existence of an atom began changing during the early nineteenth century.

*

John Dalton (1766–1844), in 1808, noticed in his book *New System of Chemical Philosophy* that on dividing a material, like an iron block, into two parts, certain properties of the material are the same for them both: the two parts can still be considered to be iron. He argued that if we were to continue dividing the material into smaller and smaller parts, then we should end up with parts having the same properties as that of the original material but these parts would be extremely tiny as compared to our usual experiences. If we were to continue dividing the material, then we would ultimately obtain the smallest part having the same properties as that of the original material. The smallest part was called an atom of that material.

Within Dalton's scheme of concepts, there should therefore be different atoms corresponding to chemical elements of different properties. The science of chemistry was considered to provide for the laws of combining atoms, to form molecules, for forming materials having mixed properties.

*

If the atom of an element were divisible, then the property of its material or the element would be lost, for the atom is defined to be that ultimate constituent of the material that has the properties of the original material. We, however, do not see within our usual experiences that a block of iron when broken into parts ceases to be iron. An atom was therefore considered indivisible, meaning unbreakable into further parts.

*

In that era, much confusion was created by misconceptions of the chemical atom being different from the physical atom. In 1811, Amedeo Avogadro (1776–1856) argued that different gases at the same temperature and pressure contain an equal number of particles when having equal volumes. Dalton as well as many other chemists did not accept this.

In 1815, William Prout (1785–1850) claimed that specific gravities of elements are integral multiples of the specific gravity of hydrogen. But this was not considered to be indicating the existence of the atom.

Joseph Louis Gay-Lussac (1778–1850) proposed that in more than one compound of two elements, the ratio of the weight of one element to the same amount of the other is a simple integer. He believed that his results were consistent with Dalton's atomic theory, but Dalton and most other chemists did not agree to this proposal.

*

On the basis of how specific chemical properties repeat themselves with the chemicals, Dmitri Ivanovich Mendeleev (1834–1907) constructed in 1865 a list of elements and its graphical representation, called the periodic table. The initial periodic table was subsequently added to and modified also with the discoveries of newer elements. This is essentially a periodic table of atoms of various elements or basic materials.

*

In the history of modern physics, the reality of atoms was established only during the early twentieth century.

Matter is made up of atoms and combinations of atoms, or molecules, of various types. Atoms of different chemical elements have different chemical properties. Most atoms are found to be stable in that they do not change themselves to other atoms over a long period of time. Ninety-two stable atoms having distinct properties have been already discovered and have been included in the modern periodic table.

Certain atoms are, however, seen to change themselves with time. These are called unstable atoms and can be extremely short lived. Atoms of the unstable type, beyond the 92 stable atoms, have to be obtained artificially. The list of atoms in the periodic table has gotten extended to 111 atoms.

During the early days of the atomic theory of matter, all elements were considered as hard and unbreakable solid balls.

However, indications existed that an unbreakable atom is not a right idea. For example, it was already well known that gases, which are generally highly insulating, become conducting at very low pressures when a high electric potential difference is applied between the electrodes of an evacuated glass tube containing a gas, or a gas discharge tube (Fig. 1.18).

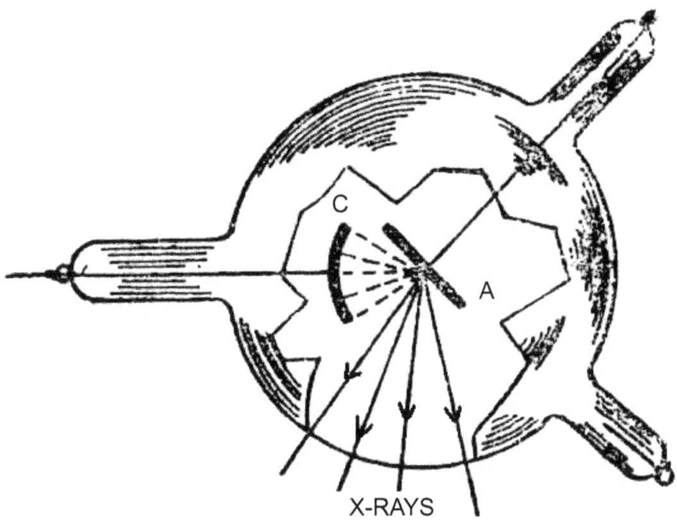

Figure 1.18 Gas discharge tube. Reprinted with permission from Ref. [22]. Copyright © 2013 by PHI Learning.

In 1896, Wilhelm Konrad Roentgen (1845–1923) discovered X-rays in an accidental manner while working with a gas discharge tube. While experimenting with the tube, Roentgen had enclosed it within an opaque shield of black cardboard. But a piece of paper painted with barium-platino-cyanide, made for a different purpose but lying on the laboratory table, showed emission of light, called fluorescence, when the tube carried current. Roentgen, a very keen observer, conducted experiments to establish that the tube was the source of highly penetrating radiation, which he called X-rays.

Hertz and some others had narrowly missed discovering X-rays, mainly because of their failure to distinguish them from other types of radiation emitted by the tube.

Roentgen discovered that X-rays penetrate the human flesh but not the bones. Interestingly, many medical men used to bring patients and crowd Roentgen's laboratory in Manchester to detect foreign bodies, like iron needles, embedded within the flesh. Such foreign bodies could be located by X-ray imaging. X-rays are a type of electromagnetic radiation.

*

In 1897, Henry Becquerel (1852–1908) announced to the world the discovery of a new phenomenon. It arose as follows:

While dealing with salts of natural elements in the laboratory, Becquerel noticed that covered photographic plates in the dark were getting exposed to radiation of some unknown kind. He discovered that the drawer containing these plates had a box of uranium nitrate in it. So he purposefully sprayed the covered photographic plate with a layer of uranium nitrate and kept it for a couple of days in the dark, isolated from all other things. On developing this photographic plate, he found it to be exposed to radiation, which could only be attributed to the uranium salt. As this source of radiation was quite different than that discovered by Roentgen, this property of the uranium salt of producing penetrating radiation was given the name *radioactivity*.

To identify the source of emission in radioactivity, Becquerel worked with a variety of salts of uranium, in different crystalline forms and in solutions, and the metal of uranium itself. In every case, he found that the intensity of radioactive emission is proportional to the concentration of uranium in the material. Becquerel also found that this proportionality between the intensity of radioactive emission and the concentration of uranium did not depend on the temperature, pressure, chemical combination of uranium with other elements, application of an electric or magnetic field, etc. This indicated the radioactivity of uranium is independent of the environment in which the atom of uranium is to be found.

*

Pierre Curie (1850–1906) and Madam Marie Curie (1867–1934) investigated an ore of uranium called pitchblende for radioactive emission. They had discovered the phenomenon of radioactivity in two new elements, now known as radium and polonium. The chemical properties of radium and polonium are quite like those of the elements of bismuth and barium.

*

Lord Ernest Rutherford (1871–1937) began his investigations of radioactivity to find that radioactive emission from atoms can be classified into two types:

- Less penetrating in the substance on which it is incident, called α particles
- Deeply penetrating in the substance on which it is incident, called β particles

When these rays were allowed to pass through a strong magnetic induction, the deflection of α particles was less compared to that of β rays. This indicated that α particles are much more massive than the β rays. The α particles deflected in a direction opposite to that of the deflection of β rays. This indicated that α particles are oppositely charged to β rays.

The α particle was identified as having a doubly charged helium nucleus. That is, if we remove the two electrons moving around the nucleus of a helium atom, then the residual nucleus is the same as an α particle. The β rays were identified as consisting of fast-moving electrons.

*

W. Villard discovered radioactive emission that is even more penetrating than α particles and β rays. This radioactive emission was called γ rays. The γ rays showed no deflection while passing through a region of strong magnetic induction, indicating them to be electrically neutral. On further investigation, the γ rays were established to be highly energetic electromagnetic radiation with a wavelength from 50 to 0.5 pm. The properties of γ rays are similar to those of hard X-rays.

As far as the ionization property is concerned, α particles produce strong ionization in the surrounding medium and have a range of a few centimeters in air in the laboratory. β rays produce much less ionization than α rays and therefore have a much larger range in air, or are more penetrating than α rays. On the other hand, γ rays produce very little ionization in air and therefore possess a much longer range, or are highly penetrating. α rays and β rays are completely absorbed by a thin metal plate, while γ rays can easily pass through such obstructions. Very thick plates of lead are needed to absorb γ rays.

*

Studies related to discharges of gases showed that positive and negative charges are always produced under the application of suitable high voltages. The negatively charged part was found to be highly mobile as compared to the positively charged part.

Different gases inside the discharge tube always led to the production of negatively charged and positively charged parts, with the negatively charged part showing quite the same properties for different gases. This indicated that the negatively charged part must be common to different gases and therefore to atoms of these elements. The positively charged parts of different gases, however, showed differences in their properties, like mobility.

This suggested that an atom (not only of the gases used in a discharge tube but also of a material of a general nature) can be split into positively charged and negatively charged parts. The negatively charged part of the atom should be much smaller in mass, as is indicated by its higher mobility, while the positively charged part should constitute the major atomic bulk being the more massive of the two parts of the atom.

For the lightest of atoms, the hydrogen atom, the positively charged part was found to be 1836 times more massive than its negatively charged part. The negatively charged part of the hydrogen atom was termed *electron*, and its positively charged part was termed *proton*. For other heavier gases, the positively charged part was termed *ion*.

*

As every atom should be electrically neutral, the number of negatively charged electrons and the number of positively charged protons in it must be equal. A natural question was then of the distribution of positive and negative charges within an atom.

Initially, it was Joseph John Thomson (1856–1940) who had proposed that an atom consists of an equal number of protons and electrons, with these charges being uniformly distributed within a small sphere of the size of the atom, typically of radius 10^{-10} m.

*

The concept of an atom representing a chemical element then became the subject of investigation in physics. Chemical elements emit light when heated. Related observations provided characteristic sharp lines of their light emission, the line spectra of light emitted by the elements. These spectra had to be explained on a suitable theoretical model of the chemical atom.

But the physical structure of the (chemical) atom was not known or was not explored then. Such a theoretical model was being developed in that era. The first such model was by Thomson.

Thomson imagined the atom as a positively charged body in which negatively charged electrons were embedded, like cherries in a pudding. Thomson's model had no experimental basis, however. Studies of gas discharge led Thomson to discover the existence of an electron. Because an electron was emitted by an atom in the discharge tube, it must be a part of that atom.

*

The stability of an atom was not an issue that was explored then. Why do the same substances possess the same properties? Why do the same crystals form out of the same substances? Why do the same chemicals form in the same chemical reactions? Why does an iron atom, for example, remain an atom of iron even after it has been subjected to a variety of external influences, like heating, immersion in a magnetic field, and immersion in an electric field? That these issues of atomic stability cannot be explained using the older physics was slowly being recognized during that era.

*

Although not all about the atom was known, the basis of Dalton's theory of the (chemical) atom was accepted by the physicists, who now needed to develop a theoretical model of the atom to answer the aforementioned questions. Thomson's pudding model of an atom was the very first such attempt. This model had to be tested for its accuracy.

To verify this model of an atom, Rutherford thought of an experiment of bombarding the atom with highly energetic α particles, which would collide with protons and receive velocities in different directions as a result of their scattering (Fig. 1.19). If the protons of the atom are uniformly distributed within their atomic sphere, then it can be expected that the scattered α particles would emerge in equal numbers in all the directions about the direction of their incidence, while simultaneously dislodging protons in the direction of their incidence. An α particle is more massive than a proton. Therefore, we should notably not observe the α particles reversing the direction of their motion in their scattering off atoms.

*

In 1911–1913, Hans Geiger (1882–1945) and Ernest Marsden performed scattering experiments at Rutherford's suggestion. Marsden was a 20-year-old student without even a bachelor's degree then.

*

Rutherford had suggested to Geiger and Marsden that they bombard with α particles from a radioactive material a thin foil of gold, whose atom is heavier than α particles. A detector surrounding the gold foil provided information about the directions in which the α particles incident on the gold foil were scattered. What Rutherford found in this experiment was that some α particles were scattered back in the direction they were incident from.

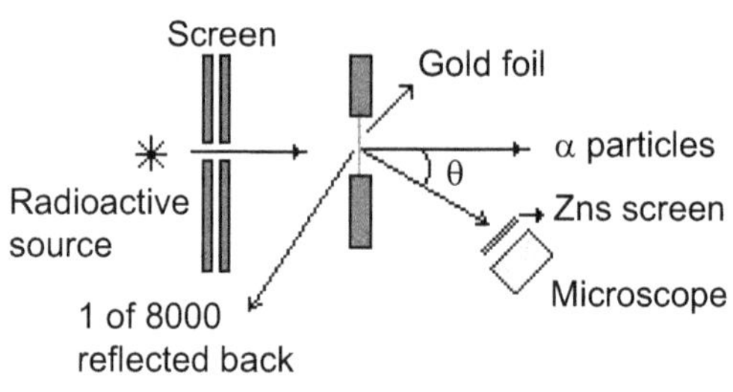

Figure 1.19 Rutherford's experiment. Reprinted with permission from Ref. [23]. Copyright © 2013 by PHI Learning.

The reversal of the direction of motion of the α particles was as if the α particles had collided with a massive object. A less massive body reverses its direction of motion upon a head-on collision with a body much more massive than it.

Considerations of elastic collision of bodies suggested that the α particles are scattered from a body much more massive than them. Thomson's picture of the uniform distribution of positively charged protons and negatively charged electrons within a sphere of atomic size cannot then be an appropriate description of the physical atom.

Experiments led Rutherford to propose that an atom has a massive nucleus, consisting of positively charged protons and negatively charged electrons moving around this nucleus. This is called a nuclear atom, with the atomic nucleus having a radius of the order of 10^{-15} m.

*

Clearly, an electron of the gold atom cannot be at rest; if it were, then it would fall onto the nucleus due to the attractive (Coulomb) force between it and the nucleus. To resolve this difficulty, Rutherford sought refuge with the planetary system of the sun. He imagined the electrons of the gold atom to be revolving around the nucleus, with the centripetal force of Coulomb attraction causing the circular motion of these electrons. This picture can then be called as that of the planetary atom.

*

Observations from the scattering experiments therefore led Rutherford to imagine a nuclear atom, that is, a nucleus (of protons and neutrons) around which an electron revolves in an orbit. This picture was already employed for the planetary system of the sun, with planets revolving in elliptical orbits around the centrally situated massive sun. This picture can therefore be called as that of the planetary atom.

In advance, the reader may note that the nuclear atom is the right concept; however, the planetary atom is not the right idea about an atom. A planetary atom would be flat like the plane of the solar system, but a real atom is not like that and behaves quite differently.

An electron in this planetary atom model is imagined to be executing a circular orbit around the positively charged centrally situated atomic nucleus consisting of protons and neutrons. An electron has an accelerated motion with an associated frequency of its circular motion.

*

There were certain marked discrepancies about the mass of a nucleus, however. The mass of the nucleus is not equal to the sum of the masses of the protons contained within it. These discrepancies were resolved with the realization that the atomic nucleus also contains an electrically neutral particle, called *neutron*, which is a little more massive than a proton. In 1932, James Chadwick (1891–1978) discovered the neutron by detecting it away from or separately from the atomic nucleus.

Chadwick's discovery of the neutron led to the nomenclature for the atom. An atom is now denoted by the following two numbers: an atomic number, denoted by the symbol *Z*, denoting the number of positively charged protons in the atomic nucleus, and an atomic mass number, denoted by the symbol *A*, denoting the total number of protons and neutrons in the atomic nucleus. Then the number of neutrons in a nucleus will be $A - Z$.

Now, the atom is denoted by writing the atomic number *Z* as the subscript prior to the chemical symbol and by writing the atomic mass number *A* as the superscript following the chemical symbol, as in $_Z$(chemical symbol)^A. For example, $_8O^{16}$ denotes the atom of oxygen, with its atomic number as 8 and its atomic mass number as 16.

*

The number of neutrons in the atomic nucleus of the same chemical element was not found to be always identical. For example, there exist three atoms of the same chemical element, oxygen: $_8O^{16}$, containing 8 neutrons; $_8O^{17}$, containing 9 neutrons; and $_8O^{18}$, containing 10 neutrons. Such atoms, having the same value for the atomic number *Z*, or the number of protons in the atomic nucleus, but different values for the atomic mass number *A*, or differing numbers of neutrons in the atomic nucleus, are called *isotopes* of the same chemical atom.

It was therefore realized that the number of protons in the atomic nucleus and therefore the number of electrons in the (electrically neutral) atom determine its chemical properties. As an example, the atom of aluminum is denoted by the symbol $_{13}Al^{27}$, showing that its atomic number (the number of protons in the nucleus of the atom of the chemical element aluminum) is 13, the number of neutrons in its nucleus is 14, and 13 electrons move around the nucleus of the atom of aluminum to make it an electrically neutral atom. The same is also true for atoms of other elements.

*

Thus, the atom considered indivisible by the then chemists was believed by the then physicists to be divisible for sure. The physicists had reasons of radioactivity to support this view of the divisible atom. Madam Curie wrote in 1900 that "atoms [of radioactive elements], indivisible from the chemical point of view, are here divisible."

Physicists established the divisibility of the atom by first discovering the subatomic constituents, such as electrons and protons. The physical atom is the chemical atom. The structure of the atom then becomes important to chemistry.

*

The discovery of radioactivity in 1896 also came as a surprise for the then well-established Newtonian physics, or classical physics. Radioactivity involves emission of particles by an atom. We would expect Newtonian laws to tell us the details of the mechanism underlying the radioactive emission of particles by an atom. For example, it should provide for the force causing the splitting of the atomic nucleus, whose fragments are emitted as the particles of its radioactivity. It should tell us the instant at which an atom disintegrates, and so on.

We may or may not be in a position to experimentally verify predictions of Newtonian mechanics for a single atom, for it is quite tiny as compared to our usual systems in the laboratory. We may therefore have to resort to a system, the statistical ensemble, of many atoms. Then the Newtonian mechanics should provide for the properties of radioactive decays of an atom, statistically averaged over many atoms. It does not!

*

Daniel Bernoulli (1700–1782) appears to have been the first to state that collisions of particles of gas with the walls of its container may be the reason for the pressure of the gas on those walls. These ideas developed into the kinetic theory of gases.

In 1857, Rudolf Julius Emmanuel Clausius (1822–1888) became the first to make the distinction between solid, liquid, and gaseous phases of matter on the basis of different types of motions of particles in them. In 1873, Maxwell calculated the statistical distribution of velocities of the particles of gas as a function of its temperature.

*

During the period 1866–1877, Ludwig Boltzmann (1844–1906) invoked statistical methods for theoretically obtaining the properties of gases as a system of many atoms. This is statistical physics. Boltzmann's works were largely neglected. In particular, his point of view that the increase of a quantity entropy is to be grasped only statistically was unusually heavily criticized in 1895 by a well-known chemist Friedrich Wilhelm Ostwald (1853–1932).

*

In a statistical treatment, we neglect the possibility of the exact measurement of the value of a variable, such as the location of a particle of the system, and ascribe its probability in a many-particle system. There, however, always exists the possibility of measuring the value of the location of a particle, or of any of these associated variables, even though the averages over many particles are computed within this treatment. Such treatment provided a statistical basis to the science of thermodynamics.

*

We could similarly expect the laws of radioactivity to be explained on the basis of Newton's laws of motion. However, the Newtonian mechanics could not be used to predict various properties of the radioactive decays of atoms, for all that we can talk about is the probability of the decay of the atom at an instant of time but cannot determine that instant.

Radioactivity provided the first instance in physics wherein the probability of atomic disintegration cannot be calculated using the usual laws (of Newtonian mechanics).

*

In 1802, William Hyde Wollaston (1766–1828) discovered that the spectrum of sunlight was crossed by seven dark lines perpendicular to its length. This observation escaped notice until it was discovered again by Joseph von Fraunhofer (1787–1826) in 1814. He had also measured the wavelengths of the strongest of the dark lines and found, for example, that the one he named by the letter D had a wavelength of 5887.7×10^{-8} m. The wavelength of the yellow light of many flames was found to be equal to this.

In 1826, W. H. Fox Talbot (1800–1877) was the first to show that "a glance at the primitive spectrum of a flame may show it to contain substances, which it would otherwise require, a laborious chemical analysis to detect." Thus had begun spectroscopic chemical analysis.

In 1852, George Gabriel Stokes told Lord Kelvin that

> I explained the connection of the bright and dark lines by the set of piano strings tuned to the same note, which if struck would give out that note, and would also be ready to sound out, to take it up in fact, if it were sounded in air. This would imply absorption of the aerial vibrations. I told Thomson, I believed there was vapor of sodium in the sun's atmosphere.

Stokes had, by then, established that the D-line, of wavelength 5887.7×10^{-8} m, was emitted by sodium when inserted to be present in a flame.

Excited, Lord Kelvin seized on these ideas and later suggested to Stokes that other vapors than sodium might also be found in the solar atmosphere and of "stars." This meant searching for substances producing, in the spectra of flames, bright lines coinciding with dark lines of the solar and the stellar spectra, other than the D-lines of sodium.

Stellar spectroscopy is the way to infer the existence of elements in a star and the composition of stellar matter.

In 1853, Anders Jons Angstrom (1814–1874) submitted a memoir to the Swedish Academy with wavelengths of dark and bright spectral lines he had measured and introduced for wavelengths the scale whose unit, 10^{-8} m or Å, is now named after him.

In 1871 G. Johnstone Stoney suggested that the structure of spectra would be best seen by mapping the spectral lines on a scale of wave numbers, that is, the reciprocal of wavelengths.

Walter Noel Hartley (1846–1913) showed that the intervals between the three spectral lines of the triplets of the zinc spectrum are the same for all the triplets when expressed in terms of their wave numbers. This discovery pointed to the addition and subtraction of wave numbers as being the key to understanding spectra. In 1871 itself, James Dewar (1842–1923) noted that certain sets of spectral lines from the spectra of alkalis and alkaline earth elements formed an evident series.

But the specific relation for the wavelengths of spectral lines forming a series was not discovered, although it did not take long to discover it.

In 1885, Johann Jacob Balmer (1825–1898) established the first numerical law relating the wavelengths of spectral lines of hydrogen forming a series, the Balmer series of hydrogen.

Many spectral series were discovered. For example, in hydrogen, we have the Paschen series of spectral lines discovered by Friedrich Paschen (1865–1947). Johannes Robert Rydberg (1854–1909) provided the unified formula connecting wavelengths of spectral lines in a given series. He had introduced a constant, now called Rydberg's constant, for this.

H. J. G. Moseley generated in 1913 the characteristic X-ray spectra of over 38 chemical elements. Characteristic X-rays are a line spectrum and characterize the target of electrons accelerated within the plates of an electron tube.

*

Just as atomists of his time believed in the indivisibility of an atom, a biologist Robert Brown (1811–1899) believed that living matter should also consist of indivisible living matter. He wanted to show that the so-called dead matter is actually not dead but is living.

To establish this doctrine, Brown collected fine dust particles of wood, which could be considered to be dead long ago. He inserted these particles in a drop of water and observed them under a microscope. The particles of dust appeared to be continuously agitated and moving about in a random fashion from place to place. Claiming that living matter that can be considered to be dead long ago is actually living, as it showed this motion, Brown published his results in 1827.

Brown's doctrine did not survive, but his observations of the agitated random motion of fine particles suspended in a liquid drop did. Physicists of later generations used Brown's observations of the motion, generally called Brownian motion (Fig. 1.20), of suspended small particles in a liquid drop to establish the existence of atoms.

In 1879, on the basis of the then young field of thermodynamics that had been developed only during the period 1830–1850, William Ramsay (1852–1916) proposed the first atomistic explanation for Brownian motion. Thermodynamics shows that the pressure of a gas or liquid arises due to randomly directed velocities of atoms of the gas or liquid. What Ramsay proposed is that the random motion of the Brownian particle is due to random motions of the atoms of the liquid in which that particle is suspended. For any such explanation, an immediate question is the following:

Figure 1.20 Brownian motion. Reprinted with permission from Ref. [22]. Copyright © 2013 by PHI Learning.

An atom can be imagined as a sphere of radius 10^{-10} m, and the typical particle of dust of wood suspended in a liquid drop is ~100 times larger in size and in mass than an atom. How can such a large-size massive particle receive motion due to collisions with tiny atoms then?

*

Avogadro had already shown that a gas contains an enormous number (10^{23}) of atoms in a small volume. So the number of atoms colliding with the Brownian particle at an instant of time is also enormous. But a particle does not receive much momentum in a particular direction on its collision with a single atom. Then how can the tiny atoms colliding with it cause the motion of a Brownian particle?

*

Here, notice that a number of persons, each exerting a small force, can move an object much more massive than each of them by applying force in the same direction so that the result is a large force capable of moving the massive object. In a similar manner, can the tiny forces due to individual atoms acting on a Brownian particle add up to a large force?

However, when a Brownian particle is introduced in a liquid, collision forces acting on it are distributed over its surface. At any instant an enormous number of these forces are acting on the surface of the Brownian particle. In a random situation, it is possible for these tiny forces to add up to a large force having a specific direction and for the Brownian particle to change its direction of motion due to this resultant force.

The direction of this large force would, however, be random, and the instant this happens would also be random. Then a Brownian particle would move in a direction a certain distance before changing the direction of its motion. After changing the direction of its motion it would move another new distance and suddenly change the direction of its motion randomly. The distance moved by the Brownian particle in each step would also be random, and the direction of its motion at each step would also be random.

A Brownian particle should then move from one place to another along a straight line some distance, change the direction of its motion there, move along a straight line again some new distance, change the direction of its motion at that new place, move along a straight line a new distance, and so on. The direction of motion of a dust particle should be changing randomly, and the amount of the distance it moves should also be changing randomly.

But this is how a Brownian particle was seen to move!

*

Statistically, the momenta of atoms of liquid are randomly directed. The average direction of these momenta is thus zero. However, when a Brownian particle is introduced in liquid, collision forces acting on it are distributed over its surface.

When the atomistic theory of Brownian motion was being developed, the physical existence of the atom could certainly not be established by directly observing an atom. However, it is noteworthy that atomistic considerations are remarkably consistent with Brownian motion. Any doubts about the physical existence of an atom should have been dispelled with the atomistic explanation of Brownian motion.

But philosophical prejudices had prevented some thinkers, like Ostwald and Ernst Mach (1838–1916), from accepting the physical existence of an atom. The doctrine, called *positivism*, of these thinkers was *what is not observed cannot exist in physical reality*.

*

The existence of atomic matter thus got established. Atomic matter interacts with radiation. The nature of this interaction needed to be analyzed. Physicists had been attempting the laws of such an interaction concurrently with the establishment of the atomic theory of matter. This history is what we consider next.

Boltzmann was the first to pioneer statistical studies in physics. On a statistical basis, he intended to explain many phenomena, in particular the thermal phenomena, as arising out of interactions of atomic matter and radiation.

1.4 Radiation and Thermodynamics

Thermodynamics developed as an experimental science, out of studies related to changes to the thermal state of a body, which can be in any state: gaseous, liquid, and solid. The unifying nature of thermodynamics is evident, for thermodynamics deals with matter in any of its forms.

As matter and radiation are basic constituents of the universe, concepts of thermodynamics are important to understanding both of them. Therefore, thermodynamics is of universal applicability in the broader sense.

*

Einstein, who greatly valued methods of thermodynamics, had said in its favor:

> A theory is the more impressive the greater the simplicity of its premises is the more different kinds of things it relates, and the more extended is its area of applicability. Therefore, the deep impression which classical thermodynamics made upon me. It is the only physical theory of universal content concerning which I am convinced that, within the framework of the applicability of its basic concepts, it will never be overthrown (for the special attention of those who are sceptics on principle).

Just as Galileo's experimental studies had led Newton to the theory of mechanics (Newton's laws of motion) on the basis of only a few postulates, certain basic axioms led us to the theory of thermodynamics. In other words, we inferred certain basic laws about the changes of the thermal state of a body from experiments and then determined or predicted the behavior of any thermal system using them.

*

Fundamental to thermal considerations are our perceptions of coldness and hotness of bodies that we may touch. It was recognized that these perceptions involve transfer of energy from one body to another—the net energy flows from a hot body to a cold body.

There are three ways in which transfer of energy takes place between locations differing in temperature. In particular, when the involved locations are in a medium, the transfer of energy must get influenced by the properties of the medium.

(i) The medium may conduct energy from one place to another; just like when a metallic solid is heated at one end, its other end gets hotter with the passage of time.
(ii) Energy may also be transferred by convection from one place to another, like that in a liquid, because of the possibility of the bulk motion of its constituents, which are free to move anywhere within the liquid.
(iii) Another manner of transferring energy occurs in the absence of a medium, that is, in vacuum, and is of radiation. A burning fire heats a substance located at a distance, even in vacuum.

When energy is transferred in any of these three ways to a body, we say that it gets hotter. It is the hotness and coldness of bodies as well as the manner of changes in them that are, broadly speaking, clarified and quantified in the science of thermodynamics. It is inherently statistical in nature, which we may wish to note in advance.

*

A system in thermal, chemical, and mechanical equilibrium has its (thermal) state completely specified by its volume, its pressure, and its temperature, which do not change with time. Such a system is said to be in thermodynamic equilibrium.

It is tacitly assumed here that every thermodynamic system possesses a well-defined temperature when in equilibrium. This is often called the *zeroth law of thermodynamics*.

Thermodynamics does not, however, tell us anything about an individual component or constituent of the system. It is to be considered as an incomplete description of the system in the sense that it provides for only the gross or average properties of the system but not for its individual constituents. This nature of thermodynamics implies its statistical basis.

Statistical fluctuations exist about the equilibrium values of the thermodynamic quantities of a thermal system. Fluctuations may contain important information about the system.

*

The principle of conservation of energy, in general terms, is that energy can be neither created nor destroyed; energy only gets converted from one form to another. The principle of conservation of energy as applied to a thermodynamic system is the *first law of thermodynamics*.

All types of nonmechanical processes by which energy can be exchanged by a thermodynamic system with its exterior are termed *heat transfer*. Heat is nonmechanical energy in the process of being transferred from one thermal system to another.

A thermodynamic system either does mechanical work or transfers heat (energy in transit) to another system. We may then think of a system that converts its heat to (mechanical) work perpetually, the so-called *perpetuum mobile*, or a thermodynamic machine causing perpetual motion.

The principle of conservation of energy, fundamental to all physical systems, indicates that it is impossible to design a (thermodynamic) machine that will create energy out of nothing and produce perpetual motion.

*

A solid melts when heated. Transition from solid to liquid is not continuous but rather quite abrupt or discontinuous. It involves the latent heat of fusion. Similarly, when heated a liquid also changes to its vapor form quite abruptly. It involves the latent heat of vaporization.

Of interest here is also the change of a solid from one crystalline form to another crystalline form, the so-called allotropic change, when heat is supplied to the solid. Allotropic change is also quite abrupt or discontinuous. During discontinuous changes, the energy of the system also changes discontinuously. The first law of thermodynamics describes these changes.

*

Even before the proper formulation of thermodynamics, Hess had stated in 1840 that if a chemical reaction proceeds directly from one state to another, and again through a series of intermediate steps between the two states, then the heat evolved in the direct change is equal to the algebraic sum of the heats of reactions in the intermediate steps, which is referred to as Hess's law of constant heat summation. It is a consequence of the first law of thermodynamics.

*

In general, the purpose of an engine is to convert the energy stored in a substance to the energy of motion. For example, a steam engine uses the expansive power of steam to cause the motion of a piston. A hot-air engine uses the expansive power of air to cause the motion of a piston. We burn some substance to generate steam or hot air. What, then, is the role of the burning substance, of steam or air, etc., for the working of an engine using which we may perform work?

Sadi Carnot showed that it does not matter which kind of working substance (coal, petrol, diesel) is used for the engine. Carnot observed that any engine extracts a certain quantity of heat from the heat source, converts part of this heat to work, and transfers the rest to the heat sink. This is essentially all that an engine does when working. This theoretical analysis formed a big step forward in the direction of the proper formulation of the general laws of thermodynamics.

The manner of the transfer of heat determines the amount of work that can be performed by an engine. If we let the heat pass from the source to the sink by conduction, then the engine will perform no work, as conduction of heat involves no mechanical motion.

*

Carnot formulated the theory of an (Carnot) engine. He imagined an ideal engine in which heat transfer occurs from one body to another, without any third body entering in this transfer. In other words, Carnot assumed that no energy is lost to the environment but is transferred to the desired body only. This is certainly an idealization of the involved situations. This idealized engine is called Carnot's engine.

Carnot proved that no engine could be more efficient than the aforementioned Carnot's engine while operating between the same two temperatures. This is Carnot's celebrated theorem on thermodynamic machines.

Concepts of reversible and irreversible processes are involved in Carnot's analysis. A process has a system and its exterior, both undergoing certain changes to their initial states. A process is said to be reversible when the system and its exterior both are brought back to their original, or initial, states. It does not matter whether we retrace the steps of the process or follow any other steps for restoring the initial states of the system and its exterior. In all other circumstances, a process is said to be irreversible. In particular, if the system is brought back to its initial state, then its exterior is not brought back to its initial state.

The irreversibility of a process refers to the impossibility of bringing back the system and its exterior both to their initial states and not to that of only one of the two. It is always possible to bring back only one of these two systems to its initial state.

*

Fundamentally, all interactions are reversible. Newton's laws show that the motion of a body can be retraced along the same path by reversing the direction of the force while keeping its magnitude the same. In dynamics, reversibility can be attained, at least in principle. That Newton's laws of mechanics are time symmetric is closely related to this reversibility.

But all natural (thermodynamic) processes seem irreversible. Then how is it that the fundamental reversible interactions of the microscopic become irreversible at the macroscopic? What is the scale of distance or of mass at which this transition from the micro to macro occurs?

*

Ludwig Boltzmann provided answers to such questions. He showed that irreversibility is the behavior of a complex structure treated as a whole and is not the feature of the individual microscopic constituents of any (thermodynamic) system. Furthermore, Boltzmann showed that the reversibility in the macroscopic domain is not really impossible but is almost infinitely improbable in the statistical sense (of probability).

Boltzmann's demonstrations made it clear that in such considerations, we neglect all measurements related to individual constituents of a thermodynamic system and focus on its average properties. In principle, we can then perform measurements on a select individual constituent. Results will not be related to the average properties, however.

*

If we consider two bodies, then the first law tells us that the heat lost by one body would be equal to that gained by the other body. However, this information does not tell us that the heat will pass from the hotter to the colder of the two bodies. We require more than just this. Concepts of hotness and coldness are needed here. Thus, we introduce the "new" concept of temperature, apart from the first law, to tell the direction of the net flow of heat from one body to another. Such a concept of temperature does not belong to the scope of the first law of thermodynamics, for it is only the principle of conservation of energy.

Carnot's considerations of the reversibility and irreversibility of thermodynamic processes, of convertibility of heat to work, etc., necessitated that a principle dealing with the impossibility of a *perpetuum mobile* be formulated. This principle is, in essence, the *second law of thermodynamics*. It embodies the concept of temperature of a thermodynamic system.

Many forms of the second law of thermodynamics were put forth in the history of these developments. Each such statement amounted to the impossibility of a *perpetuum mobile*.

*

Boltzmann focused on the atomic theory of matter during his seminal and pathbreaking works. From the atomic nature of matter and the characteristics of interatomic interactions, Boltzmann thus aimed to theoretically derive all the laws of thermodynamics. With Boltzmann's works evolved statistical thermodynamics.

The greatness of Boltzmann's works is that all statements get unified in one, single mathematical statement for the second law of thermodynamics. Boltzmann's mathematical statement of the second law of thermodynamics is then completely equivalent to a variety of earlier statements generally proclaiming the impossibility of a *perpetuum mobile*.

Boltzmann was the first to recognize that thermodynamics is statistical in character and its laws hold accordingly.

*

Statistical thermodynamics turned out to be important to further developments in physical sciences. Although Boltzmann was largely ignored and heavily criticized in Europe, the importance of his works was recognized by Josiah Willard Gibbs (1839–1903), who pursued the statistical approach.

Boltzmann committed suicide out of depression in 1905. His career remained that of a school teacher, hopping from one school to another. Only much later was he offered a position at a university, but by then he had stopped his research and was suffering from mental as well as physical illness.

Boltzmann's grave in Vienna is marked by a monument with an equation engraved on it:

$$S = k \log W$$

This equation was written first by Max Karl Ernst Ludwig Planck (1858–1947), and the constant k was also introduced first by Planck and not by Boltzmann. But Boltzmann had it implicit in his works.

*

Boltzmann provided the statistical basis to entropy. Clausius had already shown in 1857 that the entropy of a thermodynamic system can be defined as the ratio of heat extracted from the hot body to its temperature or that of the heat deposited at the cold body to its temperature, the two being equal for a reversible process. Entropy is constant for a reversible process.

Clausius had caused a paradigm shift from the "working of an engine" to a "change of the thermal state of the working substance" in that engine. This paradigm shift leads indeed to the recognition that whatever is happening with the working substance during the process of a thermodynamic transformation of its state is the appropriate or genuine reason underlying the impossibility or, more correctly, the almost infinite improbability, as was shown by Boltzmann, of a *perpetuum mobile*.

*

Caratheodory provided the proper mathematical treatment of these ideas in 1909. Caratheodory, without reference to Carnot's results, proved the existence of the entropy function. Caratheodory's principle, which is the mathematical statement of the second law of thermodynamics within this treatment, states that "within the neighborhood of any arbitrary initial state of a physical system, there exist neighboring states as close as we may desire to the initial state. But these neighboring states are not accessible from the initial state along certain sequence of the states."

*

Even if a heated body is placed in vacuum, it loses heat energy due to radiation escaping from it. During the early history of the studies related to thermal effects, it was not clear how bodies lose energy in this manner. This is because we do not see radiation as energy in motion. In other words, the radiant energy does not manifest itself during its transference but only on its absorption by matter, which gains in heat energy in the process of the absorption of radiation. Consequently, properties of radiation get inferred only on the basis of the properties of its absorption by matter.

This aspect of radiation has important consequences. In particular, on the basis of the observed properties of the absorption of radiation by matter, we may be led into ascribing certain characteristics to radiation or the radiant energy itself. Such characteristics need not be those of the radiant energy but only of the properties of its emission or absorption by matter. Such characteristics must then be fictitious characteristics of radiant energy.

When a material body absorbs energy, its momentum changes, and as energy and momentum are related, if its (kinetic) energy changes, then its momentum must also change. From this property of the absorption of radiation, we may ascribe a nonvanishing momentum to radiant energy.

But the notion of the nonvanishing momentum of a body (of radiation), whose speed is the same for all the observers, is fictitious. Notice therefore that it contradicts fundamental tenets of underlying concepts of the natural or inertial state of motion and those of change in its natural state of motion as were developed by Galileo.

We may recall here that Galileo had imagined the natural or inertial state of a body to be that of a rectilinear uniform motion. He had also realized that a body has a tendency to oppose a change in its state of uniform rectilinear motion. Galileo had thus imagined the notion of the inertia of a body: an intrinsic attribute of a body by virtue of which it opposes any change in its state of uniform rectilinear motion.

Momentum is the product of the inertia and velocity of a body. It was defined not by Galileo but by Newton. If a body has zero inertia, then its state of motion cannot be changed at all. It remains inertial, always. It thus moves with the same state of motion in relation to all observers.

Therefore, if radiation were to possess speed, which is the same for all observers, then it is to be viewed as inertialess or, equivalently, as a momentumless body consisting of only energy. Energy of a radiation body must not be thought of as its kinetic energy also.

*

Many had conceptualized about the nature of radiant energy. For example, prior to 1792, many had imagined that ice emits cold radiation, while burning wood emits hot radiation. So cold and hot radiations were also discussed. Today, these ideas sound absurd, but those discussing were accomplished thinkers talking about the nature of radiant energy.

*

In 1792, Pierre Prevost (1751–1839) proposed that the rise and fall in the temperature of a body is due to its continuous exchange of radiant energy with its environment. The modern theory of heat and thermodynamics is based on Prevost's theory of exchanges. It is assumed that at a nonzero temperature, each body emits radiant energy. When other means of transfer of heat are not dominant, the rise (and fall) of the temperature of a body is because of the absorption (or emission) of radiant energy. Radiation carries energy from one location to another as it propagates in space.

*

Maxwell's theory imagined radiation as a propagating wave of oscillating electric and magnetic vectors, which are both perpendicular to each other and perpendicular to the direction of the propagation.

*

An atomic gas has atoms moving with different speeds. During atomic collisions energy gets exchanged between atoms, and it can happen that for sufficiently high speeds of atoms, atomic electrons may receive part of the energy of their (inelastic) collision and jump to higher-energy orbits. The excited atomic electron may then emit a radiation quantum and jump back to the ground-state energy of the atom. This process is particularly important at sufficiently high temperatures of the gas. Atoms of the gas may not absorb the emitted radiation completely. The radiation that escapes unabsorbed from the gas is lost to the environment.

This is the way in which thermal energy is lost by the matter of the sun to the space external to the sun. When solar radiation reaches the earth, the matter on the earth absorbs it. On absorption of solar radiation, atoms of the matter on the earth receive energy and acquire higher kinetic energy of translational motion on the average. This means that the matter on the earth has gained in thermal energy.

Of course, if the matter of the sun and the matter on the earth were at the same temperature, then the matter on the earth would not gain in thermal energy. Radiation also serves to carry energy from one location to another and transfers energy from a hotter place to a cooler place. It therefore becomes necessary to understand how the emission of radiation varies with the temperature as well as with the properties of the body.

When operating at the same temperature, a perfectly black surface emits the maximum energy, while a polished or shinny surface emits very little energy or radiation. That is, a good emitter of radiation is also a good absorber of radiation. Substances like glass or polished and shiny bodies absorb very little of the incident radiation and are poor emitters of radiation.

*

Kirchhoff, from Heidelberg, Germany, then made the related conceptions more precise. Kirchhoff had observed, in 1859, that the dark D-lines from the solar spectrum darkened still further when solar radiation was made to pass through a sodium flame. In analyzing this observation, he led to consider a cavity enclosed by matter in thermal equilibrium with radiation, which does not escape the cavity.

Kirchhoff defined the concept of a perfectly black body, a body that absorbs all forms of radiation incident on it and converts the energy of the absorbed radiation into heat energy (Fig. 1.21).

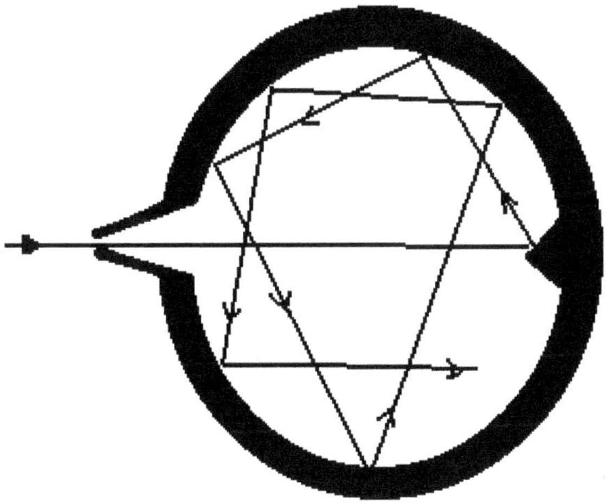

Figure 1.21 Black body. Reprinted with permission from Ref. [22]. Copyright © 2013 by PHI Learning.

It also was recognized that there exists an important difference between radiation emitted by a body at a given temperature and that emitted by a body under any other special arrangement. When we enclose both a body at a given temperature and the radiation emitted by it within a thermally isolated box, radiation is in thermal equilibrium with the body emitting it. Its characteristics do not change with time. Radiation emitted by a body at a certain temperature is therefore called *thermal radiation*. Radiation emitted by a body in all other situations is called *nonthermal radiation*.

*

Kirchhoff formulated laws about thermal radiation. In conjunction with the atomic theory of matter, Kirchhoff's law amounts to an assertion that every atom, when appropriately excited, emits radiation of a definite wavelength, which is characteristic of that atom.

Observations of the aforementioned type then led to experimental studies of elemental spectra, which aim at identifying elements by their characteristic emission and absorption lines.

This method of spectral analysis is of utmost importance in astronomy, for it is only the light received from astral bodies that provides information about their matter.

Assuming that the radiation energy absorbed by matter gets converted only into its thermal energy and a black body absorbs all radiation incident on it, Kirchhoff had thus proved that the emissive power (of a black body) is a universal function only of the temperature of the cavity and of the wavelength of the radiation contained within it.

For almost 40 years, experimental difficulties prevented data sufficient to provide this function. Basic ideas for the construction of a black body existed since Kirchhoff, however. Though Kirchhoff's proof came in 1858, a black body was actually used to verify his laws only in 1895 by Wilhelm Wien (1864–1928) and Otto Lummer (1860–1925).

*

Max Planck showed in 1901 that a specific mathematical form agrees well with experiments. Planck also provided the physical basis for the mathematical form he had proposed for the black-body spectrum. Planck had used principles of thermodynamics for this purpose.

*

Of particular significance are two approximations of Planck's aforementioned law, or formula, for the spectrum of black-body radiation.

The first approximate formula is that given by Wien, proposed in 1896, that the spectral energy density of cavity radiation, that is, energy per unit volume per unit frequency range, is proportional to the product of the cubic power and an exponential of that frequency. In 1897, Friedrich Paschen (1865–1947) verified this in the near infrared. In the far infrared, however, not only Lummer and Ernst Pringsheim (1859–1917) but also Heinrich Rubens (1865–1922) and Ferdinand Kurlbaum (1857–1927) showed that Wien's law fails in that region.

The second approximation is what Lord Rayleigh (J. W. Strutt) had established, in 1900, on the basis of the classical Maxwell–Boltzmann (statistical) law of equipartition of energy—the spectral energy density of radiation enclosed in a cavity is proportional to the square of the frequency multiplied by temperature. The same functional form was also provided by Einstein and by James Hopwood Jeans (1877–1946) in 1905. It holds where Wien's law fails.

The proportionality constant in this law was established by Einstein in 1905 and independently by Jeans in the same year. This Rayleigh–Jeans law holds for large wavelengths only, and Rayleigh was aware of this.

*

Johannes Kepler (1571–1630) had noticed that as a comet approaches the sun, the tail of the comet continuously turns around so as to be always directed away from the sun (Fig. 1.22). Why is it that the tail of a comet is always directed away from the sun?

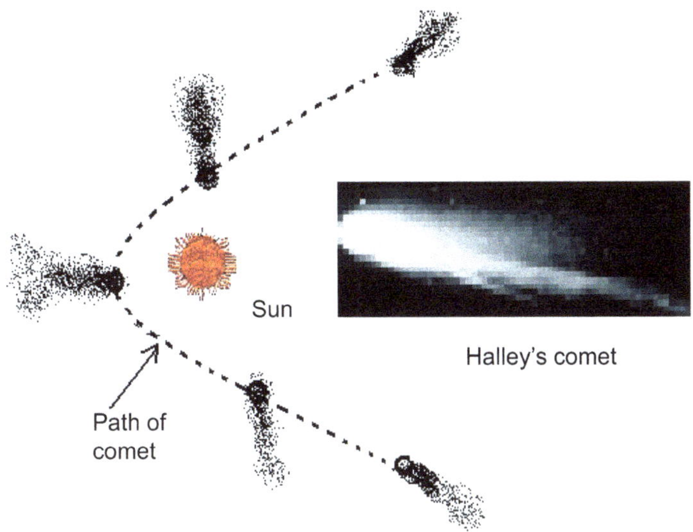

Figure 1.22 Cometary tail: always away from the sun. Reprinted with permission from Ref. [22]. Copyright © 2013 by PHI Learning.

Kepler had attempted to explain this observation on the basis that solar light exerts appropriate pressure on all bodies exposed to it, with this pressure becoming important for the motion of a body only when its size is much reduced or quite tiny.

Many experiments were performed to establish the existence of pressure by solar light, but all failed. The concept of the pressure of solar light was then forgotten for centuries.

*

An electromagnetic wave, or radiation, must exert a small but nonvanishing pressure on a material body on which it is incident.

(This is quite similar to waves in water exerting pressure on the bank, except that the constituents of water have mass. But the constituents of radiation are believed to be massless.) Maxwell theoretically established that for a parallel beam of rays of light, its pressure on a body on which the beam is incident is equal to the energy density of light in the beam.

This does not mean that the constituents of radiation possess a nonzero momentum. This is because it is the interaction of light and matter that is involved here. We will discuss this issue at a later stage.

*

We can associate the concept of pressure with radiant energy. How does the pressure of radiation arise? Is it similar to the pressure exerted by an ordinary gas on the walls of its container? The kinetic theory of ideal gases suggested a close analogy between radiation and the ideal gas—light can be thought to be consisting of individual bodies that collide with matter to impart it both energy and momentum.

Planck was the first to use this analogy with gases and to postulate that radiation consists of individual light bodies, the *quanta* of radiation. Einstein, Satyendra Nath Bose (1894–1974), and others used statistical thermodynamics to explain observed properties of radiant energy. They extended the analogy with gases to radiation and imagined radiation to be consisting of quanta, or individual components of radiation. This is the history of the grand successes of how the aforementioned analogy directed us to the quantum nature of radiation, the radical postulate by Planck, and thereby to the laws of thermal radiation.

The implications of Planck's hypothesis of a light quantum were not appreciated by the physics community at large. It was not immediately clear to many that Planck's hypothesis contradicted Maxwell's theory. Only a select few had intuitively felt its revolutionary character. Einstein was the most prominent of these select few.

*

A statement such as "Light is a wave" does not describe all the properties of light in their entirety. Planck's hypothesis of a light quantum had this important implication.

*

Hertz had accidentally observed the photoelectric effect in 1887. It consists of the emission of electrons by a material when it is irradiated. Interestingly, two high school teachers, Julius Elster (1854–1930) and Hans Geitel (1855–1923), constructed the first phototube using the photoelectric effect.

*

In 1902, Philip Lenard (1862–1947) discovered that the energy of the ejected electron does not depend on the intensity of the light incident on the material emitting that electron.

That Lenard's observations cannot be explained using Maxwell's theory of light as an electromagnetic wave was not realized immediately. It was Einstein who, in 1905, used Planck's notion of a quantum of light to explain all of Lenard's observations.

It is this interesting history underlying the idea of the quantum of radiation that we now turn to.

There were no reasons to question the wave nature of light. But no one suspected that the photoelectric effect cannot be explained on the basis of the wave theory of radiation. Einstein's explanation of this effect, postulating radiation to be made up of quanta of light propagating in space as individual bodies, came as a surprise. It baffled even the best physicists.

1.5 What Is a Quantum?

No doubt any account of the history of physical science would be incomplete without the name of Max Planck. To explain properties of thermal or black-body radiation, Planck assumed that matter absorbs and emits radiation, not in a continuous fashion but in discrete packets of energy. That is, when it absorbs or emits radiation, the energy of matter changes in discrete steps. This is the so-called quantum jump in the energy of matter.

This quantum jump in energy is similar to our exchanging currency. If we provide a person with a one rupee note, the amount of currency with that person changes by a unit, a rupee. We can count the number of rupees when exchanging, and it will be an integer multiple of the currency unit, one rupee. The denomination of exchange can also be made smaller. Nevertheless, when we exchange currency in that smaller denomination, the amount of currency with that person still changes discontinuously. Similarly with a rupee, our count will be an integer multiple of the adopted smaller denomination, that unit. And this is the basic idea of a quantum. These exchanges take place in integer multiples of a unit or a quantum; we can count them as such. And Planck had only assumed that the energy of matter changes in integer multiples of a quantum of radiant energy.

*

If things are so simple and straightforward, why was Planck's assumption of historical importance?

Well, to grasp this, we have to visit Planck's era. In that era, there did not exist, first, any (appropriate) experimental evidence to support Planck's assumption of a quantum of energy.

In science, the foremost of the criteria of verifying whether a concept is good or not, that is, whether it holds or not, is only appropriate experimental evidence. Galileo was the first one who provided us with this lesson.

In Planck's era, all experiments indicated that light was like a wave and not made up of individual objects. Second, light was at that time being imagined as a wave, similar to the disturbance that we see propagating on the surface of water. This, the so-called wave picture of light, had successfully explained many observations related to light, as we have seen before. Of significance are phenomena of interference, diffraction, and polarization of light. (Collectively, we call them the wave phenomena today.) They were explained by the assumption that light is a wave.

If light is a wave, then it cannot, simultaneously, be also corpuscular, or consisting of individually moving bodies.

If light is a wave, then the question is, what is the nature of the medium, ether, of its propagation? Only experiments related to light can answer this question, of course. We have already seen in the sequel that experiments did not detect ether. If there is no medium, how can light be a transverse wave propagating without a medium? This question bothered some, but most adopted Lorentz's idea of a field.

*

Historically, Hertz discovered the photoelectric effect: when ultraviolet light illuminates an insulated metallic plate, it becomes positively charged. This is due to the emission of electrons from the metal plate, as we know now. Further investigations by others showed that ultraviolet light has a definite role here, because photoelectron emission does not occur when red light is used, no matter how intense the red light is.

Although Hertz discovered the photoeffect, Lenard performed its detailed study. He illuminated a metal plate with light of different colors to also find the maximum velocity of the emitted electrons, which he found not to be dependent on the intensity of the incident light.

*

It was not immediately recognized that Lenard's aforementioned observations cannot be explained by the wave theory of light. The following simple considerations show that the wave theory of light is inconsistent with the observations.

*

For Maxwell's theory, light is an electromagnetic wave propagating away from its source in all directions and its strength decreases inversely with the square of the distance traveled by it in vacuum.

Consider a source isotropically emitting an amount Q of radiant energy per unit time. Then the amount of radiant energy incident on a unit area at a distance r from the source will be given as

$$Q/4\pi r^2.$$

Notice that we can "arrange" the amount of radiant energy incident on the unit area to be of any value by keeping the absorbing area at a suitable distance r from the source.

Now, the electron is assumed to be held by the atomic nucleus due to its Coulomb force of attraction. It is only on receiving energy greater than or equal to its binding energy with the nucleus that it will be ejected from the atom. The incident radiation must be adequately absorbed by the electron for it to be ejected from the atom. If A denotes the surface area of the atom, a is the absorption coefficient, and δt is the time for which the radiant energy is incident on the metallic plate, then the energy absorbed by the atom in that metallic plate is

$$QAa\delta t/4\pi r^2.$$

So if we were to irradiate the metallic plate with red light, we would observe electrons being ejected from it, provided we expose it to red light for a sufficiently long time δt. This is not observed to happen.

Moreover, we can imagine an experiment in which the intensity of ultraviolet light is diminished to such an extent that according to the earlier equation, we will have to wait for over 45 minutes to observe the photoemission of electrons. However, the photoemission of electrons is found to be instantaneous.

This contradicts the prediction of the wave theory of light. Therefore, we have to abandon the idea that the radiant energy propagates as a wave.

*

Notice that we need not characterize monochromatic radiation by frequency. As Maxwell's picture of radiation as a wave is not completely consistent with experiments, the monochromatic nature of radiation has nothing to do with the frequency of radiation imagined as a wave. The energy of the quantum of radiation is basic to our considerations, and the monochromatic nature of radiation could then mean radiation to be consisting of quanta having the same energy for them all.

*

Can properties of light then be explained by the corpuscular picture of light? Can we regard radiation to be consisting of bodies that travel through space, with each body retaining its energy content intact during the travel? These issues bring Newton's corpuscular light back into physics.

*

Following Planck's seminal works, Einstein postulated in 1905 that radiation travels in space as a parcel, that is, as a quantum of energy $\epsilon = h\nu$. Assuming this, Einstein proposed a stunningly simple explanation for the photoelectric effect as follows.

A light quantum, a parcel of energy ϵ, gives its entire energy to a single electron on their contact interaction, and this process is independent of the presence of other light quanta. An electron ejected inside a material comes to its surface after an energy loss Φ (depending on the material), which is called the work function of the material. Now, if E is the energy of the electron when this energy loss is zero, then

$$E = \epsilon - \Phi$$
$$= h\nu - \Phi.$$

The intensity of light does not affect, in any manner whatsoever, the energy of the electron in this process.

It then follows that increasing the intensity of light incident on a metallic material increases only the number of electrons being ejected by that material, but this does not increase the (kinetic) energy of any of the ejected electrons. This conclusion is in accord with Lenard's observations.

In 1906, Einstein re-examined Planck's derivation and concluded that energy of matter emitting or absorbing radiation can take on only the values that are integral multiples of $\epsilon = h\nu$, that is, in emission and absorption of radiation, the energy of matter can change only by integral multiples of $h\nu$. He concluded that radiation travels in space retaining its energy.

With these works, Einstein had essentially brought us back to Newton's corpuscular picture of light!

*

Robert Andrews Millikan (1868–1953) studied the photoelectric effect and wrote in 1948:

> I spent ten years of my life testing that 1905 equation of Einstein's and contrary to all my expectations, I was compelled in 1915 to assert its unambiguous verification in spite of its unreasonableness, since it seemed to violate everything we knew about the interference of light.

Millikan is clearly referring to the problem of explaining the wave phenomena of light on the basis of the corpuscular picture of light.

*

A lack of explanations for the wave phenomena of light using corpuscular light dated back to Newton's time. Newton's laws of motion for corpuscles of light could not explain these wave phenomena.

*

Einstein's conclusion that light travels in space as a parcel of energy $\epsilon = h\nu$ was very radical. In the history of physical science, it met with great resistance for quite a long time from even the leading physicists of the time, like Lorentz, Planck, and Niels Henrik David Bohr (1885–1962).

The problem was that most phenomena of light were explainable by assuming that light is a wave propagating as a disturbance of the electromagnetic field and therefore it could not consist of particles. That is, if light consists of particles or parcels of energy, then how do we explain the wave phenomena: interference, diffraction, and polarization?

This was the reason for the widespread impression that the corpuscular picture of light, that it is a parcel of energy, cannot be the true nature of light. This impression was not without reason. There was hardly any convincing evidence in favor of the light quantum, except only as Einstein's explanation of the photoelectric effect. On the other hand, Maxwell's wave theory of light had accumulated great experimental support.

With insights into this attitude of scientists, Abraham Pais wrote, in 1984:

> Physicists—good physicists—enjoy scientific speculation in private but tend to frown upon it when done in public. They are conservative revolutionaries, resisting innovation as long as possible and at all intellectual cost, but embracing it when the evidence is incontrovertible. If they do not, physics tends to pass them by.

In 1909, Einstein wrote:

> It is my opinion that the next phase in the development of theoretical physics will bring us a theory of light that can be interpreted as a kind of fusion of the wave and the emission theory.... [The] wave structure and [the] quantum structure are not to be considered as mutually incompatible. (See Section 1.7: Emission Origin for a Wave of Quanta.)

When Einstein wrote this in 1909, he was alone in taking the light quantum hypothesis seriously. Neither Planck nor other leading physicists of the time supported Einstein's notion of a light quantum. Notably, Einstein could easily have concluded that the light quantum of energy $\epsilon = h\nu$ has a momentum $\epsilon/c = h\nu/c$. However, he did not at all refer to the momentum of the light quantum. He only wrote that "as a consequence of fluctuations in the radiation pressure there would act on our plate only such momenta as are represented by the first term of our formula."

Nevertheless, the concept of a particle of light was his intention, for he also wrote, "the electromagnetic fields of light are linked to singular points similar to the occurrence of electrostatic fields in the theory of electrons." It seems that the question of light being particulate while also showing the wave phenomena always puzzled Einstein. This is the real puzzle, indeed, of the wave-particle duality of light radiation.

*

It was Johannes Stark (1874–1957) who, only a few months after attending the Salzburg meeting where Einstein had discussed this in 1909, explicitly introduced the momentum P_γ for a light quantum and wrote the equation

$$P_\gamma = (h\nu/c^2)c.$$

It is with this association that most physicists began associating a nonzero momentum to the light quantum.

*

Opposition to Einstein's 1905 hypothesis, that light consists of quanta or parcels of energy, had begun to disappear slowly.

On the basis of Rutherford's results of 1911 and his planetary model of atom, which imagined a nucleus (of protons and neutrons) around which an electron revolves in an orbit, Bohr provided, in 1913, explanations of atomic spectra by constructing a model of an atom.

However, this accelerated electron was assumed, flouting Maxwell's theory of electromagnetism, not to emit radiation in certain permissible orbits. Bohr related the change in the energy of atomic electron jumping from a higher-energy orbit to a lower-energy orbit with energy $\epsilon = h\nu$ of radiation it would emit. Bohr's atomic model reasonably accounted for the atomic spectra and explained the origins of Balmer's formula of 1885, Rydberg's constant, etc. This lent credence to Einstein's hypothesis of a light quantum.

*

As per Maxwell's theory, an accelerated charge radiates energy to electromagnetic waves. How does it explain the formation of atomic spectra, then? Atomic spectra cannot be explained by Maxwell's electromagnetism. This was slowly getting recognized during this era.

Atomic stability had always fascinated Bohr. We have already discussed the stability of an atom and issues related to atomic stability, and those issues were actually raised first by Bohr. He had said to Heisenberg:

> My starting point was not at all the idea that an atom is a small-scale planetary system and as such governed by the laws of astronomy, I never took things as literally as that. My starting point was rather the stability of matter, a pure miracle when considered from the standpoint of classical physics.

Bohr realized that Maxwell's classical theory of electromagnetism predicted unstable nuclear atoms. Any electron of a nuclear atom would always accelerate due to the Coulomb force of attraction between it and the central nucleus. An accelerated electron would emit electromagnetic waves and would lose energy to radiated waves. The electron would keep moving closer to the nucleus with an ever-increasing velocity, while also emitting electromagnetic waves carrying energy away. The nuclear atom would be unstable. This atomic instability is not observed in nature: have you ever seen matter emitting radiation of ever-increasing energy on its own?

The instability of the atom implied here is a direct consequence of Maxwell's theory of electromagnetism and not of the planetary model of the nuclear atom. That is to say, only Maxwell's theory of electromagnetism states that an accelerated electron radiates away energy to electromagnetic waves, but not the planetary picture of the atom. Thus, if an atom in nature is stable, then this result of Maxwell's theory need not hold in certain atomic situations, although it holds in many other situations.

Bohr then postulated that the electron can move in certain circular orbits without emitting electromagnetic radiation.

*

To any unsuspecting mind, this may not strike as any revolutionary step. So, we remind the reader that Maxwell's theory had already gathered strong experimental support.

First, all the results of electrostatics and magnetostatics are contained within the equations of Maxwell's theory, which also includes Ampere's and Biot–Savart's laws and results obtainable from such laws. Furthermore, Hertz had experimentally produced an electromagnetic wave at one place and had also detected it at another place in the laboratory. This, too, was as per the prediction of Maxwell's theory of electromagnetic radiation.

Due to overwhelming experimental support existing for it, not even Planck was ready to question Maxwell's theory of electromagnetism, in particular the part dealing with freely propagating electromagnetic waves. In a letter to Einstein written on July 6, 1907, Planck wrote disapprovingly of the light quantum, "I am not seeking the meaning of the quantum of action (light-quantum) in the vacuum but rather in places where absorption and emission occur, and I assume that what happens in the vacuum is rigorously described by Maxwell's equations."

In 1909, he again wrote, "I believe one should first try to move the whole difficulty of the quantum theory to the domain of the interaction between matter and radiation."

Even Lorentz did not want Maxwell's theory for the vacuum fields to be altered in any manner whatsoever. He, too, believed that Maxwell's theory of free radiation provided no room for modifications, in contrast with the less understood interactions between matter and radiation.

*

Historically, Bohr first realized that Maxwell's theory implies an unstable atom—that an atom would collapse to a point in no time, contrary to its observed stability. An immediate conclusion is that not all is completely right about Maxwell's theory.

Bohr's aforementioned postulate indicated that Maxwell's theory needed modifications, and that was hardly acceptable to many of the leading physicists of his time.

In 1913, Bohr provided explanations of atomic spectra using the simple planetary model of the atom, Rutherford's atomic model. But the accelerated electron of the nuclear atom was assumed by Bohr, completely against Maxwell's theory of electromagnetism, not to be emitting radiation in certain permissible atomic orbits.

Furthermore, Bohr related the change in the energy of an atomic electron jumping from a higher-energy orbit to a lower-energy orbit with the energy of the quantum of radiation it would emit. A sharp spectral line would then be seen when many atoms underwent such electronic transition from a higher- to a lower-energy orbit. Sharp spectral lines were indeed seen much prior to the development of the nuclear atom.

Bohr's atomic model reasonably accounted for the atomic spectra and explained the origins of experimentally established Balmer's formula of 1885, Rydberg's constant, etc.

*

Arnold Sommerfeld (1868–1951) developed a (planetary) model of an atom by considering special relativistic effects and by postulating elliptical paths of electrons moving around the atomic nucleus.

*

Einstein's light quantum hypothesis was then not so unreasonable a proposition as it had appeared earlier. Maxwell's theory lost its sacrosanct status eventually.

In 1950, Einstein admiringly wrote:

> That this insecure and contradictory foundation [of physics in the years from 1910 to 1920] was sufficient to enable a man of Bohr's unique instinct and tact to discover the major laws of the spectral lines and of the electron shells of the atoms together with their significance for chemistry appeared to me like a miracle - and appears to me as a miracle even today.

As was seen before, Bohr had reached his atomic model intuitively rather than by basing it on theoretically sound premises.

*

Werner Heisenberg (1901–1976) recalls:

> We had all of us learned Bohr's theory from Sommerfeld, and knew what it was about, but it all sounded quite different from Bohr's own lips. We could clearly sense that he had reached his results not so much by calculation and demonstration as by intuition and inspiration, and that he found it difficult to justify his findings.

*

Bohr's works had inspired Einstein to study atomic gases in equilibrium with a radiation bath. He noticed, in 1916, that this equilibrium requires excited atoms of the gas to spontaneously emit an energy quantum. The spontaneous emission of the light quantum is not a result of the interatomic collisions or of the interactions with another light quantum. It is as if some processes internal to the atom cause the emission of a light quantum. In the absence of detailed knowledge of these internal processes, we could assume it to be a random process, and in this manner, Einstein could estimate the chances of this happening.

With the "stimulated" emission of the quantum, that is, the one resulting from processes in which other quanta are involved apart from the atom in question, we obtain the total chances of the emission. On combining this with the chances of the absorption of light quantum, we can write an equation of a detailed balance of all the involved processes, maintaining the equilibrium of the (atomic) gas with the radiation bath at a given temperature.

The distribution of the energies of light quanta is given by Planck's formula, and that of the gas particles is given by the Maxwell–Boltzmann velocity distribution formula when the gas and radiation are in equilibrium.

Einstein essentially provided a definite link between Bohr's atomic model and Planck's formula. Einstein's works underlie the modern theory of lasers (short for *light amplification by stimulated emission of radiation*), which are quite commonly in use now. He was naturally forced to conclude also that the picture of the directed character of radiation applies to spontaneous emission of radiation, too, and that the molecule would suffer a momentum recoil of $p = h\nu/c$ in the direction opposite to that of radiation.

However, the problem was that the direction of the recoil could not be predicted. Within Einstein's theory, it could not be answered as to when the light quantum is spontaneously emitted and as to what decides in which direction it will go. Einstein then stressed that "it is a weakness of the theory that it leaves time and direction of elementary processes to chance."

He drew attention to Rutherford's theory of radioactivity proposed in 1900: the relation $\delta N = -\Lambda N\ \delta t$, where N denotes the number of disintegrating atoms and Λ is the decay constant. In Rutherford's theory also, no prediction of the time of disintegration was possible.

Einstein thus said about his theory that "it speaks in favour of the theory that the statistical law assumed for [spontaneous] emission is nothing but the Rutherford law of radioactive decay."

*

In 1923, Arthur H. Compton (1892–1962) and Peter Debye (1884–1966) independently derived the theory for scattering a light quantum off an electron. Compton also performed an experiment of scattering (K-lines) X-rays (from molybdenum) off the graphite plate to find that the theoretical relation is satisfied by the experimental data indeed. He thus asserted that "the experimental support of the theory indicates very convincingly that a radiation quantum carries with it directed momentum as well as energy."

In view of these results, completely perplexing then was the demonstration, in 1924, by Bose that Planck's radiation formula can be obtained with light as consisting of inertialess particles (quanta), using statistical techniques different from those of Maxwell and Boltzmann. In particular, this derivation assumes that the particles of light are massless, indistinguishable, and with two internal states (of polarization) and that their number is not conserved but obeys the statistical method of counting, now called Bose–Einstein statistics. Natural bodies obeying Bose–Einstein statistics are now called *bosons*.

*

Considering that probability implies incomplete information at our disposal, there should be some deterministic framework of concepts underlying these phenomena, meaning one within which the exact measurements of physical quantities are possible. Einstein's theory of radiation-matter interactions was not a deterministic theory of the involved physical phenomena.

Clearly, there was still the problem of explaining the wave phenomena displayed by light on the basis of its corpuscular picture. Neither Einstein, nor anyone else, could explain them on the basis of the corpuscular picture of light as parcels of energy. Until such satisfactory explanations were found, there seemed no option than to resort to statistical methods.

*

Noticeably, as pointed out by Pais, we always encounter in Einstein's writings the phraseology "Radiation behaves as if it consists of . . ." He was never to use any other way for describing the momentum associated with the light quantum, as remarks Pais.

A body of vanishing inertia ($m = 0$) should also have a vanishing momentum ($p = mc = 0$). Bose's statistical method, on assuming massless (and, hence, momentumless) light quanta, yields the correct Planck's law. Many experimental results are, however, explainable by assuming a nonzero momentum $P_\gamma = h\nu/c$ for the light quantum.

If a light quantum has zero inertia, then how can it have a nonzero momentum? How is it possible? Are these inconsistent results understandable? Was this what was troubling Einstein?

It is quite difficult to say conclusively what had really been troubling him in regard to the light quantum, but Einstein wrote much later, in 1951, in a letter to a friend of his childhood, Michele Angelo Besso (1873–1955), that "all these fifty years of pondering have not brought me any closer to answering the question: What are light quanta?"

*

Nevertheless, the quantum of radiation had emerged to be a strange body that has its energy as well as its momentum related to its frequency.

If the analogy of gases and radiation is complete, then we should have matter also behave similarly to radiation.

In 1924, in a genuinely daring step, Louis de Broglie (1892–1998) proposed in his doctoral thesis that matter must also show the same relationships as are shown by light with respect to its quantum and wave properties. This work implied that atoms as well as particles like electrons should show the wave phenomena like diffraction, interference, and polarization.

*

Einstein wrote to Lorentz:

> . . . de Broglie has undertaken a very interesting attempt to interpret the Bohr–Sommerfeld quantum rules. I believe it is a first feeble ray of light on this worst of our physics enigmas. I, too, have found something which speaks for his construction.

Einstein had re-examined his works of 1909 and 1917 on fluctuations of a gas around its thermal equilibrium. There existed two terms in his formula for fluctuations. The first term could be directly interpreted as being due to the fluctuations in the number of gas particles. Following de Broglie's works, he could then interpret the second term of the formula for fluctuations as being due to the wave properties of (atomic) gas particles! The same formula applies also to radiation.

*

Subsequently, Erwin Schrödinger (1887–1961), Heisenberg, Max Born (1882–1970), Bohr, Paul Adrien Maurice Dirac (1902–1993), and many others developed the theoretical framework of quantum theory. Quantum theory provided only the probability of events involving material bodies but did not talk about the material bodies themselves. Probabilistic methods of quantum theory have explained a variety of natural phenomena, like tunneling.

*

Einstein had reacted to successes of the earlier quantum ideas by saying, in 1912, that "the more successes the quantum theory has, the sillier it looks," for there was no coherent theoretical framework underlying them. It could not be called a proper theoretical framework. Even the developments following Schrödinger, Heisenberg, Bohr, Born, and others and various experimental successes of quantum theory did not make him change his attitude, which rather sharpened with time and with more successes of quantum theory, as Pais remarks.

*

Einstein did not accept the fusion of wave and particle concepts inherent in the (probabilistic) framework of quantum theory, for he believed that a deterministic framework necessarily underlies any probabilistic framework. He had begun searching for such a deterministic framework to underlie quantum theory. In 1927, he said that "what nature demands from us is not a quantum theory or a wave theory; rather, nature demands from us a synthesis of these two views which thus far has exceeded the mental powers of physicists."

To understand Einstein's vision, we need to consider conceptual difficulties associated with the concept of a quantum of radiation and the wave associated with it and that, too, from his point of view.

In what follows, we discuss these issues.

*

An atomic electron jumps in energy when it absorbs or emits a quantum of radiation. With change in its energy, it also makes a transition from one radius to another radius in an atom. But as per quantum theory, this transition in space has no associated path, that is, it does not exist in quantum theory, for its methods calculate only the probability. How do we reconcile this with our usual, day-to-day, experience?

1.6 The Mysterious Quantum

In 1901, Planck assumed that when matter absorbs or emits radiation, the energy of matter changes in discrete steps. This is the quantum jump in energy. Bohr's theory of the atom provided an atomic basis to Planck's assumption. It, however, involved the so-called quantum jump of an electron from an orbit of certain energy to an orbit of different energy.

The electromagnetic wave can impart energy to matter in a continuous manner, a result of Maxwell's theory of electromagnetism that had a wealth of experimental support in its favor. Planck's assumption had no such experimental support to begin with, except that this assumption explained the curve of the spectral energy density of black-body radiation. Even in this case there existed no theoretical compulsion for such an assumption. Planck's great intuition alone led us to the quantum.

Much mystery has also been attached in the literature to the quantum jump in energy by relating it to the issue of the quantum jump in space. This arises as follows.

We observe a ball rolling on the ground because of light received by us from it. Newton's laws of motion implicitly assume that the ball moves along a definite curve in the space, even if we do not observe it. Thus, in principle, the path of the motion of the ball exists in Newton's theory. Then we ask, does, in principle, the path of the motion of the electron as it moves from a radius r_m to a radius r_n exist in the atom?

*

Quantum theory, which developed at the hands of Bohr, Heisenberg, Schrödinger, Born, and others, provides only the probability that an electron is found at a certain location in space and at a certain instant of time. It does not tell us about the trajectory or the path of the motion of the electron when it jumps from radius r_m to radius r_n. In fact, it cannot also be assumed within this probabilistic formalism of quantum theory that the electron was at the radius r_m before it jumped to the radius r_n, as only probability can be given. In quantum theory, a quantum jump in energy is a quantum jump in space as well.

A probabilistic framework of quantum theory then led to long and involved debates, in particular, with Einstein, who believed that there is no quantum jump in space when an electron makes a quantum jump in energy. While exchanging currency, the amount of money changes by one rupee, but the currency paper has a path in space from one place to another. The quantum jump in the amount of money is unrelated to its motion in space, for the notions of money and motion of currency are unrelated. But this is not the case with an atomic electron, as per quantum theory.

*

As Pais narrates:

> It must have been around 1950, I was accompanying Einstein on a walk from the Institute of Advanced Study to his home, when he suddenly stopped, turned to me, and asked me if I really believed that the moon exists only if I look at it. The nature of our conversation was not particularly metaphysical. Rather, we were discussing the quantum theory, in particular what is doable and knowable in the sense of the physical observation.

*

Heisenberg also narrates a conversation with Einstein after the former delivered a lecture on the then new quantum theory in Berlin in 1926. After the lecture, Heisenberg recalls:

> As soon as we were indoors, he opened the conversation with a question that bore on the philosophical background of my recent work. "What you told us sounds extremely strange. You assume the existence of electrons inside the atom, and you are probably quite right to do so. But you refuse to consider their orbits, even though we can observe electron tracks in a cloud chamber. I should very much like to hear more about your reasons for making such strange assumptions."

"We cannot observe electron orbits inside the atom," Heisenberg had replied,

> but the radiation which an atom emits during discharges enables us to deduce the frequencies and corresponding amplitudes of its electrons. After all, even in the older physics wave numbers and amplitudes

> could be considered substitutes for electron orbits. Now, since a good theory must be based on directly observable magnitudes, I thought it more fitting to restrict myself to these, treating them, as it were, as representation of the electron orbits.

Einstein had then protested, "But, you don't seriously believe that none but observable magnitudes must go into a physical theory?"

Heisenberg answered:

> Isn't that precisely what you have done with relativity? After all, you did stress the fact that it is impermissible to speak of absolute time, simply because absolute time cannot be observed; only clock readings, be it in the moving reference systems or the system at rest, are relevant to the determination of time.

Einstein said:

> Probably I did use this kind of reasoning, but it is nonsense all the same. Perhaps, I could put it more diplomatically by saying that it may be heuristically useful to keep in mind what one has actually observed. But on principle, it is quite wrong to try founding a theory on observable magnitudes alone. In reality the very opposite happens. It is the theory which decides what we can observe.

Einstein continued:

> You must appreciate that observation is a very complicated process. The phenomenon under considerations produces certain events in our measuring apparatus. As a result, further processes take place in the apparatus, which eventually and by complicated paths produce sense impressions and help us to fix the effects in our consciousness. Along this whole path—from the phenomenon to its fixation in our consciousness—we must be able to tell how nature functions, must know the natural laws at least in practical terms, before we can claim to have observed anything at all. Only theory, that is, knowledge of natural laws, enables us to deduce the underlying phenomena from our sense impressions. When we claim that we can observe something new, we ought really to be saying that although we are about to formulate new natural laws that do not agree with the old ones, we nevertheless assume that the existing laws covering the whole path from the phenomenon to our consciousness – function in such a way that we can rely upon them and hence speak of observations.

*

Einstein searched for a theoretical framework in which an electron would move along a well-defined path, its actual observable nature notwithstanding. Quantum theory does not provide for a well-defined path for any transiting atomic electron. Einstein considered quantum theory to be incomplete thus, as it has a statistical basis. However, he also recognized quantum theory to be an important advance. (It may also be noted here that the mathematics of category theory did not exist during his time.)

*

The purpose of providing this discussion of the quantum jumps in energy and in space here is limited to bringing to the reader's attention some of the issues underlying quantum theory developed by Bohr, Schrödinger, Heisenberg, Born, and others.

*

Light consists of quanta, as the photoelectric effect implies. So on this basis, we should have an explanation for the wave phenomena of light. This is what we consider next.

Light is not a wave, but it consists of quanta. The quantum theory of light does not solve the puzzle of whether light is wave or particulate. Rather, it is based on it. Any solution of this puzzle must rest on showing how light quanta display the wave properties of the electromagnetic wave. We must then grasp how quanta behave like a wave.

1.7 Emission Origin for a Wave of Quanta

The atomic theory of matter established itself quite slowly. But with the establishment of the physical existence of the atom from various phenomena, like Brownian motion, it became obvious that the properties of different states of matter must be related to the atomic structure of matter. Although the atomic structure of matter developed rapidly, the details of the structure internal to an atom of matter were still being discovered.

Nevertheless, various properties of matter in different states could still be modeled on the basis of "simplifying assumptions" related to the nature of the atomic interactions.

A molecule of a monatomic gas, such as neon and helium, can be imagined to be a hard, rigid sphere of a tiny size with diameter ≈ 0.2 nm. The number of molecules of the gas is quite large, about 2.7 × 10^{19} in a cubic centimeter, at standard temperature and pressure (STP).

For our present purpose, each of these atoms is assumed to be electrically neutral, so we do not have any ions and electrons in this state. Clearly, the temperature of the gas of these atoms should then be such as to not result in ions and free electrons. Furthermore, we may also assume that the sphere does not have the structure of an electric dipole. (This assumption need not hold for real atoms. Nevertheless, at an appropriate temperature, the interactions of atomic dipoles can be neglected, and it will hold then.) Therefore, we allow only the collisions of these atomic spheres with each other and with the walls of the container of the gas of these atomic spheres.

Under the aforementioned assumptions, the momentum of any atom can change only on its collision with another atom or with the walls of the container of the gas.

For the properties of the gas, we now imagine that each of this large number of tiny hard or rigid spheres, molecules of the monatomic gas, is moving about in a random manner, with all the possible speeds being likely for it and all the possible directions also being likely for it. During the random state of their motion, these spheres then collide with each other and with the walls of their container or the vessel.

In an elastic collision of bodies, the momentum is transferred from one body to another, while also conserving the total energy. The postulate of the hardness or the rigidity of the atomic spheres now means that their collisions are elastic. Thus, we have assumed the atomic collisions to be elastic. (This assumption has limitations, and the collisions of real atoms need not be elastic at all temperatures. Nevertheless, we expect a certain temperature range within which the assumption of elastic atomic collisions would hold.) For the monatomic gas, only the kinetic energy of translational motion would then be exchanged by the colliding atoms.

Then under the aforementioned simplifying assumptions and within the range of temperatures implied by them, the average speed of a monatomic molecule of the gas will be seen to be of the order of a few kilometers per second. It will also be seen that the average distance traveled by a molecule of such a gas before its collision with another molecule is of the order of several hundred times the molecular diameter. Furthermore, the average separation between two molecules of such a gas at STP will be of the order of tens of times the molecular diameter.

Thus, the thermal state of this gas is characterized by the distribution of speeds of its molecules. It is one of the problems of the kinetic theory of gases to obtain this distribution theoretically and compare its predictions with the observed values of the pressure and temperature of a real gas. Such a distribution of molecular speeds was firstly obtained by Boltzmann and, independently, by Maxwell. It is called the Maxwell–Boltzmann distribution or the Maxwell distribution.

We have assumed the molecule of the gas to be rigid and its collision with another molecule to be elastic. So the total momentum will be conserved and the total energy will also be conserved in an intermolecular collision.

The analogy of radiation with gases played an important role in the development of the ideas related to quantum theory. Therefore we first need to consider certain important aspects of this analogy.

Now, if the walls of the container were at a temperature that was different than the temperature of the gas, there would be a transfer of thermal energy between the walls of the container and the gas. Consequently, on their colliding with any of the walls of the container, the molecules of the gas would receive energy from the walls or give energy to the walls. Since the gas is assumed to be in equilibrium in a thermal state, no change must occur in the parameters of its state. Thus, the walls of its container are also at the same temperature as the gas.

Furthermore, the wall is a collection of atoms or molecules bonded with each other. None of the molecules of the wall is free to leave its position and wander around, as is the case with a solid. Any of these molecules is then free only to vibrate about a mean position. So in a collision of a molecule of the gas with a molecule of the wall, the molecule of the wall can be considered, on average, to be not changing its kinetic energy of vibration. Therefore, on average, the molecule of the gas will not lose kinetic energy to or gain it from the molecule of the wall it is colliding with. In other words, the kinetic energy of the molecule of the gas can be assumed to be unchanged on its collision with the wall.

Thus, we will assume that the conservation principles of momentum and energy hold also for the collision of a molecule of the gas with the walls of its container. In other words, we have also assumed that the collision of any molecule with any of the walls of the container is also elastic.

*

Now, in a dynamic equilibrium situation, as is that of the gas in equilibrium, we expect the number of molecules moving in any direction to be equal, on average, to the number of molecules moving in a direction exactly opposite to it. Else, more gas will be found at one location than at another location in the container. Therefore, the total change in that (particular) direction of the momentum of all the molecules of the gas striking the unit area of the wall per unit time can be calculated.

Newton's third law of motion now requires that the gas must also exert a force on that unit area of the wall in the direction opposite to that of the motion of the gas molecules. This is nothing but the force due to pressure, according to its definition as the magnitude of the force per unit area, exerted by the gas molecules on the wall.

Our aim is to relate the mean kinetic energy of the molecules of the gas to its temperature. It is the beauty of this simple model of the kinetic theory of gases that this relation involves only the universal gas constant R and Avogadro's number N_A.

When molecules of a gas possess momentum, then only the pressure of the gas can be related to the momentum of its constituent molecules. It is crucial to recognize this.

Then if the momentum of the gas constituents is zero because their mass is zero, no relation of the pressure of the gas with the momentum of (the massless and momentumless) molecules is obtainable. In particular, this is the case for radiation enclosed in a container.

*

The constituents of radiant energy or radiation are believed to be massless, and massless bodies are momentumless as well. Such bodies are only packets or quanta of (nonkinetic) energy. There is no relationship of the pressure exerted by the gas of radiation with the momentum of its constituents. Only when a quantum of (nonkinetic) energy is absorbed by it that the energy of a molecule of the wall of the container (enclosing radiation) changes. So there is a change in its momentum, since its kinetic energy and momentum are related to each other. Such changes in the momenta of the wall molecules are then to be interpreted as the pressure of radiation on the walls.

Constituents of radiation were assumed to possess a nonzero momentum, contrary to their mass being zero. Use of these contradictory concepts is not any oversight but has origins in the deep conflict between the wave and quantum properties of radiation.

*

We recall here what Einstein had to say about such situations:

> Concepts which have proved to be useful in ordering things easily acquire such an authority over us that we forget their human origin and accept them as invariable. Then they become "necessities of thought," "given *a priori*," etc. The path of scientific progress is then, by such errors, barred for a long time.
>
> . . . It is therefore no useless game if we are practising to analyze current notions and to point out on what conditions their justification and usefulness depends, how they have grown especially from the data of experience. In this way, their exaggerated authority is broken. They are removed, if they cannot properly legitimate themselves; corrected, if their correspondence to the given things was too negligently established; replaced by others, if a new system can be developed that we prefer for good reasons. (*Phys. Zeitschr.*, **17**, p. 101, 1916)

The concept of the momentum of the quantum of radiation, a massless body, proved to be useful, but it got too negligently established as the conditions of its use do not justify or support it.

Recall Einstein's phraseology: "Radiation behaves as if it consists of . . ." He never used any other way for describing momentum, $p\gamma = h\nu/c$, associated with the light quantum. Perhaps he had in mind the conflict of the nonzero momentum of the light quantum with its mass being zero.

An inertialess or massless body cannot change its velocity with respect to any observer, for the notion of force as the cause of change in velocity is vacuous for such a body.

For radiation in a box, the light quantum cannot reverse the direction of its motion on its collision with the wall. The original light quantum moving along any direction needs to be annihilated, and a light quantum has to be created anew to travel along the changed direction of motion.

*

The conflict of a nonzero momentum of light quantum and its zero mass must be resolved while explaining its wave and quantum properties.

Recall what Einstein had said: "What nature demands from us is not a quantum theory or a wave theory; rather, nature demands from us a synthesis of these two views. . . ." Remarkable is this intuition, indeed.

Notwithstanding the successes of the (probabilistic) quantum theory, it does not resolve this conflict. It is not the synthesis of the wave and the quantum views that Einstein searched.

*

Causal synthesis of the wave and quantum views lies in the realization that the (oscillatory) motion of their emitter is the reason for the wavelike behavior of the flux of the emitted, massless and momentumless, radiation quanta. But within this causal synthesis, a single massless quantum has neither a wave nor any momentum associated with it. Therefore, it is improper to ascribe momentum $h\nu/c$ to a single quantum of light.

This emission mechanism of a wave of quanta then has the associated mental picture of the following kind, which applies also to bodies of nonzero mass, as are electrons, neutrons, etc. (Recall de Broglie's revolutionary hypothesis regarding their wave behavior.)

Thus, consider a pipe emitting a jet of water through its nozzle. Let a molecule of water move with some speed along a straight line after its emission at the nozzle. Let the speed of the molecule not change unless it collides with another object along its rectilinear path of motion. Now, water molecules get emitted from different points of the surface of the nozzle at the same instant of time. The rate of their emission determines then their flux at any location directly in front of the nozzle.

If we change the location of the nozzle, that would also change the flux of water passing at a location directly opposite the nozzle. Clearly, the oscillatory motion of the nozzle produces the oscillatory change in the flux of water passing any location. In this manner, oscillatory changes in the jet of water result because of the oscillatory changes in the location of the nozzle and we would witness a wave passing along the water jet.

This emission origin of the wavy behavior of the flux of water molecules is also applicable to considerations of the quanta, of radiation, or of matter like electrons and neutrons. Therefore, the wavy behavior of quanta of all kinds results because of the oscillatory changes in the location of their emitter.

Diffraction, interference, etc., can then be explained not only for radiation but also for electrons, neutrons, etc. The emission mechanism of a wave of quanta is then the causal synthesis of wave and quantum views.

Chapter 2

The Idea of Relativity

The theory of relativity searches for the laws of physics that hold for all observers, whatever may be the state of motion of the observer relative to others. It relates to ideas of Galileo and Newton.

2.1 Galileo and Newton

From everyday experiences, Aristotle of Athens in Greece noticed that a constant push or pull causes a constant motion of objects. For example, to cause the constant motion of a bullock cart, bullocks must exert a constant pull on the cart. If the bullocks stop pulling the cart, then the cart also stops moving. Aristotle therefore concluded that the motion of a body can only be produced when push or pull is applied to that body.

However, this conclusion reached by Aristotle is erroneous. During the seventeenth century, Galileo Galilee of Pisa in Italy performed experiments that unequivocally indicated that a body should be in uniform rectilinear motion even if no push or pull is applied to it. Galileo's conclusion, reported in a book titled *Dialogue (Fig. 2.1), shattered the strong belief in, the then more than a millennium old, conclusions by Aristotle.*

As he introduced for the first time the method of performing an experiment to verify or validate concepts and relations between

Subtlety in Relativity
Sanjay M. Wagh

ISBN 978-981-4774-44-4 (Hardcover), 978-1-351-23832-8 (eBook)
www.panstanford.com

concepts postulated about the physical world, Galileo is credited to be the father of experimental methods in physics.

Figure 2.1 Three wise men: Frontispiece of Galileo's book.

Galileo, perhaps, noticed that a body moving on a smooth surface, like that of ice, moves a larger distance as compared to one moving on the soil when both are pushed equally. He then designed smooth surfaces. When he made a body sliding down a smooth surface climb another smooth surface placed in its path, he found that body to be climbing to the very same height it was released from on the first surface.

With this observation, Galileo then devised an experiment in which a body sliding down a smooth incline climbs another smooth surface, whose inclination can be adjusted to be at different angles to the horizontal surface of the earth. He then found that for each angle of the incline of the surface the body climbed to the same height it was released from (Fig. 2.2).

Galileo then noticed that the lower is the angle of the incline made with the horizontal, the larger is the distance the body has to travel on that incline to climb to the same height it was released from.

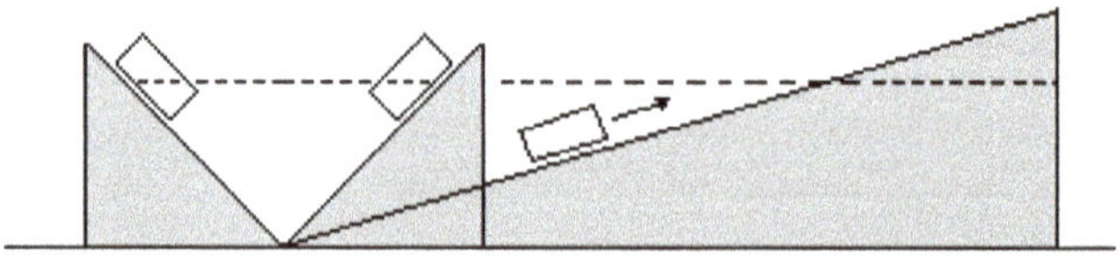

Figure 2.2 Schematic of Galileo's experiment of releasing a body on a smooth incline. Reprinted with permission from Ref. [22]. Copyright © 2013 by PHI Learning.

Galileo, a careful experimentalist, never stated it explicitly in his book that an inclined surface made horizontal can be of infinite extent so that a body can travel on it forever. Galileo was certainly aware that making any such surface would require efforts beyond possibilities within anyone's reach and critics would object to it. So, he, very simply, avoided making any statements about designing such an inclined surface.

But the reader of Galileo's book is forced to draw the conclusion that if the smooth inclined surface were made horizontal or parallel to the surface of the earth and were made of infinite extent, then the body descending down the first smooth inclined surface would reach distances far away from all the other bodies in the universe, where there would be no push or pull on the body, which would travel at a uniform velocity.

This, evidently, goes against Aristotle's doctrine.

In summary, a body that is far removed from all the other bodies in the universe and not acted upon by any push or pull of another body continues to move with a constant velocity along a straight line.

This above statement is called Newton's first law of motion because it was Newton who stated it in this form. (Galileo had not stated it in this form in his book. Galileo never concluded his Dialogue, certainly under pressure of the oppressive Roman Church. We recollect that Giordano Bruno was burned alive at the stake by the religious authorities of the Church, in 1600, for his saying that the earth revolves around the sun.)

Now, a Greek astronomer, Ptolemy imagined that bodies of the universe move around the earth in circular orbits, for we observe them to be moving from east to west. But this picture produced quite complicated patterns of motion for the planets.

It was then philosophized, by some religious authorities, that the planets had erred in performing tasks given to them by God, and therefore the Almighty had punished them by ordering them to be move along nonperfect and noncircular paths. Then stars move along perfect circular paths because they follow orders of the Lord. This view had then been favored by the religious authorities, for it suited their preaching. It was also a perfect tool in their hands to make people at large slavishly submit to their authority. It was this view of planetary motions that had then become the established doctrine or teaching.

Complicated motions of planets were increasingly becoming the subject of detailed study by many.

During the fifteenth century, Nicholas Copernicus suggested the heliocentric view of planetary motions. He imagined planets to be moving along circles around the sun. To explain the retrograde and prograde motions of planets, he further imagined them to be moving along small circular paths drawn about points of the circle centered on the Sun. Circular paths were considered by Copernicus and all to be perfect shapes.

The Copernican model opposed the then religious point of view. It was therefore criticized by many as being purely imaginary and false. Subsequently, in 1616, the Roman Church had banned the Copernican point of view from being discussed and published.

A wealthy landlord of his times, Tycho Brahe (1546–1601), too, could not believe Copernicus but for perfectly scientific reasons. The Copernican model, Brahe said, was based only on geometrical reasons. Certainly, more than just geometrical reasons are needed to think of this model as being valid. He had then built observatory to study motions of planets.

In 1610, Galileo built a telescope and had observed Jupiter's satellites, the phases of Venus, the mountains on the moon, etc. He noticed that the satellites of Jupiter revolve around it in circular orbits. This suggested to him that the Copernican view can be of merit.

From his experiments, Galileo had already concluded that a body, far removed from all the other bodies and not acted upon by pull or push of another body, would continue to move with a constant velocity, constant in speed and constant in direction, along a straight line.

A natural question for Galileo was to explain the cause for the deviation of planetary bodies from a straight line motion. He then faced these questions: Why does a body, of its own, not deviate from its uniform rectilinear motion (when it is far removed from all the other bodies in the universe and not acted upon by pull or push of other bodies)? Why does a body not follow any path whatsoever? What causes the deviation of a body from its uniform rectilinear motion, as is the case with planets, then?

The above questions amount to saying that a body has a tendency to oppose any change in its state of uniform rectilinear motion.

Galileo thus conceptualized inertia of a body: an intrinsic attribute of a body by virtue of which it opposes any change in its state of uniform rectilinear motion. Galileo could then sense a new system of philosophy about nature to be emerging out of the Copernican system or model of the solar system.

Subsequently, Johannes Kepler (1571–1630), hired by Brahe for his observatory as a mathematician, formulated the laws of planetary motions as definite mathematical relations. Kepler's laws of planetary motion are:

- Planets, including the earth, revolve around the sun in elliptical paths, with the sun being at one of the two foci of the ellipses.
- A line drawn from the sun to the planet sweeps out equal areas in equal time.
- If T is the time it takes for a planet to revolve around the sun once and ℓ is the length of the major axis of the ellipse in which the planet moves around the sun, then $T \propto \ell^{3/2}$.

Kepler's laws describe only the observationally deduced paths of planets in relation to the heliocentric frame of reference.

Kepler's laws of planetary motion were in conformity with the Copernican model of the solar system.

Prudent men of the world over had begun seeing the merits of the Copernican model.

Newton provided a mathematical foundation to Galileo's concepts and also expressed the laws of motion implicit within Galileo's book in mathematical ways. While doing so, Newton also added his own conceptions. One of these important such conceptions is gravitational mass.

Newton's first law of motion does not tell us anything about the nature of the pull or push acting on a body. However, it does tell us the conditions under which we can consider the pull or push to be acting on a body. Pull or push can then be considered to be acting on a body when its motion is not uniform rectilinear motion, that is, not of constant velocity (in magnitude and in direction both).

Not Galileo but Newton introduced the concept of force. Newton imagined the earth exerting a pull on bodies, and thus unsupported bodies fall to the earth. Galileo's point of view remains unknown.

Newton had very clearly recognized that the first law of motion does not tell us anything about the nature of the pull or push acting on bodies. He had understood that the first law of motion only tells us conditions under which we can consider a pull or push to be acting on a body: when the motion of a body is not uniform rectilinear motion. (In this context, it needs to be emphasized that the concept of motion is relative to the observer body. Then, the observer of the ball will see the motion of the person in the elevator to be accelerated and would conclude that a pull or push is acting on that person.)

Newton needed to express deviation from uniform rectilinear motion in terms of a suitable concept that would provide for the cause underlying the observed motions of planetary bodies of the solar system. He used Kepler's laws to deduce the nature of the pull acting on planetary bodies.

From Kepler's laws of planetary motions within the solar system, Newton could conclude that the pull resulting in the deviation of the planetary body from the state of uniform rectilinear motion must be directed toward the sun. He therefore attributed to the sun an appropriate value of the gravitational mass and to the planet another suitable value of the gravitational mass and postulated or assumed that there exists a pull of the sun on the planet proportional to the product of these masses.

From Kepler's laws again, Newton also concluded that the assumed pull varies inversely with the square of the distance between the sun and the planet. This is Newton's inverse-square gravitation.

Newton's inverse-square gravitation is only an assumption about the nature of the pull that is causing the deviation of a planetary body from the state of uniform rectilinear motion. This assumption says that a body like the sun acts on the planetary body situated at a distance and this force is therefore known as the action-at-a-distance force.

As the action at a distance is an assumption, no explanation can be found within the Newtonian framework of concepts for why bodies exert such a pull on other bodies.

Interestingly, Newton himself felt the need to replace the assumption of the action-at-a-distance nature of the aforementioned pull.

If we change the position of a body, then the effects of this change will have to be instantaneously felt by all the other bodies in the universe when the pull is of the action-at-a-distance type. This is quite contrary to what we observe in nature, for the effect of moving one body reaches another body situated at a distant location only after the lapse of some time.

After his general critique of Newtonian mechanics, Einstein admiringly writes: Newton, forgive me; you found the only way which, in your age, was just about possible for a man of highest thought and creative power. The concepts, which you created, are even today still guiding our thinking in physics, although we now know that they will have to be replaced by others farther removed from the sphere of immediate experience, if we aim at a profounder understanding of relationships.

Both Newton and Galileo had no reason to consider bodies of vanishing inertia, for they imagined nonzero inertia to be an intrinsic physical property of every natural body. But if a body has zero inertia, then Newton's laws of motion are not usable for it. Thus, interactions of inertialess bodies among themselves and with bodies having inertia cannot be analyzed using Newton's laws of motion. That is to say, light does not obey Newton's laws of motion if a light body has zero mass.

The concept of gravitational mass is an artifact of, that is, it arises out of, the assumption of the action-at-a-distance character for the force. However, no observation or experimental result compels us to assume such a character for the force. The concept of the gravitational mass of the body is therefore not to be considered as a fundamental physical concept.

The concept of inertia of a body is not based on any assumptions, but it is a deduction from Galileo's experiments. Therefore, in contrast to the notion of the gravitational mass of a body, its inertia is a fundamental physical concept.

Inertia was imagined by both Galileo and Newton as an intrinsic attribute of a physical body. But can we understand the origins of inertia? Below is such a thinkable possibility.

Imagine a crowded place. Let everyone of the crowd be at rest, initially at time $t = 0$. Now, introduce within this crowd a person A with nonzero velocity. Person A will collide with the nearest one in its path, and that will change the momenta of them both when the two are in contact. In this manner, changes to motion(s) take place. The directions and magnitudes of momentum (mass times velocity) of person A change, while he or she heads in a specific direction on the average.

The motion of the person A is then resisted by persons who come into contact with person A. Effectively, we may then think of one force that is an average of all the contact forces encountered by person A. Similarly, we may also effectively think of an acceleration being the average of all accelerations of person A. Then, a proportionality factor of the force and the acceleration is the inertia of the person. Denoting inertia by *m*, we may, as Newton's second law of motion, write $\vec{F} = m\vec{a}$. We have therefore explained inertia using the contact forces.

Historically, H. A. Lorentz had proposed such an explanation of inertia. However, this idea has not been worked out in detail.

No such explanation of inertia is possible on the basis of only the action-at-a-distance forces. The origin of inertia of a body cannot be explained with the assumption of action-at-a-distance forces.

Within the sphere of our usual experiences, the inertia of a body does not appear to depend on its velocity. But experiments can be performed with microscopic bodies like ions, electrons, protons, etc., by accelerating them using electric force. It was then found that the larger is the velocity of the microscopic body, the smaller is its acceleration for the same electric force. In particular, this is prominent when its velocity is $0.9c$, where $c = 3 \times 10^8$ m/s is the speed of light in vacuum.

Results of such experiments get interpreted as establishing that the produced acceleration is not proportional to force alone and depends on the velocity of the particle. We then understand these results by imagining that the inertia of a body increases with an increase in its velocity.

Galileo developed essential concepts underlying the mechanics of bodies, and Newton put those concepts on a firm mathematical footing.

Later developments, almost till the advent of the twentieth century, add only little of fundamental significance.

The word relativity became known to the world only after Einstein's revolutionary thinking of 1905. But Newton's theory is based also on four dimensions, three of the space and one for time. Newton's is also a restricted theory of relativity, as is Einstein's special theory of relativity.

2.2 What Is Relativity?

The word relativity is actually a misnomer for what we are going to describe. The aim of the theory of relativity is to determine the laws that do hold relative to every observer.

To grasp this aim, consider a tall building and a child playing with a soccer ball in a field in front of that building. Then, the child, as one observer, watches the motion of the ball.

Now, consider that a person is moving down that building by an elevator and is able to see the outside world. This person also watches the motion of the ball the child is playing with. This person is then another observer of the ball, the same object.

The aim of the theory of relativity is to determine the laws of the motion of the ball, which do not depend on who is observing, the child or the person in the elevator.

We may replace the ball in this example with any physical body. Then, the theory of relativity looks for the laws about changes of physical characteristics of a body that hold for all observers.

Therefore, the framework of a theory of relativity is that of a theory of everything of the physical world.

The direction of the motion, speed, acceleration, and other such features of the ball would just be different for the child than for the person in the elevator. How can we hope to find the same laws of the motion of the ball for both of them as observers?

To the unsuspecting mind, the aim of the theory of relativity would appear to be, simply, impossible to achieve!

Then, to understand the aim of the theory of relativity, we note that it has the following background.

Recall that we write for the position of a body in terms of the (measurable) coordinates relative to a reference body S (say, the child); for velocity relative to S, a change in each of these coordinates with time; for acceleration relative to S, a change in each component of velocity with time; etc.

Similarly, relative to another reference body S′, say, the person in the elevator, we write the position of the ball in terms of different (measurable) coordinates; the velocity of the ball as changes to these coordinates with time measured by S′; acceleration; the acceleration as changes to components of the velocity with time of S′, etc.

In other words, changes to the coordinates of the ball with time embody the entire information about its motion relative to the observer, with time also being measured by the same observer.

The observed path of motion of the ball is a curve traced by the positions of the ball relative to its observer.

One purpose of physics is to predict the path of the ball on the basis of minimal information at the disposal of the observer.

The issue is then, equivalently, that of determining quantities using which the mathematical form of the curve, as seen by the observer, can be obtained the minimal information.

Mathematical notions of a tangent to a curve and how the tangent varies along the curve are the mathematical quantities evidently useful to this purpose of physics.

The instantaneous velocity of the body (the rate of change of its distance from the observer with time at an instant) is tangent to its path; so also is the instantaneous momentum of the body (defined appropriately as a quantity characterizing the motion of that body) tangent to its path.

With insights of this kind, Newton developed corresponding mathematical techniques and realized that he needed to specify only the way the tangent varies along the path of a body to predict its motion, knowing its momentum $\vec{p}$ at any one instant. He wrote that force $\vec{F}$ is the way the tangential momentum varies along the path with time, that is, $\vec{F} = d\vec{p}/dt$. Then, on specifying the force, we can predict the path of the motion of a body!

With deep intuition, Newton used Kepler's laws of planetary motion to deduce that for a planet of the solar system, the momentum of the planet, a tangent to its path, varies in a specific (mathematical) manner. This is how he discovered the inverse-square form for gravitation.

Newton had suggested a complete scheme of quantities using which the path of the motion of a body can be predicted!

Max von Laue, admiring Newton's treatise Principia, wrote: In this treatise we find the two pronouncements that the rate of change of (linear) momentum of a mass point per unit time equals the force acting on it, and that the forces between two mass points are equal and opposite (equality of action and reaction). It follows immediately that the interaction of an arbitrary number of mass points never changes their total momentum, but that this total momentum is constant for any system not subject to external forces. Newton calculates the total momentum as a vector quantity, by adding vectorially the individual momenta. His formulation of the first law above appears almost prophetic: He equates the force not to the product of mass and acceleration, but to the rate of change of momentum, even though both formulations are equivalent if the momentum is assumed to be the product *mv*.

Furthermore, on relating the coordinates relative to S to those S′,we can relate the motion of the ball relative to the observer S to that relative to the observer S′.

The mathematical relation of the coordinates relative to S determining the coordinates relative to S′ is called a coordinate transformation from the coordinates of S to those of S′.

For uniform relative motion of observers S and S′, the involved coordinate transformation is, then, linear in coordinates. (No squares or higher powers of coordinates appear in the transformation.)

Notably, Newton's equation, that force $\vec{F}$ is equal to the rate $d\vec{p}/dt$ of change of momentum with time, retains its form in the coordinate systems of the two observers when the two are in uniform relative motion, when it has this form in relation to one of the observers, importantly.

We say that Newton's equation is invariant under linear or Galilean coordinate transformations.

Now, when the relative motion of the two observers is not uniform, but involving acceleration, then Newton's equation, of the motion of a body the two observers are observing, acquires, with respect to one of them, certain terms that are to be interpreted as forces acting on that body.

But these forces do not arise due to the interaction of a body with other bodies. We say that these forces do not possess causal origin in a physical body. Therefore, these forces are to be considered as fictitious forces. In other words, these forces are arising due to the nonuniform relative motion of the two observers.

We then say that Newton's equation is *not* invariant under general coordinate transformations, as the motion with acceleration is more general than uniform motion.

The question then arises, how do we decide which of the involved observers has nonuniform motion? This is the question Newton had faced while proposing his theory of mechanics.

Notice in this context that a curve transforms to another curve under a general transformation of coordinates. The tangent and how the tangent changes along the curve are not notions that generally remain invariant. The characterization of a curve in terms of the tangent and how it varies along that curve has these features.

We may now consider that all the paths of a body for which Newton's second law of motion, that is, the equation $\vec{F} = d\vec{p}/dt$, holds with force $\vec{F}$ possessing causal origin in some or the other physical body, define one type of curves, paths of its motion, in the space. For any other curve in space, fictitious forces then arise in this equation, and these curves can thus be said to define other types of curve in the space.

To answer this question we have to tell using some principle as to how one type of curves is preferable to other type of curves in the space. Notably, such a principle must be based only on the notion of the tangent to the curve and how it varies along the curve, since Newton's laws of motion use only these concepts.

But this is circuitous as the tangent to the curve and how it varies along the curve are to be characterized using the same notions. Therefore, we need to involve something beyond these notions to settle the above issue. This is what Newton did when he proposed Absolute Space as that something relative to which acceleration is to be defined.

However, this is unsatisfactory as the Absolute Space has no physical significance as a physical body. Newton was aware of this difficulty with his resolution of the question that he himself had raised.

Within Newton's scheme, no satisfactory answer can be found to the above question. Many, including Newton, felt uncomfortable with fictitious forces arising in Newton's formulation of mechanics, because force should have a causal origin in a physical body.

We will not discuss this issue further, because it only shows that the characterization of path in terms of tangent and how it changes along the path has this undesirable feature. We may then think of some different mathematical characterization of the path.

We thus say that Newton's equation is not invariant under the general coordinate transformations; as the motion with acceleration is more general than the uniform motion.

The aforementioned are the origins of the idea of relativity. We say that Newton's is a theory of restricted relativity, since its laws retain their form under restricted coordinate transformations, the linear or Galilean transformations of coordinates.

Historically, we note that these were recognized only after Einstein had proposed his special theory of relativity.

Notably, Newton assumed that time difference measured by the observer S and that measured by the observer S′ are equal. In other words, time runs at the same rate for the two observers. Time running at the same rate for all observers is called as the Absolute Time, then.

Newton's theory is then a 4D description of the mechanics of physical bodies; three dimensions of the space of motion and one dimension of the (absolute) time.

Generalization of the following kind is now permissible for a Newtonian law of physics, which can be a statement other than Newton's laws of motion, in restricted Newtonian relativity, now.

Permissible physical law of Newtonian relativity is then a mathematical relation of quantities that retains its form under the restricted or the Galilean transformations of space coordinates.

Now, Einstein stated the principle of relativity as: If a coordinate system S is chosen so that physical laws hold good in their simplest form with respect to this frame then, the same laws also hold good in relation to other coordinate system S′ moving in uniform translation relatively to S. This is the special principle of relativity. Together with the principle of the constancy of the speed of light in vacuum, it leads to the special theory of relativity.

Einstein's special principle of relativity is the same as the principle of relativity of Newton's theory. The word special indicates that the principle is restricted to the case of uniform translational motion of observers and does not extend to their nonuniform motion.

Of definite importance now is the assumed constancy of the speed of light, c, for the observers. Application of the Newtonian law of addition of velocities leads to the result that the speed of light is $c - v$ if the observer is receding from and $c + v$ if the observer is approaching the source of light. In other words, the velocity of light should depend on the observer.

As seen before, many experiments—those of Fizeau, Michelson, Michelson–Morley, and others—can be interpreted to mean that the speed of light does not vary in the way predicted by Newtonian law of addition of velocities. That is to say, these experiments indicated that the speed of light in vacuum is the same for all observers.

Newton's theory needed modification, then!

Newton's laws do not apply to bodies of zero mass, but only to those of nonzero mass. The concept of force is vacuous for a body of zero mass. Hence, the Newtonian laws, in particular, the law of addition of velocities, cannot be applied to Light, if its body has zero mass. Newton's theory does not apply to Light, then. This was missed during that era, however.

Much literature is available on how Einstein embarked on a revolutionary path by postulating that time runs at different rates for observers in uniform relative motion while proposing the special theory of relativity. Einstein's notion of relative time had caused consternation with the scientific community, initially, and later with the public at large.

We thus avoid discussing Einstein's path, here, as our main purpose is to go beyond Einstein in our analysis.

For consistency of the restricted principle of relativity and the constancy of the speed of light for (inertial) observers, Einstein had introduced the concept of relative time in 1905.

But, is time relative, for sure?

2.3 Problem of the Constancy of the Speed of Light

The problem of the constancy of the speed of light in vacuum for different observers is then as given next.

Consider a source of light at a location S and its detector at a location A, both at rest relative to an observer O and separated by distance L as measured by that observer. Let the source of light at S emit light at time = 0 as measured by that observer. Let that light reach A and get detected at A at time $t = T$ as measured also by the same observer. Therefore, the speed of light as measured by that observer is $c = L/T$.

Now, consider another observer O′ moving with uniform velocity u along the line SA and in the direction of A. Newtonian law of addition of velocities implies, now, that the velocity of light with respect to observer O′ is $c - v$. Also, if the observer O′ were moving in the direction of S along the line SA, then the velocity of light would be $c + v$ relative to O′.

But we require the speed of light to be the same, having the value c, for both the observers, O and O′.

This is the problem of the constancy of the speed of light for different observers.

2.4 Einstein's Relative Time

During his era, Einstein proposed the best possible solution to the problem of the constancy of the speed of light in vacuum. His solution was based on Fitzgerald's idea of length contraction—the length L contracts to length $L' = L\sqrt{1-(u^2/c^2)}$ as measured by the observer O′ and time dilation, as used by Lorentz and that time T dilates to $T' = T/\sqrt{1-(u^2/c^2)}$, as measured by the observer O′. The length contraction and time dilation could then explain experimental results due to Michelson–Morley and others. This much was already known before Einstein.

In 1905, Einstein elevated observations of Fitzgerald and Lorentz to the status of coordinate transformations, the Lorentz transformations. He showed that Newton's law of addition of velocities gets modified under Lorentz transformations in such a way that the ratio of the speed of body with respect to the observer and the speed of light always lies between 0 and 1. Then, the constancy of the speed of light and the restricted principle of relativity are consistent.

But Einstein's time is relative (T and T' differ in value) in contrast to Newton's absolute time (T and T' have the same value).

2.5 Measuring the Speed of Light

We note that when Einstein proposed, in 1905, special theory of relativity using the Lorentz transformations, the quantum nature of light was being discovered. It was slowly being discovered that the mass of this quantum needs to be zero.

Newton's laws of motion are vacuous for understanding the motion of a body of zero mass. It had slinds that Newton's laws of motion cannot be applied to light if the mass of a light quantum is zero. No physical meaning can then be attached to the coordinates of light quantum, for we cannot make any measurement of these coordinates.

What do we mean by the speed of light, then?

To grasp the meaning of what we understand by the speed of light, consider a source S of light as in Fig. 2.3.

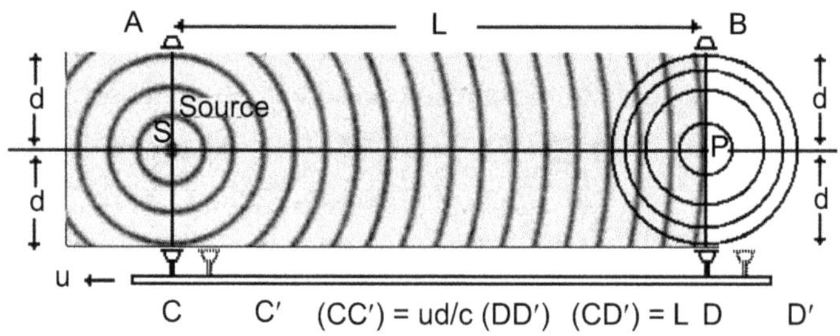

Figure 2.3 Measuring the velocity of light.

We have four detectors of light, at locations A, B, C, and D, and a reflector cum splitter at P, such that distance SP = AB = CD = L. Then, light from S takes time L/c to reach P, while it takes time d/c to reach A (and C), where d is the distance SA = SC.

The reflector at P sends light from S to detectors at B and D, with that light taking time d/c to reach them from P, since PB = PD = d.

Let the source S emit light at time $t = 0$ and let that light reach the detectors at A and C after time d/c. Light from S then reaches the reflector at P after time L/c, and on reflection at P the light reaches the detectors at B and D after time $L/c + d/c$.

Thus, the time difference between detection of light in detectors at A and B is L/c, and that between detection at C and D is also L/c.

Knowing distance L and the time difference(s) in the detection of light in the detectors, we measure the speed of light in such an experiment.

We have described the above experiment to emphasize the following: Recall that the path of light body cannot be changed as the concept of force is vacuous for a body of zero mass and thus it cannot be observed. Then, light that is reflected at location P is not the one traveling along SP, but the original light gets destroyed and new light gets created by the material of the reflector at P to travel along the paths PB and PD. The measurement of the speed of light in any experiment has the above features.

By the speed of light, we then mean the ratio of the distance between the emitter and the detector of light and the difference between the time of its detection and emission, even when we cannot observe light during its motion from emitter to detector.

2.6 Time Is Absolute

Now, we need to emphasize at this stage that the detectors at A and C detect the light simultaneously relative to the source S, and the same holds also for detectors at B and D.

Let us arrange the detectors at C and D to be on a platform and move that platform with uniform speed u relative to the source, as has been indicated in Fig. 2.3. Then, the simultaneity of detections of light at A and C is maintained relative to the source S, if the detector to be at C were at point C″, located at distance ud/c.

Similarly, for the simultaneity of detection of light with that at B, the detector, which is to be at location D, is needed to be at point D′, which is at distance ud/c from D.

Then, the distance from C′ to D′will be L, measured by the moving observer; and only then the speed of light as measured by that observer will be c, the same as that measured by the observer at rest with the source.

If the platform were moving with uniform acceleration, then the detector at point C′ will have to be located at distance $ad^2/2c^2$ from point C, with its motion starting from rest. It will also be the case with the location of the detector at point D′ from point D. Again, the distance from C′ to D′ = L, and only then the speed of light measured by the observer of the platform undergoing uniform acceleration will be c.

In general, for the simultaneity of detections at A and C, the detector (to be at C) will be located at point C′ which is at a suitable distance from C. Similar will be the case with the detector to be at D. The distance from C′ to D′ will be equal to L, notwithstanding the nature of the motion of the platform. It is only then that the speed of light measured by the observer on the moving platform will always be c.

As the events of detection of light at A and C and at B and D are arranged to be simultaneous in the rest frame of the source, the time difference of events at A and B must be the same as that of events at C and D, notwithstanding the nature of the motion of the platform with detectors. This means that time runs at the same rate for the observer of the platform and that of the rest frame of the source S of the experiment. That is, relative time is not implied by the constancy of the speed of light in vacuum.

Einstein's idea of relative time is untenable because the aforementioned discussion of the measurement of the speed of light shows that $T = T'$, that is, time runs at the same rate for all observers.

During his era, Einstein suggested the best possible path for the progress of our understanding of natural phenomena. His ideas of special and general relativity were the intellectual feats of a true genius, his attempts to conceptually grasp nature.

But it was generally missed during the era following Einstein's special relativity that the results of Newton's theory cannot be applied to light if it has zero mass. As seen here, the nonobservable nature of the path of a light quantum has the discussed consequences for the measurement of its speed. Thus, time cannot be relative, contrary to what Einstein presumed for his special and general relativity.

Nevertheless, we remain deeply grateful to Einstein, for he alone laid the initial steps for us to the right path. Recall Einstein's phraseology about radiation, recall his attempts to arrive at the synthesis of the wave and the quantum views of nature, and recall his letter to Besso.

> Einstein was first to arrive at the general principle of relativity. His preliminary attempt to formulate a theory based on this principle is also his theory of gravitation.

2.7 General Principle of Relativity

Now, the special principle of relativity is an extension of Newton's principle to incorporate the laws of light. It achieves this extension by acknowledging that in our day-to-day experiences, we use light to observe.

But it, too, is beset with problems of being restricted only to uniform motion between observers. There is therefore a definite theoretical need to extend the special principle of relativity.

Einstein extended the special principle of relativity on the basis of Mach's reasoning, as given next.

Mach's reasoning concerns the following situation: Consider two identical fluid bodies so far from each other and from other material bodies that only the self-gravity of each one needs to be considered. Let the distance between them be invariable, and in neither of them let there be internal motions with respect to each other. Also, let either body, as judged by an observer at rest relative to the other body, rotate with a constant angular velocity about the line joining them. This is, importantly, a verifiable relative motion of the two identical fluid bodies.

Now, using a surveyor's instruments let an observer at rest relative to each body take measurements of the surface of that body. Let the revealed surface of one be spherical and of the other be an ellipsoid of revolution.

The question then arises of the physical reason behind this difference in the shapes of these two bodies.

No answer is to be considered satisfactory unless the given reason is observable, for the law of causality has significance only when observable effects ultimately appear as causes and effects.

As seen before, Newton's theory (also special relativity) requires fictitious or pseudo forces to provide an answer to this issue. The reason given by these two theories is unsatisfactory as pseudo forces are unobservable.

Any cause within the system of these two bodies alone will not be sufficient. An internal cause will not be in conformity with the law of causality, because the two bodies are identical and only in relative rotation.

The only satisfactory answer is that the cause must be outside this physical system, and that must be referred to the real difference in motions of distant material bodies relative to each fluid body.

Then, the frame of reference of one fluid body is equivalent to that of the other for describing the motions of other bodies. Mach concluded that no observable significance can be attached to the cause of the difference in the shapes of the two bodies without this equivalence.

Laws of physics must then be such that they apply to systems of reference in any kind of motion (without the introduction of any fictitious forces). This is then the extended or general principle of relativity.

Tt is clear that even the reference frames must be constructed out of material bodies and motions of other material bodies must also affect the constructions of the reference frames. Therefore, the general principle of relativity also means that the laws of physics must be general enough so as to incorporate even these situations in their entirety.

Equally important to Einstein's thinking is the fact that physical time must undergo appropriate changes when the above is implemented. In particular, time must depend on acceleration and velocity both, in contrast to Newton's theory or the special theory of relativity.

Einstein connected the general principle of relativity with the observation that a possible uniform gravitation imparts the same acceleration to all bodies. This insight led Einstein to the equivalence principle.

Let K be a Galilean frame of reference relative to which a material body is moving with uniform rectilinear motion when far removed from other material bodies. Let K'be another frame of reference that is moving relatively to K in uniformly accelerated translation. Then relatively to K', the same material body would have an acceleration that is independent of its material content as well as of its physical state.

The observer at rest in frame K'can then raise the question of determining whether frame K' is really in an accelerated motion, that is, whether this is the only reason or the cause for the acceleration of bodies being independent of their material contents.

Now, let various bodies of differing material contents and of different inertias fall freely under the action of the earth's gravity after being released from the same distance above the ground and at the same instant of time. Galileo had, supposedly at the Leaning Tower of Pisa, observed that these bodies reach the ground at the same instant of time and had thereby concluded that these bodies fall with the same accelerations.

Hence, the decrement in distance between material bodies displaying only the phenomenon of gravitation is uniquely characterized by the fact that the acceleration experienced by material bodies occupying a sufficiently small region of space near another material body of a large spatial dimension is independent of their material content and their physical state. Here, the gravitational action of the larger material body can then be treated as being that of uniform gravitation.

Therefore, the answer to the aforementioned question raised by the observer at rest in the frame K' is in the negative, since there does exist an analogous situation involving the phenomenon of uniform gravitation in which material bodies can possess acceleration, independent of their material content and the physical state.

In other words, the observer at rest in the frame of reference K' can alternatively explain the observation of the "acceleration being independent of the physical state or the material content of bodies" on the basis of the phenomenon of uniform gravitation.

The mechanical behavior of material bodies relative to the frame K' is then the same as that in the frame K, supposedly being considered as special as per the special principle of relativity. We can therefore say that the two frames, K and K', are equivalent for the description of the facts under consideration. Clearly, we will then have to extend the special principle of relativity to incorporate the accelerated frames of reference.

Borrowing Einstein's words, this situation is suggestive of the fact that the systems K and K' may both, with equal right, be looked upon as stationary, that is, they have an equal title as systems of reference for the physical description of phenomena. (Note the word "suggestive".)

Now, the equivalence of inertial and gravitational masses of a material body refers to the equality of corresponding qualities of a body. But this is permissible only in a theory that assumes force as an external cause of motions of bodies. The concept of the gravitational mass is, however, irrelevant when the concept of force is abandoned, but not the concept of the inertia of a body.

What, then, is the status of the general principle of relativity in a theory that abandons the concept of force? Does it still hold?

Evidently, the general principle of relativity stands even when the concept of force is abandoned, for it only deals with the observable concept of the acceleration. It rests only on the observation that uniform gravity imparts the same acceleration to all the bodies. In other words, the concept of force is not at all essential to establish this fact.

Now, it is crucial to recognize that the equivalence principle establishes only the consistency of the phenomenon of gravitation with the general principle of relativity. The equivalence principle is not logically equivalent to the general principle of relativity.

As noted earlier, Einstein had, certainly, been quite careful to use the word "suggestive" in stating the relation of these different principles. He wrote:

> . . . in pursuing the general theory of relativity we shall be led to a theory of gravitation, since we are able to produce a gravitational field merely by changing the system of coordinates.

A theory of gravity will be improper without a proper general theory of relativity. From this discussion, it should be equally clear that the general principle of relativity can be reached from more than one vantage issue. Each such issue can then indicate only that some physical phenomenon related to that issue is consistent with this principle of relativity. The mutual consistency of the general principle of relativity and various physical conceptions then becomes the requirement of a satisfactory theory.

Therefore, the general theory of relativity, a theory explicitly based on the general principle of relativity, will not be just a theory of gravitation but, of necessity, also the theory of everything. In other words, its physical and mathematical framework must be applicable to all physical bodies. Also, the physical construction of the frames of reference is the primary requirement of a satisfactory theory based on the general principle of relativity—the satisfactory general theory of relativity.

Remarkable is Einstein's vision about the general principle of relativity, which is scattered over many of his writings done over a period of more than a few decades. He was never satisfied with the field equations of his theory of gravitation, which represents gravity by the curvature of the space-time geometry, for other forces of nature, in particular polar forces of electricity and magnetism, cannot be represented in this way.

Einstein had, in 1928, concluded that his equations of general relativity as proposed in 1916 were not a satisfactory formulation of the general principle of relativity. He therefore began his attempts at numerous formulations for a satisfactory unified field theory.

Wolfgang Pauli was then led to criticize, in a published review of these unified field theories, by sarcasm, that

> [Einstein's] never-failing inventiveness as well as his tenacious energy in the pursuit of [unification] guarantees us in recent years, on the average, one theory per annum . . . It is psychologically interesting that for some time the current theory is usually considered by its author to be the "definitive solution" . . .

Pauli had already demanded to know from Einstein, in a letter dated December 19, 1929, as to "what had become of the perihelion of Mercury, the bending of light, and the conservation laws of energy-momentum."

Einstein had no proper answers to these questions within his proposed formulation of unified field theory. However, he was not overly concerned about the issues raised by Pauli. In a letter on January 1, 1930, to W. Mayer he, however, complained, "Nearly all the colleagues react sourly to the theory because it puts again in doubt the earlier general relativity."

He attempted some satisfactory unified theory of all fields. In 1931, in a note he also admitted that his earlier attempts at a unified field theory constituted a wrong direction to follow.

A true genius, he was an open-minded thinker, always ready to dump his ideas if found unsatisfactory. He had said, "Anyone who has never made a mistake has never tried anything new."

> The general principle of relativity does not consider the structure of the frame of reference of the observer. It also remains silent about how a physical body is to be represented mathematically.

2.8 Universal Principle of Relativity

We have, now, contemplated on the general principle of relativity in sufficient detail to conclude that a satisfactory theory based on this principle will, necessarily, be a theory of everything.

We have also seen that the proper physical construction of the frames of reference, the physical coordination of the physical space, is also a requirement of a satisfactory general theory of relativity. Therefore, unless all the effects of physical phenomena on the frames of reference are properly incorporated within the framework of the theory, it cannot be considered to be based on the general principle of relativity.

A crucial issue is now of the manner in which we can account for effects of physical phenomena on the construction of reference frames.

This issue is that of a proper mathematical framework (accounting for these effects) alone and not that of the physical nature of these effects. The following discussion aims at clarifying this.

The purpose of a frame of reference is of ascertaining a location of space by assigning measurable coordinates to it. When the frame of reference gets affected by physical phenomenon, the assignment of measurable coordinates to locations gets affected also.

We may then demand that the genuine laws of physics be such as to allow for or incorporate the above situation. That is to say, the mathematical statement of the law of physics needs to be such as to hold irrespective of the assignment of measurable coordinates changing arbitrarily due to the frame of reference getting affected by *any* physical phenomenon.

This demand can be called as that of the principle of the free reference frame structure of the laws of physics.

Mach's logical arguments were used by Einstein to arrive at the general principle of relativity, which he had clearly recognized as only a point of view and not any specific principle or law of nature.

Then, any body is to be "equivalent as a frame of reference" with any other physical body, whatever may be its state of motion relative to any other body. Then, when a reference body is changed, the coordination of the space also changes as a result. That is, the coordinates of any location of the space are then different for the changed reference body. An arbitrary or general transformation of coordinates is then implied here.

Therefore, Einstein had concluded that the laws of physics must be such that they hold as equations having the same form under a general transformation of coordinates.

Einstein was the first to recognize that the genuine laws of physics must be generally covariant, that is, the laws of physics must retain their mathematical form (be invariant), under general transformations of coordinates. This is the principle of general covariance.

Recall that Einstein invoked the equivalence principle, which demands that the results of special relativity hold good locally. Invariance of the equations of special relativity under the Lorentz transformations could have played an important role for his further thinking in this context. Therefore, for Einstein, *time* was also one of the coordinates to be transformed in a general manner, as were the three spatial coordinates to be transformed, when the frame of reference is changed.

Einstein's field equations of his theory of gravitation were the result of his preliminary attempts to grasp the implications of the general covariance of physical laws, with his versions of the unified field theory being more of such (unsuccessful) attempts at a field theory.

Einstein's principle of general covariance is technically different than the principle of the free reference frame structure. The former principle needs to be consistent with the latter principle always. This consistency requirement arises in the following manner.

Newton's theory shows us that the simplest body is a point of the (3D) Euclidean space. It is assigned inertia, and Newton's laws then provide us with the means to describe its motion. To this end, we postulate a force acting on it, as the cause of its motion, relative to a Galilean frame of reference, another point in that space.

Our postulate of the force may require that we assign additional physical qualities (gravitational mass, electric charge, etc.) to the (point) body whose motion is being described.

To describe motion, we require coordination of the space, which is achieved using the Euclidean formula for the distance between two points of the 3D space. The (point) body describes a curve of the Euclidean space during its motion, and its motion is characterized by its velocity, momentum, acceleration, etc.

Physical bodies are not point-like, however; they possess extension in space. In other words, a physical body is thinkable as a subset of the points of the Euclidean space. And the distance between two physical bodies is distance between two subsets of the Euclidean space.

Newton's theory does recognize this issue; and shows that the motion of the extended physical body is representable as being that of its center of mass—a point at which the entire mass of the extended body can be considered to be concentrated.

As a consequence, we can continue to use the coordination of space by the Euclidean distance with impunity and replace a physical body by its center of mass, a point body, to study its motion using the Newtonian laws. In Newton's theory, the spin of the extended body is then the rotational motion of its constituent points about its center of mass.

Newtonian theory is an internally consistent framework of concepts, as Einstein had termed it.

In Newton's theory, coordination of the space by Euclidean distance is, in principle, also measurable using measuring rods.

Effects of physical phenomena on the frames of reference are accountable in Newton's theory in this manner in terms of changes any extended body being used as a reference frame.

So Newton's theory satisfies the principle of the free reference frame structure. However, as discussed earlier, the laws of Newton's theory are not covariant under general coordinate transformations.

Appealing is the way Newton's theory treats the issue of an extended body. Thus, we may search for the generally covariant laws, while maintaining some of the features of Newton's theory within a new framework of concepts constituting a generally covariant formalism.

However, a new aspect related to representation of a physical body emerges with quantum theory. It is of the representation of a body by the probability distribution, not as a point or an extended body. Schrödinger's (or Pauli's or Dirac's) equation of quantum theory represents a body as a probability distribution over the whole space.

For example, an electron gets represented by a spinor (a mathematical object). When free, an electron is likely to be found anywhere in the space, with the probability of its occurrence being equal for any two locations.

The addition of another electron (or another body interacting with the electron) to the space changes this probability distribution, in accordance with the laws of quantum theory.

The spin of an electron is an eigenvalue of the spin operator and can take on only discrete values: $+1/2$ or $-1/2$ in suitable units. (This situation is quite unlike the spin of an extended body in Newton's theory.) The eigenvector of the operator provides for the probability and has all the information about the electron.

In general, an observable quantity in quantum theory is an operator (in a definite mathematical sense) and its measurable value is an eigenvalue of that operator. The corresponding eigenvector provides for the probability of the measurement of that observable yielding that value.

Notably, the framework of the usual quantum theory is not generally covariant. That is, the equations of the usual quantum theory are not generally covariant equations.

These aspects of the representation of a physical body then indicate, in particular, that a body can be represented differently and the knowledge of the physical world can be obtained differently.

A question, however, arises: Which one of these different ways, formalisms, is more fundamental than the others?

Logically, one of these formalisms cannot be considered to be basic or fundamental than any another.

But a formalism can be obtained by generalizing common features of those that use a representation of a physical body. A generalized formalism will not depend on the representation of the physical body. A formalism independent of the representation of a physical body will be more fundamental than those depending on such a representation.

The question is, therefore, of finding the formalism that does not depend on the representation of a physical body. It will also be universal of applicability, for all the features of formalisms depending on representation of a physical body will be obtainable within it.

We have reached the premise of the universal principle of relativity at this point of our considerations. It states that physical laws are independent of the representation of physical body by any mathematical structure and are such that the principle of the free reference frame structure and the general principle of relativity both are satisfied, too.

What, then, is the mathematical formalism suitable for the universal principle of relativity? The following considerations now show that the most general mathematics of category theory is the basis to implement this most general physical principle mathematically.

Now, set theoretical aspects treat a point of 3D space as a *singleton* member of the collection of all points of the 3D, Euclidean, space. Thus, for set theory, Newton's theory treats a body as a singleton of the 3D Euclidean space.

Similarly, quantum theory represents a physical body by a singleton of the Hilbert space, in general.

We are therefore led to consider the representation of a physical body by a singleton of the collection of mathematical structures of any chosen type, whatsoever it may be. (It is, of course, open as to what physical picture we may associate with the chosen mathematical structures representing a physical body. But this issue is of no direct concern to us here.)

The motion of a physical body from one to another location in the space, as defined by these bodies only, is then to be looked upon as a transformation of the mathematical structure representing it.

Considerations of the aforementioned kind lead us to the framework of category theory as the mathematical basis of the universal principle of relativity. The most general physical theory therefore "meets" here the most general mathematical theory, for category theory is also the mathematics of mathematical theories.

> Light quantum cannot be observed during its motion, if it has zero mass or inertia. The quantum of light reveals itself only during its interaction with bodies of nonzero inertia.
>
> The quantum world has also revealed an important principle to us, Heisenberg's uncertainty principle.

2.9 A Light Quantum in Universal Relativity

When quantum theory developed, probability was not new to physics. Statistical techniques in physics were already developed by Maxwell, Boltzmann, Gibbs, and others. These techniques let us forgo the information on individual physical bodies and deal with their ensemble properties.

Information about an individual constituent of an ensemble is, however, obtainable in such considerations, in principle.

We refer to these statistical techniques to constitute an incomplete theory, which has an underlying complete theory in the form of

Newton's theory that provides, in principle, for the information that has been forgone by these (classical) statistical methods.

The question arises of the completeness of the theory based on the chosen representation of the physical body. Einstein held that if probability is the only knowledge associated with a representation, then the theory based on it is incomplete in the aforementioned sense. He believed that there should be another representation of a physical body underlying the probabilistic quantum theory, which provides for a complete theory.

Specifically, Einstein held the view that there must be another complete framework of concepts underlying quantum theory.

Quantum theory challenged Einstein's views as its methods assign probability to a single body, like an electron. These methods do not provide for the exact information of the location of electron in a measurement but only the probability for the value of location in any attempted such measurement.

A line was drawn between Einstein and other physicists (like Bohr, Heisenberg) over this issue. Einstein had attempted to show that quantum theory is incomplete in the sense already described, while others (Bohr, Heisenberg) argued against Einstein's views.

Heisenberg showed that the location and momentum of a body cannot be measured simultaneously to arbitrary precision—Heisenberg's uncertainty principle. After opposing it during the initial stages of the development of quantum theory, Einstein agreed to the correctness of this important principle of the quantum world. Nevertheless, he did not accept the probabilistic nature of quantum theory as being that of any final theory and searched for a deterministic framework to underlie quantum theory.

This interesting debate had not concluded within the lifetimes of these physicists. During this debate, Einstein put forward the ensemble interpretation of quantum theory, saying that the probability associated with a single body is to be understood on the basis of measurements performed on individual members of an ensemble of identical such bodies, that, too, under the same conditions of measurement.

To understand Einstein, consider an atom of hydrogen; and let us locate its electron in relation to its nucleus. To this end, we shine a light quantum on the atom so that the electron reflects it and we "see" where the reflected quantum emerges from within the atom to ascertain the location of that electron. One such reflection of light provides us one location of the electron in relation to the nucleus of the hydrogen atom.

Either by allowing the same hydrogen atom to settle to its original state or by taking another hydrogen atom for this reflection experiment, we repeat this measurement many times.

In this way, we obtain a picture of the locations of electron in the hydrogen atom. This picture will be that of dots, the locations of electrons, distributed around the atomic nucleus.

Dots will be distributed in a spherically symmetric way around the nucleus, as is the prediction of quantum theory. There will be dots close to the nucleus also, but much less in number than dots around the Bohr radius of the hydrogen atom. As we go further and further away from the nucleus, the number of dots will also go on decreasing.

This is the structure of the s shell of the electron in a hydrogen atom (Fig. 2.4).

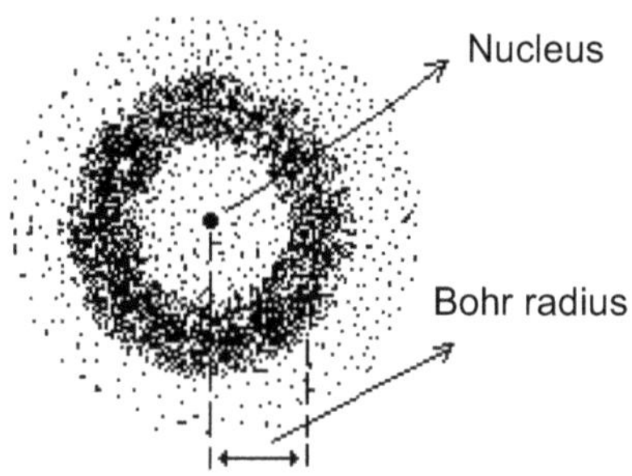

Figure 2.4 The s shell of a hydrogen atom.

Einstein's point of principle is that this picture will not be obtained in just one, single, reflection of the light quantum off an electron of the hydrogen atom. Only the superposition of many measurements of the location of the electron in the atom provides this picture.

When we observe it, the electron is seen only at one location within the space around the atomic nucleus. But we see it at a different location when we observe it next when the atom settles in the same state or when using another member of the ensemble.

We are considering an ensemble of hydrogen atoms and ascertaining the location of the electron in each member of this ensemble to arrive at the above picture, of location(s) of electron(s) dot(s).

For Einstein, dealing with an ensemble meant the use of probability while forgoing information (about the location of an electron) within individual members of this ensemble.

Many may consider all this as only polemics, playing with words! But be warned that many questions now arise: Is the electron this s shell? Does the path of the motion of the electron exist within the atomic space or not? Is the hydrogen atom spherical in shape?

Bohr argued, even after Einstein's death, as if Einstein was alive to respond to his arguments. Bohr would pose these kind of questions and would attempt to seek answers to them from his point of view.

Surely, Bohr had taken Einstein seriously. Most others, on the other hand, engaged themselves in exploring the consequences of, the then new methods of, quantum theory, justifiably leaving the issues of the aforementioned kind to more competent people like Bohr.

Many texts have been written on quantum theory, never dealing with such issues. These texts led to the widespread impression that the hydrogen atom is spherical in shape.

But if we shine many light quanta reaching the hydrogen atom at the same instant, then only one of them will reflect off the electron of the atom and all the other will pass undisturbed through it. This would not happen if that atom had a spherical shape.

As Bohr had put it, it is the only right thing to say that the hydrogen atom behaves as if it is spherical in shape, not that it is spherical in shape! Bohr had grasped Einstein's point: In a single measurement of the position of its electron, the shape of the hydrogen atom will not be seen as a sphere. Always, it will just be a dot away from the nucleus of the hydrogen atom, and nothing else!

Einstein as well as Bohr had been great thinkers of their era. They both were very careful indeed to state in words what they wanted to convey to others. Bohr had embarked, in fact, on a project of creating a proper language for conveying results of the probabilistic quantum theory to anyone, specifically students!

Einstein had raised the issue with Abraham Pais: if I really believed that the moon exists only if I look at it. He had also raised the issue with Heisenberg: You assume the existence of electrons inside the atom . . . But you refuse to consider their orbits . . . He referred to these issues often.

To put them in the present context, we may ask, "If we do not observe an atomic electron by shining light on it, then does that electron exist inside the atom or not? Does the path of motion of that electron exist inside the atom or not? That is, if we do not observe some body, then does it exist or not? Does the path of its motion exist or not? Does "the physical reality" exist independently of the act of observation or not?"

Quantum theory provides us with the means of calculating the probability of the location of the electron within the atomic space but not about its unobserved path. Does is exist or not?

Einstein firmly believed in reality as it exists independently of the act of observation. He thus did not believe that the theory of the final nature, a theory of everything, which is based on the general principle of relativity, will be probabilistic in character. He relentlessly searched a theory for the reality of his firm conviction.

Indeed, the universal theory of relativity, a theory based on the universal principle of relativity and using category theory as its mathematical basis for formulating physical laws, does not deal with probability as its basic or fundamental concept. Rather, probability is a derivable notion within its formalism, which has absolute or universal time.

The universal theory of relativity recognizes that a light quantum is massless and its unobservable path of motion is always a straight line in the 3D Euclidean space. A light quantum is therefore treated as only a momentumless energy quantum. The emission wave mechanism is then the reason for the wave properties of light quanta, not of a single light quantum. Thus, within the framework of the universal relativity, a single quantum of light has no wave associated with it.

Also, for other bodies like electrons, protons, atoms, etc., no wave is associated with an individual body, their single quantum. Such individual bodies follow their inertial straight line motion unless they encounter a contact with other bodies in the course of their motion.

A formalism of universal relativity then says that the light quantum, which is momentumless, is completely absorbed or is completely emitted by a physical body of nonvanishing inertia and it changes the energy of the latter body in the process. The energy and momentum of a body of nonzero inertia are related to each other. Therefore, a change in the energy of such a body is also a change in its momentum. This is how the pressure of radiation in the framework of universal relativity is explained.

Within universal relativity, every body has a well-defined path of motion, even if we do not observe that path. Thus, its framework describes the reality as it exists independently of the act of observation.

> The Lord is not just subtle, but much more subtle indeed than what we had imagined Him to be -

Problems and issues of various natures are still there to be solved! We are yet to answer many questions within the framework of universal relativity. For example, when we attempt to see it, how is it that the electron in an atom of hydrogen is most likely to be found at a distance of the Bohr radius away from the atomic nucleus?

A clarification seems essential for those who may think that the answer is already provided by quantum theory.

To deal with this problem, quantum theory assumes spherical symmetry with the atomic nucleus at the center and the electron in its Coulomb potential. A solution of Schrödinger's equation then provides for the probability distribution of the electronic location.

But we need to answer not in terms of these assumptions but in terms of how, in the first place, quanta mediating the interaction of the nucleus (its Coulomb potential) influence the electron! (One may recall that Hertz wanted to free physics from the concept of potential.)

Chapter 3

Doppler Effect

The Doppler shift of a spectral line provides information about the motion of its source relative to the observer. The Doppler effect is important to astronomy and cosmology, therefore.

3.1 Doppler Effect in Light

As seen in Chapter 1, the finite velocity of propagation of light leads to the effect discovered by Christian Doppler, who showed that the motion of a source of light relative to an observer modifies the period or frequency of oscillations of the light received by the observer.

In essence, Doppler's discovery is that if a source of wave is in motion relative to an observer, then the wavelength of this wave as observed by that observer must be different than that it is emitted by the source with. The Doppler shift (in wavelength or frequency of the wave) results from only the relative motion of the source and the observer. From the Doppler shift of a source, the observer can then infer the nature of its motion.

In this context, we also note that the same Doppler shift must result when the observer is moving and the source of light is at rest. Otherwise, the observer will be able to infer its own motion contrary to its right of considering it to be at rest. This is as per the principle of relativity.

Subtlety in Relativity
Sanjay M. Wagh

ISBN 978-981-4774-44-4 (Hardcover), 978-1-351-23832-8 (eBook)
www.panstanford.com

The Doppler effect is vital for astronomy, as light is the only available source of our information about astronomical objects. It therefore pays to consider the derivation of the Doppler shift in detail here.

3.2 Derivation of the Doppler Shift

Figure 3.1a shows stationary source emitting the first pulse at time t_1 and the second pulse at time t_2, the time measured in the frame of the source. The frequency of emission is then $1/(t_2 - t_1)$. The initial line of sight (SO) is shown making an angle of $\bar{\theta}$ with the Y axis. When the first pulse reaches the observer at O, the time elapsed is $T_1 = t_1 + \text{SO}/c$.

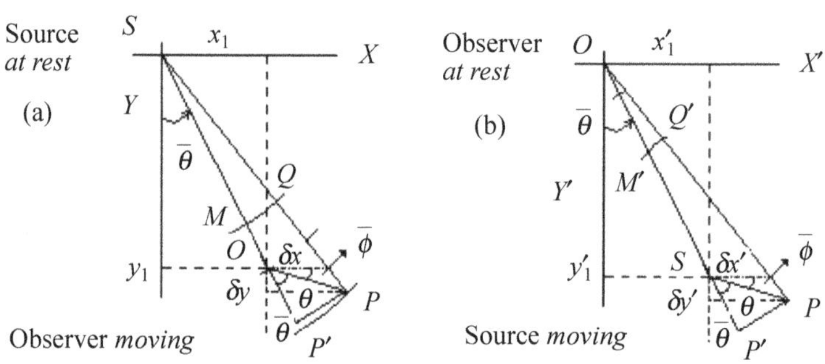

Figure 3.1 Geometry of the Doppler effect.

Let the observer move along the line OP, making an angle $\bar{\phi}$ with the X axis, with P being the point at which the second pulse reaches the observer at time $T_2 = t_2 + \text{SP}/c$. Then

$$T_2 - T_1 = t_2 + \text{SP}/c - t_1 - \text{SO}/c. \tag{3.1}$$

With $\text{SO} = \sqrt{x_1^2 + y_1^2}$, $\text{SP}^2 = \sqrt{(x_1 + \delta x)^2 + (y_1 + \delta y)^2}$, $x_1 >> \delta x$, $y_1 >> \delta y$, we get

$$T_2 - T_1 \approx t_2 - t_1 + \cos\bar{\theta}\delta x + \sin\bar{\theta}\delta y\,. \tag{3.2}$$

But $\delta x = \text{OP}\cos\bar{\phi}$ and $\delta y = \text{OP}\sin\bar{\phi}$; and we get to first order

$$T_2 - T_1 \approx t_2 - t_1 + \frac{\text{OP}}{c}\cos(\bar{\theta} + \bar{\phi}) \equiv t_2 - t_1 + \frac{\text{OP}}{c}\cos\theta$$

$$c\Delta T = c\Delta t + (\text{OP})\cos\theta, \tag{3.3}$$

where θ is the angle of motion with the line SO. Notice that Eq. 3.3 holds for any type of motion of the observer, generally.

(If the line of motion OP of the observer in Fig. 3.1a makes an angle θ with the initial line of sight SO, then the line of motion SP of the source (as seen by the observer) in Fig. 3.1b also makes an angle of θ with the initial line of sight OS (see Fig. 3.2). Thence, we do not need to distinguish the angle θ in the two situations.)

Figure 3.2 Symmetry of the Doppler effect.

If we consider observer to be stationary and source to be moving, similar considerations applied to Fig. 3.1b then yield to first order:

$$T_2' - T_1' \approx t_2' - t_1' + \frac{\text{SP}}{c}\cos(\bar{\theta} + \bar{\phi}) \equiv t_2' - t_1' + \frac{\text{SP}}{c}\cos\theta$$

$$c\Delta T' = c\Delta t' + (\text{SP})\cos\theta. \tag{3.4}$$

In this case, we also indicate the time measured by the observer with an overhead prime to distinguish it from the case of Fig. 3.1a. Equation 3.4 also holds generally. The frequency of the wave as measured by the observer is then $1/(T_2' - T_1')$.

It may be emphasized, once again, that we are not required to specify how the source travels the distance SP or the observer travels the distance OP, whether with uniform velocity or with acceleration. Therefore, Eqs. 3.3 and 3.4 are basic to any analysis of the Doppler effect, whatever may be the nature of the relative motion of the source and the observer.

Now, the motion of the observer in Fig. 3.1a or that of the source in Fig. 3.1b is to be considered[1] as taking place during successive detections of two consecutive wave crests.

[1]While deriving the Doppler shift, a subtle error was committed in the past by setting SP = $u(t'_2 - t'_1)$. As a consequence, all observational results using the consequently derived erroneous Doppler shift formula are invalid. Such erroneous observational results include those important to quasars, active galaxies, and cosmology. We also stress that time is absolute, contrary to that assumed for the corresponding special and general relativistic models invoked to explain these erroneous observational results.

Therefore, in Fig. 3.1a, the motion of the observer takes place for the duration $T_2 - T_1$, while that of the source takes place for the duration $T_{2'} - T_{1'}$ in Fig. 3.1b.

Then when the observer is moving with a uniform velocity u and the source is at rest, for Fig. 3.1a, we have

$$\mathrm{OP} = u(T_2 - T_1) \equiv u\Delta T, \tag{3.5}$$

and when the source is moving with a uniform velocity u and the observer is at rest, for Fig. 3.1b, we have

$$\mathrm{SP} = u(T'_2 - T'_1) \equiv u\Delta T'. \tag{3.6}$$

Then for uniform relative motion, with $\beta = u/c$, we have

$$\Delta T = \frac{\Delta t}{1-\beta\cos\theta} \quad \text{and} \quad \Delta T' = \frac{\Delta t'}{1-\beta\cos\theta}. \tag{3.7}$$

3.3 Difficulty with Special Relativity

As noted earlier, a subtle error was committed in the past by setting $\mathrm{SP} = u(t'_2 - t'_1)$. The following are its consequences.

In place of Eq. 3.7, we then obtain

$$\Delta T' = (1+\beta\cos\theta)\Delta t' \;\Rightarrow \frac{1}{\Delta T'} \equiv \nu_{\mathrm{o}} = \nu_{\mathrm{e}}\frac{\sqrt{1-\beta^2}}{1+\beta\cos\theta}, \tag{3.8}$$

where we have used $\Delta t' = 1/\nu_{\mathrm{e}}\sqrt{1-\beta^2}$, and $\beta = u/c$, with ν_{e} being the frequency of the emitted wave in the rest frame of the source. Equation 3.8 is the standard formula for the Doppler shift.

On using the inverse Lorentz transformation, we also obtain Eq. 3.8. This is the manner in which special relativity ensures that the observer does not detect self-motion using the Doppler effect then.

However, the distance SP is not $u\Delta t'$ but $\mathrm{SP} = u\Delta T'$. The source (or observer), if at all it does so, moves only during the period $\Delta T' = T'_2 - T'_1$ (or ΔT) and not during $\Delta t' = t'_2 - t'_1$ (or Δt).

Then using $\mathrm{SP} = u\Delta T'$, the only correct expression, we obtain

$$\Delta T' = \frac{\Delta t'}{1-\beta\cos\theta} \;\Rightarrow \nu_{\mathrm{o}} = \nu_{\mathrm{e}}(1-\beta\cos\theta)\sqrt{1-\beta^2}. \tag{3.9}$$

The same equation, Eq. 3.9, is also obtained using the inverse Lorentz transformation. Consequently, it is therefore ensured that the observer does not detect self-motion using the Doppler effect.

In terms of wavelengths, we obtain from Eq. 3.9

$$\lambda_o(\theta) = \frac{\lambda_e}{(1-\beta\cos\theta)\sqrt{1-\beta^2}}. \tag{3.10}$$

On calculations for $\theta = 0$ and $\theta = \pi$, we then obtain for the first and second order wavelength shifts from the above as

$$\Delta^{(1)}\lambda = \lambda_e\beta, \quad \Delta^{(2)}\lambda = \frac{3}{2}\lambda_e\beta^2 \quad \Rightarrow \quad \Delta^{(2)}\lambda = \frac{3}{2\lambda_e}\left[\Delta^{(1)}\lambda\right]^2. \tag{3.11}$$

Clearly, Eq. 3.11 provides the parabolic relation of the first order and second order wavelength shifts, but with three times the coefficient, as is implied by the standard formula, Eq. 3.8, for the Doppler shift.

Recall that the Ives–Stilwell experiment shows that

$$\Delta^{(2)}\lambda \approx \frac{1}{2\lambda_e}\left[\Delta^{(1)}\lambda\right]^2. \tag{3.12}$$

Equation 3.11 is ruled out by the Ives–Stilwell experiment. *So **a theoretically correct result of special relativity as contained in Eq. 3.9 is inconsistent with** the results of the Ives–Stilwell experiment.* Also, many other experiments supporting the standard formula, Eq. 3.8, of the Doppler shift rule out Eq. 3.9.

Special relativity gets experimentally rejected then.

As seen in Chapter 2, the nonobservational nature of the path of motion of a light quantum, having zero rest mass, is consistent with time running at the same rate for all observers. That is to say, time is absolute. Thence, we have $\Delta T - \Delta T'$ and $\Delta t - \Delta t'$ in Eqs. 3.3 and 3.4. Then the same Doppler shift results for the two situations of Fig. 3.1.

Henceforth, we will not make any distinction between the times measured by the two observers. Therefore, $\Delta T = \Delta T'$ and $\Delta t = \Delta t'$, in general; and $\nu_e = 1/\Delta t$ is the frequency of radiation emitted by the source and $\nu_o = 1/\Delta T$ is the frequency of that radiation as observed by the observer.

3.4 Doppler Shift Formulas with Acceleration

For the case of uniform acceleration a, we have

$$\mathrm{SP} = u(T_2 - T_1) + \frac{1}{2}a(T_2 - T_1)^2 \equiv u\Delta T + \frac{1}{2}a(\Delta T)^2. \tag{3.13}$$

We then obtain using Eq. 3.4

$$\frac{a}{2c}\cos\theta(\Delta T)^2 - (1-\beta\cos\theta)\Delta T + \Delta t = 0\,. \tag{3.14}$$

We have to solve this quadratic to get a relation of ΔT and Δt to obtain the Doppler shift in this case. We have the observed frequency as

$$\nu_{\text{o}} \approx \nu_{\text{e}}\left[1-\frac{\beta^2}{2}\sin^2\theta - \beta\cos\theta\right] - \frac{a}{2c}\,. \tag{3.15}$$

But we do not encounter uniform acceleration in any natural situations with astronomical bodies. So the case of variable acceleration is better suited for astrophysical situations.

But an important question is, what kind of variable acceleration can we expect with astronomical bodies?

For the nature of variable acceleration in many astronomical situations, analogy of the following kind suggests itself.

Consider an engine pulling or pushing a train of railway carriages that are attached to each other by links. The engine pulls a link to the first carriage that moves, and in its turn that carriage pulls its link to set the next one in motion. This is how the train gets set in motion.

With pull acting on it at one end, a link becomes taut and pulls along the carriage attached at the other end. But when the carriage accelerates due to the link pulling it, that link slackens as the carriage pulls back. The slackened link becomes taut again when the engine pulls it next.

When in motion, due to friction of its wheels with the rails, a carriage slows down in speed till its link with the engine becomes taut to pull it again. This is the case with every carriage of the train.

Even if the train were to move with a constant average velocity, the movement of its carriage is an accelerated motion. It would be erroneous to treat this motion as a case of uniform velocity evidently.

The aforementioned description of the motion of a train indicates that it is a case of quasi-periodic acceleration. There is a definite period for variation of acceleration as the engine starts pulling the train. The period of variation of acceleration changes, however, depending on various factors.

In many astronomical situations, we encounter situations analogous to that of the motion of a train.

A primary mover (e.g., the pressure of radiation from a source) pushes matter. The pushed matter may then encounter other matter in its path of motion, with that matter getting pushed in that encounter. This must slow the speed of the initial blob of matter. But the prime mover may push it again. This is also the case of quasi-periodic acceleration.

In what follows, we therefore consider the case of periodic acceleration of matter emitting spectral lines. Quasi-periodic changes can be accounted for by changing the frequency or the period.

Let us assume that acceleration is changing with time as

$$\frac{da}{dT} = a_0 \sin \omega T \,, \tag{3.16}$$

where a_0 is constant. Then on integration, we obtain for the acceleration

$$a = -\frac{a_0}{\omega} \cos \omega T + k_1 \,. \tag{3.17}$$

On further integration, we have

$$\mathrm{SP} = \frac{a_0}{\omega^3} \cos(\omega \Delta T) + \frac{1}{2} k_1 (\Delta T)^2 + k_2 (\Delta T) + k_3 \,, \tag{3.18}$$

where k_1, k_2, and k_3 are integration constants. When $\omega \Delta T << 1$, we have

$$\mathrm{SP} \approx \frac{a_o}{\omega^3} + \left(k_1 - \frac{a_0}{\omega} \right) \frac{1}{2} (\Delta T)^2 + k_2 (\Delta T) + k_3 \,. \tag{3.19}$$

We may set $k_1 = 0$ as a nonzero value corresponding to constant acceleration, $k_2 = \beta c$ as the initial velocity, and $k_3 = 0$. Then we obtain the quadratic equation

$$\frac{a_0}{2c\omega} \cos\theta (\Delta T)^2 + (1 - \beta \cos\theta)(\Delta T) - \Delta t - \frac{a_0}{c\omega^3} \cos\theta = 0 \,. \tag{3.20}$$

On solving it, we get the Doppler shift in this case. On manipulating the terms, the observed frequency ν_o is obtained as

$$\nu_o \approx \left[1 - \frac{\beta^2}{2} \sin^2\theta - \beta \cos\theta \right] \nu_e - \frac{a_o}{c\omega^3} \nu_e^2 \cos\theta + \frac{a_o}{2c\omega} \cos\theta \,. \tag{3.21}$$

From Eq. 3.21, it follows that the source parameters β and θ, the acceleration parameter a_o, and ω determine the redshift $z = (\nu_e - \nu_o)/\nu_o$ of the spectral line from the source.

But notice that, in terms of the redshift z, we have

$$\frac{1}{1+z} = f - \left(\frac{a_0}{c\omega^3}\right)\nu_e \cos\theta + \frac{a_o}{2c\omega\nu_e}\cos\theta\,, \tag{3.22}$$

where we have set

$$f(\beta,\theta) = 1 - \frac{\beta^2}{2}\sin^2\theta - \beta\cos\theta\,. \tag{3.23}$$

Clearly, the redshift does not determine the velocity β and the acceleration a uniquely. Thence, we can only constrain the involved source parameters from the redshift of the spectral lines emitted by the source.

Now, consider the ratio ν_o/ν_e. The order of the first term that is independent of ν_e, that is, the term $f(\beta, \theta)$, is unity. As the order of ν_e is 10^8, the order of its coefficient can be set to be 10^{-8}. Thus, the acceleration a_0/ω is of the order $10^{-8}\omega^2$, which yields a reasonable value.

Then the last term of Eq. 3.21 is negligible.

It is now important to realize that Doppler broadening of the line follows the standard quantum treatment, since the atom is not accelerated when the emission or absorption of photon occurs. (The atom of the gas moves with a uniform velocity after an elastic collision with another atom.)

Therefore, atomic acceleration due to the emission or absorption of a photon is irrelevant to Doppler broadening of a spectral line. In other words, the width of the spectral line emitted by a source is provided by the standard formula obtained from quantum theory. Therefore, apart from its line broadening effect, the motion of the atom in the rest frame of the line-emitting matter is not relevant to our analysis of the Doppler effect.

The overall acceleration of the source, the line-emitting matter, is therefore the factor a_0/ω^3 in Eq. 3.21.

Now, observationally, we know only the redshift of the spectral line, from the laboratory frequency ν_e and the observed frequency ν_o. Then the important questions are: How do we determine the source parameters? And how do we even estimate the values of these source parameters?

The following suggests itself as such a method for determining the source parameters from the observed redshift of spectral lines from that source. We term it as Doppler analysis with acceleration.

3.5 Doppler Analysis with Acceleration

Neglecting the last term of Eq. 3.21, we can write

$$z^{\star} \equiv \frac{\nu_{\rm e}-\nu_{\rm o}}{\nu_{\rm e}} = \frac{\beta^2}{2}\sin^2\theta + \beta\cos\theta + \left(\frac{a_0}{c\omega^3}\right)\nu_{\rm e}\cos\theta$$

$$= \frac{\beta^2}{2}\sin^2\theta + \beta\cos\theta + \left(\frac{a_0}{\omega^3}\right)\frac{1}{\lambda_{\rm e}}\cos\theta . \tag{3.24}$$

Notice that

$$z^{\star} = \frac{\lambda_{\rm o}-\lambda_{\rm e}}{\lambda_{\rm o}} = \left(\frac{\lambda_{\rm e}}{\lambda_{\rm o}}\right) z \equiv \frac{z}{1+z}, \tag{3.25}$$

where z is the redshift of the spectral line. Thus, when the spectral line is red-shifted, $z^* < z$; and when it is blue-shifted, $z^* > z$. We will refer to the quantity z^*, in the text, as the scaled redshift.

Clearly, if the source matter were moving with a velocity β, experiencing an acceleration parameter a_0 and moving in the direction θ, and if it were emitting spectral lines (of a different rest wavelength $\lambda_{\rm e}$), then the plot of the scaled redshift z^* versus $1/\lambda_{\rm e}$ will be a straight line for it with a slope

$$S = (a_0/\omega^3)\cos\theta \equiv \alpha\cos\theta \tag{3.26}$$

and having an intercept

$$\kappa = (1-f) = \frac{\beta^2}{2}\sin^2\theta + \beta\cos\theta . \tag{3.27}$$

Then if we can infer the angle θ of motion of matter emitting spectral lines from some independent considerations, we will be able to infer its velocity and acceleration from the intercept and slope, respectively.

Of course, two or more spectral lines are required for the aforementioned method to yield results. Recall that for linear regression, we fix a point and fit the line passing through it.

Now, consider that line-emitting matter has an angular spread in motion. This is, for example, the case for spherical distributed matter and axisymmetric distributed matter, as in a torus. Then we cannot determine its velocity and acceleration using slope and intercept, since we do not know the angle θ of its motion. However, the following strategy emerges.

Notice that the magnitude of velocity and that of acceleration of matter are identical for its parts in these situations, except that the direction of motion is differing for them. Then for the linear regression of data of the spectral lines, we will obtain a different slope and intercept when we vary the fixed data point and fit a line passing through it.

Then, of the obtained values of the slope, a maximum value S_{max} exists. Notice that we have

$$\theta = \cos^{-1}\left(\frac{S}{\alpha}\right). \tag{3.28}$$

Thus, the minimum value α_{min} of the acceleration term α is the maximum value S_{max} of the slope.

If we can estimate the angle θ of motion of line-emitting matter from some independent considerations, then we can estimate the value of the acceleration term α as $S_{max}/\cos\theta$.

Now, the intercept κ, from linear regression analysis, provides us with a quadratic equation in velocity β. On solving it, we obtain

$$\beta = -\frac{\alpha S}{\alpha^2 - S^2} + \frac{\alpha S}{\alpha^2 - S^2}\sqrt{1 + \frac{2\kappa}{S^2}(\alpha^2 - S^2)}\,. \tag{3.29}$$

When we know the values of α, S, and κ, we can estimate the value of the velocity β from this equation.

We emphasize here that the frequency ω of variations of acceleration is not determinable, or is a free parameter of above considerations. These considerations are, therefore, applicable to considered situations irrespective of the rapidity of these variations. We emphasize that this situation is very much similar to that of a railway train.

Furthermore, we emphasize that the spatial extent of the line-emitting system under consideration is also not determinable within these considerations of the Doppler effect. (This is similar to the length of the railway train not being relevant to the variation in the acceleration of its carriage.) For this reason, these considerations are applicable to a laboratory system also.

Temporally (sinusoidally) variable acceleration thence allows, in general, theoretically proper and realistic estimates for the magnitude of the velocity and acceleration of the source emitting spectral lines.

Part II

Category Theory and Universal Relativity

Chapter 4

What Is a Category?

4.1 Introduction

Category theory begins with the concept of an arrow. A category is a collection of arrows satisfying certain natural axioms regarding their existence and their composition.

Whenever we collect or gather or accumulate elements, we need to be careful about certain paradoxical situations. One such famous paradox relevant to us here is Russell's paradox.

Russell's paradox arises when we attempt to form any gathering, collection, or accumulation of elements all of which have been chosen according to say a property P and when we furthermore have that this gathering also has the same property P.

To see this, let us intuitively define a set. We shall call a gathering of elements collected on the basis of a well-defined property a set. Then being a member of a set is also a well-defined property that underlies this notion of a set. So any set S is a member of itself: $S \in S$.

Now, being a set is also a well-defined property. Hence, we have that a gathering u of all the sets is also a well-defined set. However, for set $S = \{A | A \in u \text{ and } A \notin A\}$, we have that $S \in S$ if and only if (iff) $S \notin S$! Set S, even when it is a gathering of the elements based on a well-defined property of being a member of a set or not, is not a member of u since all the subsets of u have the property that they

Subtlety in Relativity
Sanjay M. Wagh

ISBN 978-981-4774-44-4 (Hardcover), 978-1-351-23832-8 (eBook)
www.panstanford.com

are the members of themselves. This paradox arises mainly because we have encountered here a situation in which the gathering u has the property of being a set and all of its members also satisfy this property.

A way out of this paradoxical situation is to define a set in such a way that the gathering of all the so-defined sets is *not* a set, unambiguously. This can be achieved by appealing to properties of the operation of the gathering of elements, like certain natural notions of being able to form a larger gathering by the concatenating of smaller gatherings.

Notice, however, that we are calling only a particular type of collection or gathering of elements as a set here. So any collection that does not satisfy the criteria stipulated by us is not called a set. Nonetheless, we are free to form a collection that is not a set.

However, it is not that category theory cannot be understood without these set theoretical foundations, which have a secondary place in category theory. Still, these foundations are important, for we deal with a collection of arrows (and of objects) in this theory.

Category theory deals with a collection of arrows. Thus, we begin with the concept of an arrow in what follows.

4.2 An Arrow

An arrow, commonly also called a morphism or a map or a transformation, is assumed to have associated with it a source and a target. If S is the source and T is the target of an arrow f, we use the following notation to denote arrow f:

$$f: S \to T \text{ or } S \xrightarrow{f} T$$

Both the source and the target of an arrow will be called objects. We could say that an arrow then connects two objects, or originates from one object and ends at another object.

Two arrows f and g are said to compose to give another arrow h, provided that the target of f is the source of g. This can be called the matching condition for the composition of arrows. That is, if we have arrows $A \xrightarrow{f} B$ and $B \xrightarrow{g} C$, with the target B of f being the source of g, only then do we have an arrow $A \xrightarrow{h} C$, and we denote it by $h = g \circ f$. (There, of course, can exist other arrows, like $A \xrightarrow{\alpha} C$

, but one of them is arrow $h = g \circ f$.) The composition of arrows is, evidently, not defined then if the above matching condition for their composition is not satisfied.

By appropriately supplying the objects for the arrows under consideration, it can be easily seen that the following are equivalent (TFAE) matching conditions:

- The compositions $g \circ f$ and $h \circ g$ are defined.
- The composition $h \circ (g \circ f)$ is defined.
- The composition $(h \circ g) \circ f$ is defined.

Now, if we are given three arrows that can be composed as $f \circ g$ and $g \circ h$, that is, three arrows f, g, and h satisfying the above matching conditions, then we assume that $(f \circ g) \circ h = f \circ (g \circ h) \equiv f \circ g \circ h$. That is, we assume that the composition of arrows is associative. Then any number of arrows can be composed with each composition obeying the matching condition.

Next, we will assume the existence of an identity arrow corresponding to every object, that is, an (identity) arrow from a given object A to the same object A and denote it as $A \xrightarrow{\mathrm{id}_A} A$.

But every endo-arrow $A \xrightarrow{f} A$ need not be an identity arrow! For example, consider the permutation group $\vec{G}$ of set $G = \{1, 2, 3, 4\}$. We imagine that it is just a single object G and that the group elements are the arrows with their source and target being that single object G. Then only the group identity e, that is, the identity permutation $1 \to 1, 2 \to 2, 3 \to 3, 4 \to 4$, corresponds to the identity arrow $G \xrightarrow{\mathrm{id}_G} G$, while all the other group elements do not. Other group elements are, nonetheless, arrows from object G to object G, that is, of the form $G \xrightarrow{f} G$.

Therefore, an identity arrow $A \xrightarrow{\mathrm{id}_A} A$ is one of the many permissible endo-arrows $A \xrightarrow{f} A$, and we assume that there exists, at least, such an identity arrow for every object that we may collect. Then other arrows of the type $A \xrightarrow{f} A$ may or may not exist in our collection.

Moreover, we also assume that the identity arrow for any object * serves also as the identity with respect to the composition of arrows. That is to say, we assume that compositions $* \xrightarrow{\mathrm{id}_*} * \xrightarrow{f} B = * \xrightarrow{f \circ \mathrm{id}_*} B = * \xrightarrow{f} B$ and $A \xrightarrow{f} * \xrightarrow{\mathrm{id}_*} * = A \xrightarrow{\mathrm{id}_* \circ f} * = A \xrightarrow{f} *$ are always defined.

Often, we will omit the symbol $\circ$ of composition and write $h = gf$. Note also the reverse order of writing the composition of arrows. This notational practice is only a convenience that has no fundamental significance. Note that the notation for an arrow, $f: A \to B$ or $A \xrightarrow{f} B$, is reminiscent of that of functions; but arrows are, in general, *not* functions.

4.3 Partial Binary Algebra of Arrows

We are now almost ready to formalize a precise mathematical definition of what we shall call a category. Before proceeding to do so, we will, however, require a few definitions.

The first is the following definition of a partial binary algebra. It formalizes the composition of arrows as a binary operation for a pair of arrows.

Definition 1 A partial binary algebra is a pair $(X, \circ)$ consisting of a collection X and a partial binary operation $\circ$ on X, that is, a binary operation defined on a subcollection of $X \times X$. We denote the value of $\circ(x, y)$ by $x \circ y$.

Notice at this point that the subcollection of $X \times X$ can be empty. An empty subcollection of $X \times X$ does not cause any problems because the partial binary operation $\circ(-, -)$ yields an empty arrow as the composition of two empty arrows and the result of this composition is also as well defined as that of others.

Moreover, we use the word "collection" to generically mean the operation of accumulation of elements without the associated paradoxical situations arising in its considerations.

The next definition characterizes the existence of identities ($\mathrm{id}_* \circ f = f$ and $f \circ \mathrm{id}_* = f$) for the above partial binary operation. We shall, in conformity with the relevant mathematical literature, call this identity also as the unit of the partial binary algebra.

Definition 2 If $(X, \circ)$ is a partial binary algebra, then an element $u \in X$ is called the unit (or identity) of $(X, \circ)$ provided that $x \circ u = x$ whenever $x \circ u$ is defined and $u \circ y = y$ whenever $u \circ y$ is defined, where $x, y \in X$ are arbitrary elements of X.

4.4 Definitions of a Category

We now define a generic category in the following manner:

Definition 3 (object-free) A generic category $\mathbb{C}$ is a partial binary algebr $\mathbb{C} = (\mathcal{A}, \circ)$ where the elements or the members of collection $\mathcal{A}$ are called the arrows and the $\circ$ is a partial binary operation defined on $\mathcal{A}$ such that the following are satisfied:

- **Matching conditions:** For any three arrows f, g, and $h \in \mathcal{A}$ TFAE:
    - The compositions $g \circ f$ and $h \circ g$ are defined.
    - The composition $h \circ (g \circ f)$ is defined.
    - The composition $(h \circ g) \circ f$ is defined.
- **Associativity condition:** For any three arrows f, g, and $h \in \mathcal{A}$ satisfying the matching conditions, we have $(h \circ g) \circ f = h \circ (g \circ f)$.
- **Unit existence condition:** For every arrow $f \in \mathcal{A}$, there exist units u_c and u_D of the partial binary algebra $(\mathcal{A}, \circ)$ such that $u_C \circ f (= f)$ and $f \circ u_D (= f)$ are defined.
- **Smallness condition:** For any pair of units u_1 and u_2 of the partial binary algebra $(\mathcal{A}, \circ)$, the following is a collection (for which any paradoxical set theoretical situations do not arise), called a hom-collection: $[u_1, u_2]_\mathbb{C} = \{f \in \mathcal{A} \mid f \circ u_1 \text{ and } u_2 \circ f \text{ are defined.}\}$

We also denote $[u_1, u_2]_\mathbb{C}$ by $\mathbb{C}(u_1, u_2)$ or by $\hom_\mathbb{C}(u_1, u_2)$.

The above definition is evidently an object-free definition of a generic category as it uses only the notion of an arrow that we began with.

Objects within the context of the above object-free definition of a generic category can now be defined using the units of its partial binary algebra. To this end, we index the collection of all the units of the generic category $\mathbb{C}$ with another collection and call that indexing collection as the collection of objects of the generic category $\mathbb{C}$. Such a collection of objects of the generic category $\mathbb{C}$ will be denoted by the symbol $\mathrm{Ob}(\mathbb{C})$. There exists therefore a one-to-one (1-1) correspondence between the objects and the units or the identity arrows of a generic category.

We shall now state some equivalent definitions of a generic category using objects. Remember, however, that these definitions are equivalent without the equivalence being an isomorphism of the categories. The correspondence between these definitions is neither injective (1-1) nor surjective (onto). But the same category theory would result since it is an equivalence. These words would be formalized later using functors.

An object-dependent definition of a category involves two collections—the first that of the objects and the second that of the arrows connecting the objects so collected. The first object-dependent definition of a category can be framed as follows.

Definition 4 A category is a quadruple

$$\mathbb{A} = (\mathrm{Ob}(\mathbb{A}), [-, -]_{\mathbb{A}}, \mathrm{id}, \circ)$$

consisting of

- A collection $\mathrm{Ob}(\mathbb{A})$, whose members are called objects
- For each pair (A, B) of $\mathbb{A}$-objects, a collection $[A, B]_{\mathbb{A}}$, whose members are called the $\mathbb{A}$-arrows from A to B

  (The collection of all $\mathbb{A}$-arrows is defined to be the union of all the collections $[A, B]_{\mathbb{A}}$ in $\mathbb{A}$ and is denoted by $\mathrm{Mor}(\mathbb{A})$.)
- For each $\mathbb{A}$-object A, an arrow $\mathrm{id}_A : A \to A$, called the $\mathbb{A}$-identity on A
- A composition law associating each $\mathbb{A}$-arrow $A \xrightarrow{f} B$ and each $\mathbb{A}$-arrow $B \xrightarrow{g} C$ with an $\mathbb{A}$-arrow $A \xrightarrow{g \circ f} C$, called the composite of f and g

These data are then subject to the following conditions:

(i) The composition is associative, that is, for $A \xrightarrow{f} B$, $B \xrightarrow{g} C$, and $C \xrightarrow{h} D$, $h \circ (g \circ f) = (h \circ g) \circ f$ holds.

(ii) $\mathbb{A}$-identity arrows act as identities with respect to the composition, that is, for $\mathbb{A}$-arrows $A \xrightarrow{f} B$, we have $\mathrm{id}_B \circ f = f$ and $f \circ \mathrm{id}_A = f$.

(iii) The collections $[A, B]_{\mathbb{A}}$ are pairwise disjoint.

(The condition (iii) above is for technical convenience only. It can always be imposed when all the other conditions are satisfied by defining each arrow $f \in [A, B]_{\mathbb{A}}$ to be a triple (A, f, B), with $f \subseteq A \times B$ such that for each $x \in A$ there exists a unique

$y \in B$ with $(x, y) \in f$. When verifying that an entity is a category or not, we neglect therefore the condition (iii).)

(The composition $\circ$ is a partial binary operation on the collection More($\mathbb{A}$).)

It may now be noticed that the matching conditions of definition 3 and the collections $[u_A, u_B]_{\mathbb{C}}$ can be used to define a function as in the following object-dependent definition of a category. Still another way to define a category is then the following.

Definition 5 A category $\mathbb{C}$ is a collection $\mathbb{C}$, together with another collection Mor($\mathbb{C}$), which is a disjoint union of the form

$$\mathrm{Mor}(\mathbb{C}) = \bigcup_{(A,B)\in\mathbb{C}\times\mathbb{C}} [A,B]_{\mathbb{C}},$$

where each $[A, B]_{\mathbb{C}}$ is a collection. Furthermore, to each triple (A, B, C) of the members of $\mathbb{C}$, we have a function from $[B, C] \times [A, B]$ to $[A, C]$. The image of the pair (f, g) under this function is the composition, $f \circ g$, of an arrow f by an arrow g; $f \in [B, C]$ and $g \in [A, B]$. The members of $\mathbb{C}$ are the objects, and those of Mor($\mathbb{C}$) are the arrows of category $\mathbb{C}$.

The composition functions are further subject to the following axioms:

- **Associativity:** Whenever the compositions are defined, we have $(fg)h = f(gh)$.
- **Existence of identities:** For each $A \in \mathbb{C}$, there exists an element $1_A \in [A, A]$ such that $1_A f = f$ and $g 1_A = g$ whenever the compositions are defined. 1_A is unique.

We can then prove the following proposition as an example of the relation between these definitions of a category:

Proposition 1 If $\mathbb{A} = (\mathrm{Ob}(\mathbb{A}), [-, -]_{\mathbb{A}}, \mathrm{id}, \circ)$ is a category defined as per definition 4, with Mor($\mathbb{C}$)as the collection of its arrows, then

- An $\mathbb{A}$-arrow is an $\mathbb{A}$-identity iff it is a unit of the partial binary algebra (Mor($\mathbb{A}$),$\circ$).
- (Mor($\mathbb{A}$),$\circ$) is an object-free category as per definition 3.

Proof (Mor($\mathbb{A}$),$\circ$) is clearly a partial binary algebra, where $f \circ g$ is defined iff the matching conditions are satisfied. Hence, each $\mathbb{A}$-identity is a unit of (Mor($\mathbb{A}$),$\circ$).

If $u : A \to B$ is a unit of $(\mathrm{Mor}(\mathbb{A}),\circ)$, then $u \circ \mathrm{id}_A = u$ and $u \circ \mathrm{id}_A = \mathrm{id}_A$, where the first equality holds since id_A is an $\mathbb{A}$-identity and the second equality holds since u is a unit of the partial binary algebra. This establishes both parts of the proposition that follows from it.

With every category as per definition 4, we can therefore associate an object-free category as per definition 3. This association, of pair $(\mathrm{Mor}(\mathbb{A}),\circ)$ and quadruple $(\mathrm{Ob}(\mathbb{A}),[-, -]_{\mathbb{A}},\mathrm{id}, \circ)$, is neither injective (1-1) nor surjective (onto); still it is an equivalence of two definitions of a category, as will be defined later using functors.

In what follows, we provide still another object-dependent definition of a generic category that uses functions from the collection of arrows to the collection of objects as its basis.

Definition 6 A category $\mathbb{C}$ is a quintuple

$$=\{\mathrm{Ob}(\mathbb{C}),\mathrm{Mor}(\mathbb{C}), \mathrm{dom}, \mathrm{cod}, \circ\},$$

where

- $\mathrm{Ob}(\mathbb{C})$ is a collection whose members are called $\mathbb{C}$-objects.
- $\mathrm{Mor}(\mathbb{C})$ is a collection whose members are called $\mathbb{C}$-arrows.
- dom, called a domain function, and cod, called a codomain function, are functions from $\mathrm{Ob}(\mathbb{C})$ to.
- $\circ$ is a function from the collection
  $D = \{(f, g) | f, g \in \mathrm{Mor}(\mathbb{C}) \text{ and } \mathrm{dom}(f) = \mathrm{cod}(g)\}$
  into $\mathrm{Mor}(\mathbb{C})$, called the composition law of $\mathbb{C}$ with the value of $\circ\,(f, g)$ being written as $f \circ g$ and $f \circ g$ being defined iff $(f, g) \in D$ such that the above data satisfies the following conditions:
  - **Matching condition:** If $f \circ g$ is defined, then $\mathrm{dom}(f \circ g) = \mathrm{dom}(f)$ and $\mathrm{cod}(f \circ g) = \mathrm{cod}(g)$.
  - **Associativity condition:** If $f \circ g$ and $h \circ f$ are defined, then $h \circ (f \circ g) = (h \circ f) \circ g$.
  - **Identity existence condition:** For each $\mathbb{C}$-object A, there exists a $\mathbb{C}$-arrow e such that $\mathrm{dom}(e) = A = \mathrm{cod}(e)$ and
    (i) $f \circ e = f$ whenever $f \circ e$ is defined and
    (ii) $e \circ g = g$ whenever $e \circ g$ is defined.
  - **Smallness condition:** For any pair (A, B) of $\mathbb{C}$-objects, any paradoxical set theoretical situations do not arise for the collection
    $[A, B]_{\mathbb{C}}\ \{f | f \in \mathrm{Mor}(\mathbb{C}), \mathrm{dom}(f) = A \text{ and } \mathrm{cod}(f) = B\}$.

Like before, we, once again, can notice that the association of a category as a pair Mor($\mathbb{A}$),$\circ$) with a category as a quintuple {Ob($\mathbb{C}$),Mor($\mathbb{C}$), dom, cod, $\circ$} or that of a category represented as a quadruple (Ob($\mathbb{A}$),$[-,-]_{\mathbb{A}}$,id, $\circ$) with a category represented as a quintuple {Ob($\mathbb{C}$),Mor($\mathbb{C}$), dom, cod, $\circ$} is neither injective (1-1) nor surjective (onto). However, these associations are also an equivalence of the involved two categories.

In the early days of the developments of category theory, the partial binary algebra structure behind it was not clear. It was then necessary, and had also been quite a common practice, to supply definitions of various terms in the definition of what was called a category.

For the sake of providing an example of the above and also as an example of a simple categorical proposition, we now note the following result in the form of a proposition:

Proposition 2 Let $\mathbb{C}$ be a category, and let A be any of its objects. Then there exists one and only one arrow $e : A \rightarrow A$ in $\mathbb{C}$ satisfying the identity existence conditions of definition 6.

Proof Let e and e' be two endo-arrows of that object A satisfying the stated conditions. Then from identity existence condition (i), we have $e' \circ e = e$, and from identity existence condition (ii), we have $e' \circ e = e$. Hence, $e = e'$.

That is why we have often provided the following definition:

Definition 7 (identity arrow) For each object A of a category $\mathbb{C}$ of definition 6, there exists a unique $\mathbb{C}$-arrow $\mathrm{id}_A : A \rightarrow A$ satisfying the identity existence conditions of that definition. This unique arrow id_A is called the identity of object A. It is alternatively also denoted by 1_A.

One purpose behind providing various object-dependent definitions of a category was to provide a reader with a flavor of notations that, quite frequently, occur in the literature on category theory. As had been stated before, the same category theory results on using any of these definitions. Consequently, we may work with any one of these definitions of a category to obtain results within the framework of category theory. Because of the aforementioned equivalence of these definitions, we can also freely use the objects and arrows in a combined situation with impunity.

After Eilenberg and McLane pioneered the very first definition of a category, category theory developed in many ways at the hands of many people. A category is therefore presented in the literature in many different ways. Often, a category is presented in the form of two collections, that of its objects and that of the hom-collection of its arrows, as $\mathbb{A} = (\mathrm{Ob}(\mathbb{C}), [A, B]_{(A, B) \in \mathbb{C}\times\mathbb{C}}, \circ)$. When this is done, however, generally speaking, we require the hom-collections to be pairwise disjoint. Otherwise, an arrow does not have a unique source and a unique target. This minor difficulty is then resolved by considering another hom-collection of the form

$$\overline{\mathrm{hom}}(A.B) = \{(A, f, B) \mid f \in \mathrm{hom}(A, B)\}.$$

Quite often, a reader may have to keep this in mind when working through the examples of categories.

Attention: A hom-set is a hom-collection that is a set. A category in this case is called a locally small category, as is defined in the next section. In the older literature on category theory, it is usual to talk about a hom-set rather than a hom-collection, even when it is a class. With the old literature, a reader may have to keep track of whether the hom-collection is a set or not. It would be advisable to first check whether the hom-collection is a set for a given category or not.

Now, we would like to emphasize that a category can also be defined in the following way:

Definition 8 (collection category) A (collection) category is a triple of the form $\mathbb{C} = (\mathrm{Ob}(\mathbb{C}), U, \mathrm{hom})$, where

- $\mathrm{Ob}(\mathbb{C})$ is a collection whose members are called objects of category $\mathbb{C}$
- $U : \mathrm{Ob}(\mathbb{C}) \to \mathfrak{C}$ is a collection-valued function, where $\mathfrak{C}$ is the collection of all the collections such that for each $\mathbb{C}$-object A, there exists $U(A)$, called the underlying collection of A
- $\mathrm{hom} : \mathbb{C} \times \mathbb{C} \to \mathfrak{C}$ is a collection-valued function, where for each pair (A, B) of $\mathbb{C}$-objects, $\mathrm{hom}(A, B)$ is called the collection of all $\mathbb{C}$-arrows with source (domain) A and target (codomain) B, such that the following are satisfied:
    - For each pair (A, B) of $\mathbb{C}$-objects, the collection $\mathrm{hom}(A, B)$ of $\mathbb{C}$-arrows is a subcollection of the collection $U(B)^{U(A)}$— the collection of all functions from $U(A)$ to $U(B)$.

- For each $\mathbb{C}$-object A, the identity function $1_{U(A)}$ is a member of hom(A, A).
- For each triple (A, B, C) of $\mathbb{C}$-objects, $f \in$ hom(A, B), $g \in$ hom(B, C) implies that arrow $g \circ f \in$ hom(A, C), where $\circ$ denotes a law of the composition of functions.

Definition 9 (collectable category) A category presentable as a collection category will be called a collectable category.

Remember that by "collection" we mean a gathering or accumulation of elements for which set theoretical paradoxes do not arise. Therefore, in the above definition, we only require that the collection $\mathfrak{C}$ of all the collections is not of the type of the collections that it is a collection of. Since we can consider collections (that are not sets) and functions defined on such collections, the above definition is meaningful.

If we replace the words "collection-valued" by "set-valued" and replace the word "collection" by "set" in the above definition, we obtain the often-quoted definition of a concrete category. A concretizable category is a category that can be presented as a concrete category. (This is analogous to the pair of a measure space and a measurable space.) Note that every category is not concretizable.

Nonetheless, we then propose a conjecture (without proof here) that every category of definition 3 is a collectable category.

4.5 Some Standard Nomenclature about Categories

Let us now mention some standard nomenclature about generic categories that is quite often used in the existing vast literature on category theory.

- When the collection A of arrows in $\mathbb{C}$ is a small set, we refer to the generic category $\mathbb{C}$ as a small category. (In this case, the collection Ob$(\mathbb{C})$ of objects in $\mathbb{C}$ will also be a small set, since there exists a 1-1 correspondence between the identity arrows and the objects.)
- When the collection $[u_1, u_2]_{\mathbb{C}}$ is a set, we refer to the generic category as a locally small category. (The collection A or Ob$(\mathbb{C})$ is, however, not required to be a set.)

- When the collection $[u_1, u_2]_{\mathbb{C}}$ is a class, we will refer to the generic category as a large category.
- When the collection $[u_1, u_2]_{\mathbb{C}}$ is a conglomerate, we will refer to the generic category as a quasi category.

  (A reader may refer to the appendix for the definitions of a set, a class, and a conglomerate.)

It will then be easily seen that we cannot form a quasi category of all the quasi categories. Any such attempt will lead to paradoxical set theoretical situations. However, we can, as examples, form a category of all the small categories as well as the quasi category of all the categories. These are legitimate entities as per the adopted definitions and are amenable to treatment similar to any other category. This often-quoted fact should no longer be surprising to the reader since it should be clear that the same partial binary algebra structure underlies any of them.

Here, we also note the following:

- A category $\mathbb{A}$ is small provided that its collection of objects, $\mathrm{Ob}(\mathbb{A})$, is a set. When this is not the case, that is, when $\mathrm{Ob}(\mathbb{A})$ is not a set, a category is large.

  When $\mathrm{Ob}(\mathbb{A})$ is a set, so is $\mathcal{A}$ a set and, therefore, category $\mathbb{A}$ must also be a set.
- Since each small category is a set, the collection of all the small categories is a class.
- A collection $\mathbb{C}\mathrm{at}$ of all the small categories then has as members small categories. However, this collection $\mathbb{C}\mathrm{at}$ is itself not small since the collection of all sets is not a set but a proper class.
- We therefore need to free the concept of a category from these set theoretical restrictions if we desire to consider many such mathematically legitimate collections as the collection of all categories. This is precisely what is achieved with the notion of a quasi category.

The above nomenclature should now be of historical significance only as it is the partial binary algebra structure that can be seen to be fundamental to category theory. That is why, we will, most often, simply refer to a generic category as only a category and will leave it to the reader to figure out the (aforementioned) type of that (generic) category. Still, one must guard against the set theoretical paradoxes from arising.

Now, the collection $\mathcal{A}$ of arrows can be empty. We will then have an empty category. This causes no more problems than the existence of an empty set of the usual set theory because an empty collection $\mathcal{A}$ causes no problems with the definition of the partial binary algebra or of a category.

Notice also that the collection $[u_1, u_2]_{\mathbb{C}}$ can be empty. This also causes no problems with the definition of a category.

4.6 Some Examples of Categories

Below, we provide examples of categories. An advanced reader would then notice that this list of examples is not quite exhaustive, and we do not claim its completeness either, since categories really abound in mathematics!

We have already considered one such example of a category in the form of a category of a group (of permutations of a finite set). We will also mention it in the list that is provided below.

Note that many of the categories mentioned below are structured sets (also called constructs) and structure-preserving functions between them. Category theory then enables us to retain this information by providing some means for a formal definition of a construct.

Example 1 $\mathbb{S}$et: Whose Ob($\mathbb{S}$et) is the collection (class) of all sets; $[A, B]$ is the set of all functions from A to B, id_A is the identity function on A, and $\circ$ is the usual composition of functions.

Example 2 $\mathbb{R}$ – $\mathbb{M}$od: For every ring R, a category $\mathbb{R}$ – $\mathbb{M}$od can be constructed whose collection of objects is the collection of all left-modules and $[A, B]$ is the set of all module homomorphisms or the R-linear transformations from A to B. The composition of arrows is the usual composition of homomorphisms. This category is called the category of left R-modules.

Example 3 In a way quite similar to the above construction of the category of left R-modules (we will list only the one pair, objects and arrows), we can construct the following categories:

- $\mathbb{M}$od – $\mathbb{R}$: Right R-modules and module homomorphisms; category of right R-modules
- $\mathbb{R}$ng: Rings and ring homomorphisms
- $\mathbb{F}$ield: Fields (for which $0 \neq 1$) and field homomorphisms
- $\mathbb{POS}$: Partially ordered sets and monotone functions
- $\mathbb{L}$at: Lattices and lattice homomorphisms
- $\mathbb{B}$oo$\mathbb{A}$lg: Boolean algebras and Boolean homomorphisms
- $\mathbb{G}$: A group as a category with a single object G and arrows as group elements
- $\mathbb{M}$: A monoid (i.e., a semigroup with an identity) as a category with a single object M and monoid elements as arrows
- $\mathbb{A}$b: Abelian groups and group homomorphisms
- $\mathbb{SG}$rp: Semigroups and semigroup homomorphisms
- $\mathbb{M}$on: Monoids and identity preserving semigroup homomorphisms
- $\mathbb{R}$ – $\mathbb{A}$lg: R-algebras and R-algebra homomorphisms where R is a commutative ring
- $\mathbb{T}$op$_2$: Hausdorff spaces and continuous functions
- $\mathbb{CR}$eg$\mathbb{T}_2$: Completely regular Hausdorff spaces and continuous functions
- $\mathbb{C}$omp$\mathbb{T}_2$: Compact Hausdorff spaces and continuous functions
- $\mathbb{T}$op$\mathbb{G}$rp: Topological groups and continuous homomorphisms
- $\mathbb{L}$in$\mathbb{T}$op: Linear topological Hausdorff spaces and continuous linear transformations
- $\mathbb{NL}$in$\mathbb{S}$p: Normed linear spaces and bounded (=continuous) linear transformations
- $\mathbb{B}$an$\mathbb{S}$p$_1$: Complex Banach spaces and bounded linear transformations
- $\mathbb{B}$an$\mathbb{S}$p$_2$: Complex Banach spaces and norm-decreasing linear transformations
- $\mathbb{CB}$an$\mathbb{A}$lg: Commutative complex Banach algebras with unit and norm-decreasing algebra homomorphisms
- $\mathbb{C}$*– $\mathbb{A}$lg: Complex Banach algebras with involution * satisfying $||a * a|| = ||a||^2$, and norm-decreasing, involution preserving algebra homomorphisms

Example 4 The following constructs, that is, categories of structured sets and structure-preserving functions between them, $\circ$ will always be the composition of functions and id_A will always be the identity function on A:

- $\mathbb{V}$ec: All real vector spaces and all linear transformations between them
- $\mathbb{G}$rp: All groups and all homomorphisms between them
- $\mathbb{T}$op: All topological spaces and all continuous functions between them
- $\mathbb{R}$el: With objects all pairs (X, ρ), where X is a set and ρ is a (binary) relation on X Morphisms $f: (X, \rho) \to (Y, \sigma)$ are relation-preserving maps, that is, maps $f: X \to Y$ such that $x\rho x'$ implies $f(x)\sigma f(x')$.

Example 5 For constructs, it is often clear what the morphisms should be once the objects are defined. However, this is not always the case. For instance:

- There are at least three natural constructs each having as objects all metric spaces, namely
  - $\mathbb{M}$et: With morphisms all nonexpansive maps (=contractions)
  - $\mathbb{M}\mathrm{et}_u$: With morphisms all uniformly continuous maps
  - $\mathbb{M}\mathrm{et}_c$: With morphisms all continuous maps
- There are at least two natural constructs each having as objects all Banach spaces, namely
  - $\mathbb{B}$an: With morphisms all linear contractions
  - $\mathbb{B}\mathrm{an}_b$: With morphisms all bounded linear maps (=continuous linear maps = uniformly continuous linear maps)

Example 6 $\mathbb{M}$at: With objects all natural numbers, and for which $[m, n]$ is the set of all real $m \times n$ *matrices,* $\mathrm{id}_n : n \to n$ is the unit diagonal $n \times n$ matrix and the composition of matrices is defined by $A \circ B = BA$, where BA denotes the usual multiplication of matrices. (The objects and morphisms are not structured sets and structure-preserving functions in $\mathbb{M}$at.)

Example 7 $\mathbb{A}$ut: With objects all (deterministic, sequential, Moore) automata. Objects of $\mathbb{A}$ut are sextuples $(Q, \Sigma, Y, \delta, q_0, y)$, where Q is the set of states; Σ and Y are the sets of input symbols and output symbols, respectively; $\delta : \Sigma \times Q \to Q$ is the transition map; $q_0 \in Q$ is the initial state; and $y : Q \to Y$ is the output map. Morphisms from an automaton $(Q, \Sigma, Y, \delta, q_0, y)$ to an automaton $(Q', \Sigma', Y', \delta', q_{0'}, y')$ are triples (f_Q, f_Σ, f_y) of functions $f_Q : Q \to Q'$, $f_\Sigma : \Sigma \to \Sigma'$, and $f_y : Y \to Y'$, satisfying the following conditions:

- Preservation of transition: $\delta'(f_\Sigma(\sigma), f_Q(q)) = f_Q(\delta(\sigma, q))$
- Preservation of outputs: $f_Y(y(q)) = y'(f_Q(q))$
- Preservation of the initial state: $f_Q(q_0) = q_{0'}$

The objects and morphisms of $\mathbb{A}$ut are not structured sets and structure-preserving functions.

Example 8 Classes as categories: Every class X gives rise to a category $C(X) = (O, \text{hom}, \text{id}, \circ)$—the objects of which are the members of X and whose only morphisms are identities—as follows:

$$\mathcal{O} = X, \text{hom}(x,y) = \begin{cases} \varnothing & \text{if } x \neq y, \\ \{x\} & \text{if } x = y, \end{cases} \quad \text{id}_x = x, \text{and } x \circ x = x.$$

$C(\phi)$ is called the *empty category*. $C(0)$ is called the *terminal category* and is denoted by 1.

Example 9 Preordered classes as categories: Every preordered class, that is, every pair $(X, \leq)$ with X a class and $\leq$ a reflexive and transitive relation on X, gives rise to a category $C(X, \leq) = (O, \text{hom}, \text{id}, \leq)$—the objects of which are the members of X—as follows:

$$\mathcal{O} = X, \text{hom}(x,y) = \begin{cases} \{(x,y)\} & \text{if } x \leq y, \\ \varnothing & \text{otherwise,} \end{cases} \quad \text{id}_x = x$$

and $(y, z) \circ (x, y) = (x, z)$.

Example 10 Monoid as a category: Although we have already mentioned monoid as a category earlier, we provide some more details here. Every monoid $(M, \bullet, e)$, that is, every semigroup $(M, \bullet)$ with a unit e gives rise to a category $C(M, \bullet, e) = (O, \text{hom}, \text{id}, \circ)$—with only one object—as follows:

$$O = M, \text{hom}(M, M) = M, \text{id}_M = e, \text{and } y \circ x = y \bullet x.$$

Example 11 $\mathbb{S}\text{et} \times \mathbb{S}\text{et}$: As objects all pairs of sets (A, B), as morphisms from (A, B) to (A', B'), all pairs of functions (f, g) with $A \xrightarrow{f} A'$ and $B \xrightarrow{g} B'$, identities given by $\text{id}_{(A, B)} = (\text{id}_A, \text{id}_B)$, and composition defined by $(f_2, g_2) \circ (f_1, g_1) = f_2 \circ f_1, g_2 \circ g_1$. Similarly, for any categories

$\mathbb{A}$ and $\mathbb{B}$ one can form $\mathbb{A} \times \mathbb{B}$, or, more generally, for finitely many categories $\mathbb{C}_1, \mathbb{C}_2, \ldots, \mathbb{C}_n$, one can form the product category $\mathbb{C}_1 \times \mathbb{C}_2 \times \ldots \times \mathbb{C}_n$.

Example 12 Category of sets and injective (surjective, bijective) functions: Objects are sets and $[A, B]$ is the set of all injective (surjective, bijective) functions from set A to set B.

Example 13 Category of topological spaces and open functions: Objects are topological spaces, and arrows in $[A, B]$ are all the open functions from a topological space A to topological space B.

Example 14 $p\mathbb{S}$et: A category of pointed set has as objects pairs of the form (A, a), where A is a set and $a \in A$, and as arrows elements of set $[(A, a), (B, b)] = \{f | f : A \to B \text{ and } f(a) = b\}$. This is called a category of sets with base point or a category of pointed sets. By a similar construction, we can obtain $p\mathbb{T}$op as the category of pointed topological spaces. We can also construct a category of bipointed sets with objects as (A, a, b) and functions preserving the bases a and b.

Enough of these examples. Even a cursory glance at this (incomplete, unordered) list of examples should be sufficient to convince anyone that the concept of a category has indeed captured the heart of many of the diverse branches of mathematics. Apart from providing standard notations, to provide the uninitiated with a view of the expanse of the canvass of category theory was indeed one of the purposes of giving this list here.

In these examples, we began with a known mathematical structure, then collected objects similar to it, and then associated with members of that collection arrows forming a partial binary algebra to construct the associated category. It should thus be apparent to the reader that such a construction of a category involves identifying appropriate arrows to correspond to chosen objects for any mathematical study. Choosing objects here requires prior knowledge of their mathematical properties.

Then one may begin to wonder as to what use category theory can be put to if the construction of a category involves prior knowledge of objects and arrows of quite specific kind to correspond to those objects. Is it that category theory is just a way of recasting old mathematical structures in terms of a new language, with that new language not possessing any more power than those old mathematical methods?

Well. For anyone, the above are naturally arising, legitimate questions. In response to them, we note that the major goal of a category theorist, after the initial developments, has always been to arrive at a categorical procedure using which one can tell which mathematical object (ring, field, monoid, group, topological space, metric space, Banach space, vector space, automata, etc.) is the subject of study only on the basis of the properties of an arrow or those of their categorical collection. Then, provided, of course, that this categorical procedure is at hand or available, one would obviously not need any kind of prior knowledge of the mathematical nature of the objects and of the corresponding arrows to study these structures.

The above hunch of a category theorist, that the arrows "know" about the objects which they connect, rests on the fact that an object-free definition of a category is equivalent to any of its object-dependent definitions in that the same category theory is obtained by adopting any of these definitions. Thence, the object part is completely determined by the arrow part—and emphasizing this is the main thrust of the present discussion. Consequently, we think of characteristics of only the arrows using which we can tell mathematical nature of the object that is the subject of our study.

The question is, what exactly are these characteristics of the arrows? Over the years, category theorists have established many general categorical procedures, for example, that of describing the adjunction between two categories, which we will discuss in the chapters to come. However, at the present moment, we simply do not know whether any of these general procedures has succeeded in achieving the above goal.

We also discuss additive structures definable over a category. In general, a category is not additive. We then define the notion of measures in the categorical context. Measures are additive at a fundamental conceptual level and are special functors from any category to an additive category. Using categorical measures, we obtain a variety of mathematical structures over any category. For example, we are able to obtain a metric space structure or a vector space structure.

With measures, we may have achieved the goal of freeing category theory altogether of the necessity of possessing some prior knowledge about the mathematical structure of the objects and arrows for the construction of a category. One may, however, be able to recast the construction of measures in terms of some earlier categorical procedure. (As it is presently unclear as to whether this is so, our earlier comment about the ignorance regarding the achievement of the goal of category theory by any of the earlier categorical means appears justifiable.)

As a policy comment, we note here that the arrows, in the most general categorical context, are to be treated as the only basic constituents of a category, and not the objects. Therefore, to construct any general mathematical structure over the arrows of a generic category, we need to freely reorganize these arrows and construct another category whose arrows exhibit that structure. As we are only reorganizing the arrows of a general (generic) category for this construction of another category, the smallness condition of definition 3 gets automatically satisfied by the constructed category. We therefore need only to ensure that the reorganization of arrows results in a partial binary algebra and that the matching, the associativity, and the unit existence conditions of definition 3 are obeyed. Clearly, such structures are then constructible for any category.

Within the categorical context, these can be considered to be the origins of different mathematical structures. Is a hierarchy of such mathematical structures obtainable from a generic category? Many such issues will be the subjects of our discussion at a later stage.

4.7 Duality Principle

With our definition of a category, an interesting question emerges: if we reverse the directions of all the arrows in the collection $\mathcal{A}$ of arrows in a category $\mathbb{A}$ and with that reverse also the order of all the compositions that we can form from those arrows, do we still obtain a category? The answer to this question is not only in the affirmative but also an important aspect of category theory. The new category so obtained is called the dual category or the opposite category to category $\mathbb{A}$ and is defined formally as:

Definition 10 (dual or opposite category) For any generic category $\mathbb{A} = (\text{Ob}(\mathbb{A}),[-,-]_{\mathbb{A}},\text{id}, \circ)$, the dual or opposite category is defined to be category $\mathbb{A}^{\text{op}} = \left(\text{Ob}(\mathbb{A}),[-,-]_{\mathbb{A}^{\text{op}}},\text{id},\circ^{\text{op}}\right)$, where we have $[A,B]_{\mathbb{A}^{\text{op}}} = [B,A]_{\mathbb{A}}$ and $f \circ^{\text{op}} g = g \circ f$.

It is easy to check that $\mathbb{A}^{\text{op}}$ as defined above is indeed a category. In particular, we note that the units of the partial binary algebra of the arrows of $\mathbb{A}$ are also the units of the partial binary algebra of $\mathbb{A}^{\text{op}}$, since the objects of $\mathbb{A}$ and of $\mathbb{A}^{\text{op}}$ are the same.

A dual category has an important role within category theory. To understand this role of the dual category, we first need to understand what a statement in category theory is.

In general, a statement related to a category will consist of objects, arrows connecting objects, and the compositions of arrows. Any composition of arrows leads ultimately to some arrow within the same category. A statement or a property related to a category then consists only of arrows and objects within that category. We could define any categorical property as:

Definition 11 (property) Property P involves objects and arrows of a category. Hence, $P = P_{\mathbb{A}} = (A, B, \ldots, f, g, \ldots)$.

A property can be dualized by reversing the direction of each arrow and the order in which the arrows are composed. Such dual property will be denoted by P^{op} and we will refer to P and P^{op} as properties dual to each other. If for some property P we have that $P = P^{\text{op}}$, then we call the property P as being a self-dual property.

Notice now that any statement $S_{\mathbb{A}^{\text{op}}}(A) = (A, f)$ about an object A of category $\mathbb{A}^{\text{op}}$ is translatable into a logically equivalent statement $S_{\mathbb{A}}^{\text{op}}(A) = (A, f^{\text{op}})$ concerning the (same) object A of category $\mathbb{A}$, where by f^{op}, we mean an arrow f with its direction reversed.

Therefore, if we begin with any property $P_{\mathbb{A}}(X)$ about objects X in a category $\mathbb{A}$, then we can construct a dual property, $P_{\mathbb{A}}^{\text{op}}(X)$, also about objects X in the same category $\mathbb{A}$ as follows:

- In a property $P_{\mathbb{A}}(X)$, replace all the occurrences of $\mathbb{A}$ by $\mathbb{A}^{\text{op}}$ and call the resultant property $P_{\mathbb{A}^{\text{op}}}(X)$.
- Translate $P_{\mathbb{A}^{\text{op}}}(X)$ into a logically equivalent statement for

objects X in category $\mathbb{A}$ by reversing the directions of all the arrows involved in it. This is the dual property $P_{\mathbb{A}}^{\text{op}}(X)$.

This strategy can also be used to translate any property $P_{\mathbb{A}}(f)$ about an arrow f in $\mathbb{A}$ into its dual. Notice that $P_{\mathbb{A}}(X)$ and $P_{\mathbb{A}}^{\text{op}}(X)$ need not be the same statements in category $\mathbb{A}$.

For gaining clarity, consider the following concrete example of this procedure using a category $\mathbb{A}$.

Consider the following statement $P_{\mathbb{A}}(X)$: There exists in category $\mathbb{A}$ an object X with the property that there is one and only one arrow f from it to any other object A in $\mathbb{A}$, that is, $X \xrightarrow{f} A$ is the only arrow.

We first replace the occurrences of $\mathbb{A}$ in this statement by $\mathbb{A}^{\text{op}}$ and obtain the statement $P_{\mathbb{A}^{\text{op}}}(X)$ as follows: There exists in category $\mathbb{A}^{\text{op}}$ an object X with the property that there is one and only one arrow f from it to any other object A in $\mathbb{A}^{\text{op}}$, that is, $X \xrightarrow{f} A$ is the only arrow.

Next, we convert this statement $P_{\mathbb{A}^{\text{op}}}(X)$ to a logically equivalent statement $P_{\mathbb{A}}^{\text{op}}(X)$ in category $\mathbb{A}$: There exists in category $\mathbb{A}$ an object X with the property that there is one and only one arrow f to it from any other object A in $\mathbb{A}$, that is, $A \xrightarrow{f} X$ is the only arrow.

Thus, the statement $P_{\mathbb{A}}(X)$ and the statement $P_{\mathbb{A}}^{\text{op}}(X)$ are quite different and evidently involve entirely different objects X of the same category $\mathbb{A}$.

As an example that the statements $P_{\mathbb{A}}(X)$ and $P_{\mathbb{A}}^{\text{op}}(X)$ involve different objects of the same category $\mathbb{A}$, consider category $\mathbb{S}\text{et}$. Then the aforementioned statement $P_{\mathbb{S}\text{et}}(X)$ is a statement holding only for an empty set since it is the only object in category $\mathbb{S}\text{et}$ for which there is one and only one function (arrow) from it to any other set (another object of $\mathbb{S}\text{et}$). On the other hand, the aforementioned statement $P_{\mathbb{S}\text{et}}^{\text{op}}(X)$ holds only for a singleton set since it is the only object in category $\mathbb{S}\text{et}$ for which there is one and only one function (arrow) to it from any other set (another object of category $\mathbb{S}\text{et}$).

Now, we say that whenever $P_{\mathbb{A}} = (A, B, \ldots, f, g, \ldots)$ holds for all the objects in $\mathbb{A}$ and also for all the arrows in $\mathbb{A}$, then $\mathbb{A}$ has the property P or that $P(\mathbb{A})$ holds.

On the basis of the above, we can then prove that for all the categories $\mathbb{A}$ and all the properties P, the following hold:

- $(\mathbb{A}^{op})^{op} = \mathbb{A}$
- $P^{op}(\mathbb{A})$ holds iff $P(\mathbb{A}^{op})$

The following duality principle for the categories is an immediate consequence of the above:

Definition 12 (duality principle) Whenever a property P holds for all the categories, then the property P^{op} holds for all the categories.

Often the dual concept P^{op} of a concept P is denoted as co-P. A concept P is said to be self-dual iff $P = P^{op}$.

As a consequence of the aforementioned duality principle, we can also establish:

Proposition 3 If S is a categorical statement, then S holds for all the categories satisfying property P iff S^{op} holds for all the categories satisfying the dual property P^{op}.

In category theory, each result therefore has two equivalent forms. Notably, if we begin with one of the two equivalent forms, then the other form may appear to be a very different statement! However, it should be clear now that only one of these two forms actually needs to be proved; the other follows from the aforementioned duality principle. That is why duality is an important principle of category theory for us. It essentially halves our load of proving results in the categorical context!

4.8 Subcategory

Obviously, any arbitrary subcollection of the collection $\mathcal{A}$ of arrows in a category $\mathbb{A}$ need itself not form a partial binary algebra of arrows. Therefore, any arbitrary subcollection of $\mathcal{A}$ cannot, in its own rights, be considered as a category by itself.

Then a natural enquiry is about the conditions under which any subcollection of the arrows of a category can also be considered as a (sub) category. The following rules define this notion of a subcategory.

Definition 13 A category $\mathbb{A} = (\mathcal{A}, \circ^S)$ is said to be a subcategory of another category $\mathbb{B} = (\mathcal{B}, \circ)$ provided:

(i) $\mathcal{A} \subseteq \mathcal{B}$

(ii) For each $A, A' \in \mathbb{A}$, $[A, A']_{\mathbb{A}} \subseteq [A, A']_{\mathbb{B}}$

(iii) For each $\mathbb{A}$-object A, the $\mathbb{B}$-identity of A is also the $\mathbb{A}$-identity of that object A

(iv) The composition law $\circ^S$ in $\mathbb{A}$ is the restriction of the composition law $\circ$ in $\mathbb{B}$ to the arrows of $\mathbb{A}$

Notice that conditions (i), (ii), and (iv) of the above definition of a subcategory do not imply the condition (iii). Then we need to specify not only the objects but also the collection of arrows for a general subcategory of any category.

However, in the case of a special subcategory, we will get away by specifying only its objects, that is, its identity arrows. This is when each hom-collection of the subcategory is the same as the corresponding hom-collection of the category whose subcategory it is. Such a special category is the following.

Definition 14 A category $\mathbb{A}$ is called a full subcategory of $\mathbb{B}$ provided that in addition to the conditions of definition 13 of a subcategory, for each $A, A' \in \mathbb{A}$, $[A, A']_{\mathbb{A}} = [A, A']_{\mathbb{B}}$.

Because of the nature of full subcategories, a full subcategory of a category $\mathbb{B}$ can be specified by merely specifying its objects within $\mathbb{B}$.

4.9 Generating New Categories from Those Existing

A category is a partial binary algebra of arrows satisfying certain naturality axioms. On the basis of some property, we could form bunches of arrows of an existing category in such a manner as to generate also a partial binary algebra for the bunches so generated. If, in addition, we also manage to satisfy other conditions in definition 3 of a category, then we will have generated a new category from the old one. Note that within this procedure of generating a new category, the smallness condition of definition 3 is automatically satisfied because it is satisfied by the existing category.

4.9.1 Quotient Category

Recall that an equivalence relation R on a collection X is a transitive (xRy and yRz imply xRz for all $x, y, z \in X$), reflexive (xRx for all $x \in X$), and symmetric (xRy implies yRx for all $x, y \in X$) relation. Under any equivalence relation R on X, the collection X also gets partitioned into mutually disjoint or mutually distinct equivalence classes (bunches).

For the purpose of generating a new category from an existing category by forming appropriate bunches of its arrows, we may then consider an equivalence relation on the collection of arrows $\mathcal{A}$ of a category $\mathbb{A}$ and call it a congruence of arrows under certain conditions as follows.

Definition 15 An equivalence relation $\sim$ on the collection $\mathcal{A}$ of a category $\mathbb{A}$ is called a congruence on $\mathbb{A}$ provided:

- Every equivalence class under the relation $\sim$ is contained within $[u_C, u_D]_{\mathbb{A}}$ for some $u_C, u_D \in \mathcal{A}$.
- Whenever $f \sim f'$ and $g \sim g'$, we have that $g \circ f \sim g' \circ f'$, provided that the compositions are defined.

Notice that the first of the above conditions ensures matching conditions for the arrows in $\mathbb{A}$, while the second of these conditions is the compatibility of the binary operation $\circ$ or the composition law of the arrows in $\mathbb{A}$ and $\sim$.

Now, we can define a composition law $\tilde{\circ}$ for the equivalence classes (themselves considered as arrows) as follows.

Definition 16 If $\tilde{g}$ is to denote an equivalence class of arrow $g \in \mathcal{A}$ of category $\mathbb{A}$ under the congruence relation $\sim$, then

$$\tilde{g} \,\tilde{\circ}\, \tilde{f} = \widetilde{g \circ f}$$

defines the composition law for the equivalence classes.

By noting that the equivalence classes of the collection $\mathcal{A}$ under the equivalence $\sim$ are pairwise disjoint and, hence, distinct, it is easy to verify that the collection of equivalence classes $[\tilde{*}]$ of $\mathcal{A}$ under the equivalence relation $\sim$ together with the composition $\tilde{\circ}$ forms a category.

Notice also that every such category has the same collection of the units of the partial binary algebra and, hence, has the same objects as that of $\mathbb{A}$. This means that such a category is not obtained if the equivalence relation $\sim$ identifies two or more units of the partial binary algebra of $\mathbb{A}$.

Formally, we therefore have:

Definition 17 (quotient category) If $\sim$ is a congruence relation as per definition 15 on the collection $\mathcal{A}$ of the arrows of a category $\mathbb{A}$, then the pair $(\mathcal{E},\tilde{\circ})$, where ε is the collection of all the equivalence classes of $\mathcal{A}$ under the congruence relation $\sim$ and $\tilde{\circ}$ is the composition law for the equivalence classes as per definition 16, is a category that is called the quotient category of $\mathbb{A}$ with respect to $\sim$ and is denoted by $\mathbb{A}/\sim$.

4.9.2 Slice or Comma Category

There exists another way of generating a category from a given category $\mathbb{A}$, and it is that of considering all the arrows into a specific or a fixed object B of category $\mathbb{A}$. That is to say, we consider all the arrows of the form $X \xrightarrow{f} B$ in $\mathbb{A}$, where X in any object in $\mathbb{A}$.

This collection of all the arrows $* \xrightarrow{\#} B$ in $\mathbb{A}$ is often denoted by $\mathbb{A}/B$ and is called the slice of category $\mathbb{A}$ by its object B.

Consider now two arrows $X \xrightarrow{h} B$ and $Y \xrightarrow{g} B$ from the collection $\mathbb{A}/B$ and an arrow $f: X \to Y$ in category $\mathbb{A}$ such that $h = gf$. We then have the following commuting diagram:

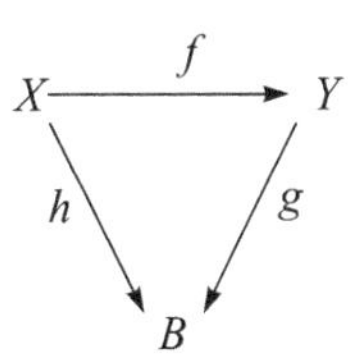

Effectively, arrow f, usually called the arrow over object B in $\mathbb{A}$, connectsarrows h and g.

The collection $\mathbb{A}/B$ as a collection of objects and the collection of all the arrows over object B in $\mathbb{A}$ both together form a category called the slice category of $\mathbb{A}$ over its object B. It is sometimes denoted by the same symbol $\mathbb{A}/B$. We will, however, use a different symbol below.

Formally, we therefore have the following definition:

Definition 18 (slice category of category $\mathbb{A}$ over its object B) It is a category, denoted by $\mathbb{A} \downarrow B$, whose objects are those $\mathbb{A}$-arrows having a target B and whose arrows from $A \xrightarrow{f} B$ to $C \xrightarrow{g} B$ are those $\mathbb{A}$-arrows $h : A \to C$ for which the triangle

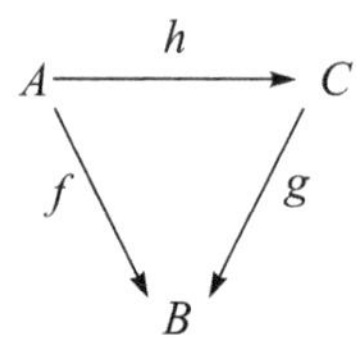

commutes. The composition of arrows in $\mathbb{A} \downarrow B$ is the same as the composition of arrows in $\mathbb{A}$.

A slice category of $\mathbb{A}$ over its object B is also called in the literature as a comma category of category $\mathbb{A}$ over its object B.

A reader may also have noticed another way of obtaining a new category—the one using all the arrows originating from a specific or a fixed object B of category $\mathbb{A}$. That is to say, we consider all the arrows of the form $B \xrightarrow{f} X$ in $\mathbb{A}$, where X is any object in $\mathbb{A}$. Formally, we have:

Definition 19 (slice category of object *B* over its category $\mathbb{A}$) It is a category, denoted by $\mathbb{A}\ B\ \mathbb{A}\ B$, whose objects are those $\mathbb{A}$-arrows having a source B and whose arrows from $B \xrightarrow{f} A$ to $B \xrightarrow{g} C$ are those $\mathbb{A}$-arrows $h : A \to C$ for which the triangle

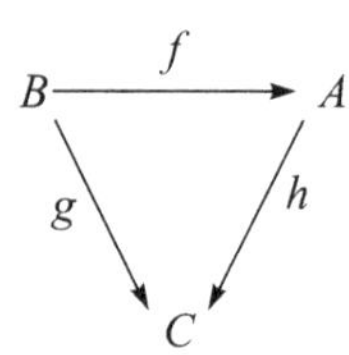

commutes. The composition of arrows in $\mathbb{A} \uparrow B$ is the same as the composition of arrows in $\mathbb{A}$.

A slice category of object B over its category $\mathbb{A}$ is also called in the literature a comma category of object B over its category $\mathbb{A}$. A reader may also have noticed the duality involved in these aforementioned two constructions.

4.9.3 Arrow Category

We may now think of pairing of arrows to generate a category from the given category $\mathbb{A}$, as is done in the following example of an arrow category of category $\mathbb{A}$, and is usually denoted by $\mathbb{A}^2$ or as $\mathbb{A}\mathrm{r}(\mathbb{C})$.

Definition 20 (arrow category of category $\mathbb{A}$) If $\mathbb{A} = (\mathcal{A}, \circ)$ is any category, then the arrow category of category $\mathbb{A}$, denoted by $\mathbb{A}^2$, has $\mathcal{A}$ as its class of objects and for which a $\mathbb{A}^2$-arrow from $A \xrightarrow{f} B$ to $C \xrightarrow{g} D$ is a pair (a, b) of $\mathbb{A}$-arrows such that the following square commutes:

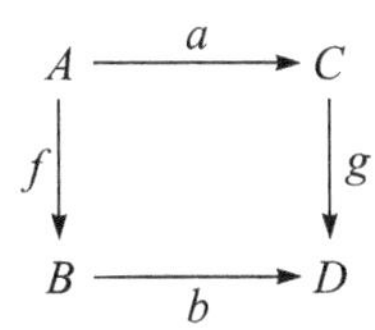

The composition law for the $\mathbb{A}^2$-arrows is defined as

$(a, b) \circ (c, d) = (a \circ c, b \circ d)$.

It is easy to verify that the above is indeed a category. In particular, notice that the $\mathbb{A}$-identities provide the $\mathbb{A}^2$-identities.

4.9.4 Triangle Category

Another construction of a category is that of a triangle category for category $\mathbb{A}$. It is denoted by $\mathbb{A}^3$. A formal definition of this category goes as:

Definition 21 (triangle category for category $\mathbb{A}$) If $\mathbb{A}$ is any category, then the triangle category for $\mathbb{A}$, denoted by $\mathbb{A}^3$, whose objects are precisely all the commutative triangles of $\mathbb{A}$ and for which an $\mathbb{A}^3$-arrow from

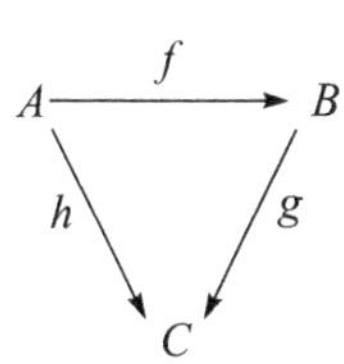

to

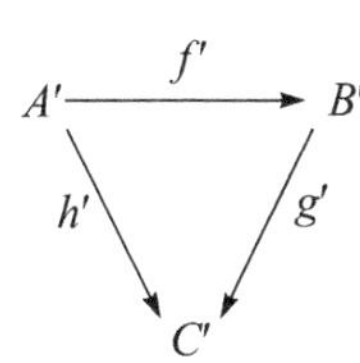

is an ordered triple (a, b, c) of $\mathbb{A}$-arrows $A \xrightarrow{a} A'$, $B \xrightarrow{b} B'$, and $C \xrightarrow{c} C'$ such that each square in the following diagram commutes:

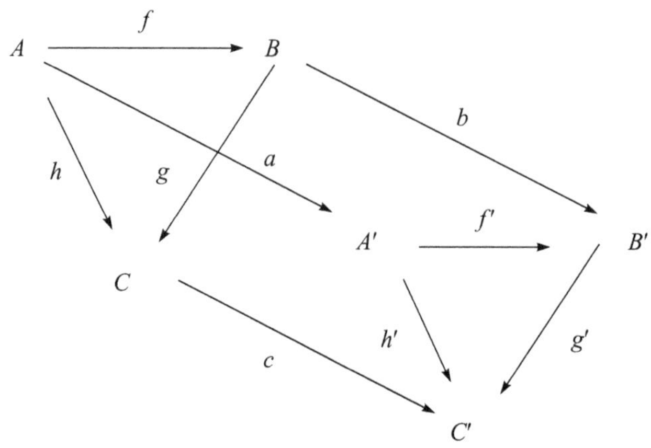

The law of the composition of $\mathbb{A}^3$-arrows is defined as

$$(a, b, c) \circ (a', b', c') = (aa', bb', cc').$$

4.9.5 Product and Sum Categories

At the next level of our constructions of new categories from the existing ones, we consider two categories $\mathbb{A} = (\mathcal{A}, \circ)$ and $\mathbb{B} = (\mathcal{B}, *)$. Since the cartesian product $\mathcal{A} \times \mathcal{B}$ of the collections $\mathcal{A}$ and $\mathcal{B}$ is a well-defined concept, we can consider a pair of arrows (f, g) with $f \in \mathcal{A}$ and $g \in \mathcal{B}$. We can define the composition of such pairs of arrows by

$$(f_1, g_1) \circ^p (f_2, g_2) = (f_1 \circ f_2, g_1 * g_2)$$

whenever, and iff, the composition $f_1 \circ f_2$ is defined in $\mathbb{A}$ and $g_1 * g_2$ is defined in $\mathbb{B}$, that is, the left hand side is defined only when the compositions on the right hand side are defined.

It is then easy to verify that $(\mathcal{A} \times \mathcal{B}, \circ^p)$ defines a category, called the product category, denoted as $\mathbb{A} \times \mathbb{B}$, of category $\mathbb{A}$ and category $\mathbb{B}$.

Evidently, we can extend this procedure to any number of categories to construct their product category.

On the other hand, a disjoint union, $\mathcal{A} \uplus \mathcal{B}$, of the collections $\mathcal{A}$ and $\mathcal{B}$ is also well defined. Consequently, we can define an arrow for the disjoint union as a pair (f, i) where i is the index; we can, for example, have $i = 1$ for $\mathcal{A}$ and $i = 2$ for $\mathcal{B}$. The composition operation on these arrows is then defined as

$(f, i) \circ^S (g, j) = (f \circ g, i)$ if and only if $i = j$.

It is then easy to verify that $(\mathcal{A} \uplus \mathcal{B}, \circ^S)$ defines a category, called the sum category, denoted as $\mathbb{A} \coprod \mathbb{B}$, of categories $\mathbb{A}$ and $\mathbb{B}$.

Evidently, we can also extend this procedure to any number of categories to construct their sum category.

4.9.6 Category of Families of Objects of a Category

Still another way to generate a new category from the existing one is the following construction that uses families of objects of a category $\mathbb{C}$. Notice that one has to be careful about the smallness condition here.

Let us denote by A_I the family $(A_i)_{i \in I}$ of objects in an arbitrary category $\mathbb{C}$, indexed by some collection I. Category $\mathbb{F}$amily($\mathbb{C}$) of the families of objects in $\mathbb{C}$ can then be defined in the following manner:

- Objects in category $\mathbb{F}$amily($\mathbb{C}$) are the families A_I, B_J, C_K, . . . of objects in $\mathbb{C}$
- An arrow from A_I to B_J in category $\mathbb{F}$amily($\mathbb{C}$) then involves a map of indexing collections I and J to associate index $i \in I$ to index $j \in J$ and a family of arrows from $\mathrm{A}_i \in \mathrm{A}_I$ to those $B_{j(i)} \in B_J$ for each $i \in I$. An arrow from A_I to B_J is then a pair $(f, \mathfrak{f})$, with $f : I \to J$ as a map of collections and $\mathfrak{f}$ as a family $\mathfrak{f} : \left(A_i \xrightarrow{\mathfrak{f}_i} B_{j(i)} \right)_{i \in I}$ of arrows in category $\mathbb{C}$.
- For any two arrows $(f, \mathfrak{f}) : A_I \to B_J$ and $(g, \mathfrak{g}) : B_J \to C_K$ in category $\mathbb{F}$amily($\mathbb{C}$), the composition $(g, \mathfrak{g}) \circ (f, \mathfrak{f})$ is defined as arrow $(h, \mathfrak{h}) : A_I \to C_K$ such that $h = gf$ and $\mathfrak{h}_i = \mathfrak{g}_{f(i)}\mathfrak{f}_i$.

An identity arrow for A_I in $\mathbb{F}$amily($\mathbb{C}$) is the identity map $\mathrm{id}_I : I \to I$ and the family, $\{\mathrm{id}_{A_i}\}_{i \in I}$, of identity arrows for $\mathbb{C}$-objects A_i, $i \in I$. It is easy to verify that $\mathbb{F}$amily($\mathbb{C}$) is indeed, as stated above, a category by checking that the identity arrows of $\mathbb{F}$amily($\mathbb{C}$) act also as identities with respect to the above composition, which is easily seen to be associative. Notice that the smallness condition in the definition of a category can also be satisfied in this construction. Note also that a family can be empty.

Now, since we can regard objects in $\mathbb{C}$ as one-member families, category $\mathbb{C}$ is a full subcategory of $\mathbb{F}$amily($\mathbb{C}$).

It will turn out that every object of category $\mathbb{F}\text{amily}(\mathbb{C})$ is a coproduct of some objects of category $\mathbb{C}$. (A product and a coproduct are defined in the next chapter.) This feature of the (freely constructible) category $\mathbb{F}\text{amily}(\mathbb{C})$ will be be seen to be useful at a later stage to define measures in the categorical context.

4.10 Categories Presented as Diagrams

Firstly, we note the following terminology:

Definition 22 (discrete category) Any category $\mathbb{A}$ is called discrete provided that all of its arrows are the identities of the corresponding partial binary algebra of arrows.

As per the above definition of a discrete category, its hom-collection is given as

$$[A, B] = \begin{cases} \emptyset & \text{for } A \neq B \\ \text{id}_A & \text{for } A = B. \end{cases}$$

Definition 23 (connected category) Any category $\mathbb{A}$ is called connected iff for every pair (A, B) of its objects, the collection $[A, B]_{\mathbb{A}} \neq \emptyset$. That is to say, none of the hom-collections is empty in a connected category $\mathbb{A}$.

It is easy to see that the only category that is a discrete as well as a connected category is the empty category.

Sometimes, a category has so few objects and arrows that it is convenient to present it in terms of a diagram that has all the objects represented as dots, •, and nonidentity arrows as arrows. Then the following diagrams

 and 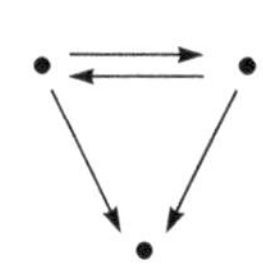

are categories.

However, every such diagram is not a category; for example, the following diagrams are not categories:

 and 

In such diagrams, the composition of the presented arrows is not any of the presented arrows. This violates the corresponding condition in the definition of a category and, hence, such diagrams are not categories.

We note also that categories corresponding to ordered sets ϕ, {0}, {0, 1} . . . of natural numbers (with the usual order of the natural numbers serving for the arrows of the corresponding categories) are often denoted by the symbols 0, 1, 2, ...

Diagrammatically, we then have

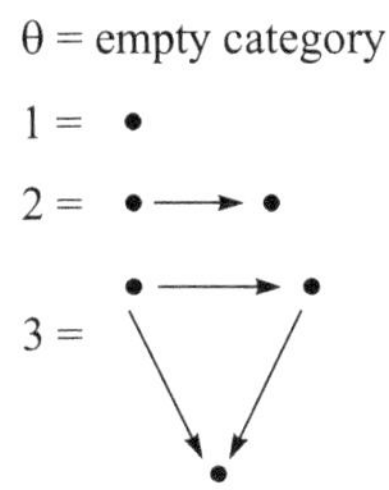

and so on.

4.11 Summary

One should now note that the role of any identity arrow in the collection of arrows in a category is, in fact, double: it is an arrow, and it has a unique label that we call as its corresponding object. An identity arrow essentially identifies a unique corresponding object for us. Therefore, many would consider an identity arrow to be equivalent or even identical to a corresponding object in its category. That is, no distinction is made between an identity arrow and the corresponding object in a category.

As we have seen, an object-free definition of a category, after all, exists. It is indeed a useful practice therefore to not distinguish identity arrows from the corresponding objects.

To conclude this chapter, we note that there exist many other ways of constructing categories from the existing ones apart from those that we considered in this chapter. Some of these ways involve now the properties of the arrows in the existing category.

Moreover, when we consider structure-preserving functions from the collection of arrows of one category to that of another category, we will have arrows from one to another category. These are called functors.

Using functors as arrows and categories as objects, we can therefore construct a category of all the categories, subject of course to an important caveat that we do not violate the smallness condition in the definition of a category in any such construction.

We will now turn to properties of arrows and objects in a category. It may then be noted that every category need not possess any arrows or objects with such properties. Still, these properties are important to the analysis of mathematical structures that form categories.

Chapter 5

Properties of Arrows and Objects

5.1 Introduction

Some arrows and some objects in a category can possess certain specific properties that are, many times, of importance to the analysis of a mathematical structure forming a category.

We have already considered two examples of such properties of objects in Section 4.7, Chapter 4. As an example of a categorical property of an object, we considered that there exists in category $\mathbb{A}$ an object X with the property that there is one and only one arrow from it to any other object A in $\mathbb{A}$, that is, $X \xrightarrow{f} A$. Such an object is called an initial object of category $\mathbb{A}$. We also saw that, in category $\mathbb{S}$et of the sets and functions, there does exist such an object, the empty set, from which there is one and only one function (arrow) to any other set.

Another example is the dual of the above statement: there exists in category $\mathbb{A}$ an object X with the property that there is one and only one arrow to it from any other object A in $\mathbb{A}$, that is, $A \xrightarrow{f} X$. Such an object is called a terminal object. We had then seen that in category $\mathbb{S}$et of sets and functions, there does exist such an object, the singleton set, to which there is one and only one function (arrow) from any other set.

Subtlety in Relativity
Sanjay M. Wagh

ISBN 978-981-4774-44-4 (Hardcover), 978-1-351-23832-8 (eBook)
www.panstanford.com

However, there exist categories in which neither an initial object nor a terminal object exists. One example of such a category is provided by a partially ordered set (poset) considered as a category, denoted as pos. Category pos of a poset is quite often used as a test bed for checking the use of the conceptions of category theory.

Any poset $(P, \leq)$ gives rise to a category for which the elements of the poset P are the objects and there exists one and only one arrow from p to q if and only if (iff) $p \leq q$ with $p, q \in P$. The arrows of this category, pos, are therefore pairs (p, q) such that $p \leq q$, and the domain and the codomain operations of definition 6, Chapter 4, of a category are the first and second projection operations for this pair.

Identity arrows of objects arise uniquely from the reflexivity of the partial order. The law of the composition of arrows also arises uniquely from the transitivity of the partial order of a poset.

A poset P is complete iff every subset of P has a lowest upper bound (denoted as lub, sup, or join) and a greatest lower bound (denoted as glb, inf, or meet).

Then unless there exists a smallest element in the poset, there is no initial object, and unless there exists a largest element in the poset, there is no terminal object in category pos. Hence, pos in general, has neither an initial object nor a terminal object. The initial and terminal objects both simultaneously exist only for a complete poset.

Nonetheless, when existing in a category, an initial object and a terminal object play quite important roles in the analysis of the properties of that category.

Such properties of the arrows in a category are described in this chapter. But a term to be defined first.

5.2 Diagram and Commuting Diagram

Whenever we have three arrows f, g, and h forming the following type of a triangle and for which $h = gf$,

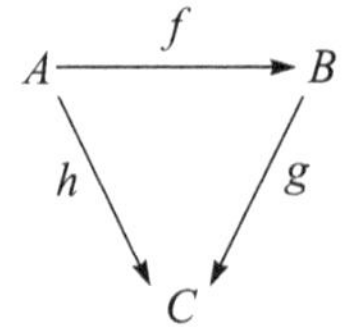

we say that arrow h factors through B or that the above triangle commutes. (Remember that h (= gf) is a specific arrow from A to C, and for another arrow $k : A \to C$ the above triangle need not commute.)

We have already used this concept in Chapter 4. A general definition of a commuting diagram is then:

Definition 24 (diagram and commuting diagram) By diagram in category we mean any collection of its objects together with any collection of arrows connecting various of these objects.

Any diagram is said to commute when, given any two objects in the diagram and any two arrows between them, such arrows obtained by the composition of all the arrows in the diagram, these two arrows are equal.

Commuting diagram is an important concept of category theory, since its statements essentially amount to saying that a certain diagram commutes in the category. It is no exaggeration that the aim of a category theorist is to always state any of the statements of category theory in terms of a commuting diagram. Also, many of the commuting diagrams may then involve arrows of a specific kind and, hence, such diagrams may, in general, not exist within a generic category.

5.3 Section and Retraction Arrows

Definition 25 (section) An arrow f in category $\mathbb{A}$ is called an $\mathbb{A}$-section provided there exists some arrow g in $\mathbb{A}$ such that $gf = u_C$ for some u_C. Arrow f is then said to have a left-inverse. Diagrammatically, the following commutes:

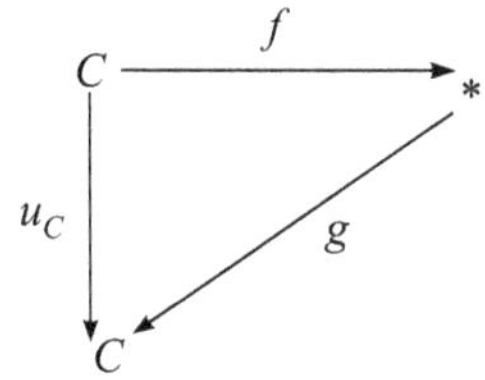

The left-inverse of a section need not be unique.

Proposition 3 If arrows f and g in a category $\mathbb{A}$ are $\mathbb{A}$-sections, then so is gf an $\mathbb{A}$-section whenever the composition is defined.

Proof Let $h \in \mathbb{A}$ be such that $hf = u_c$ and $k \in \mathbb{A}$ be such that $kg = u_D$, since f and g are given to be $\mathbb{A}$-sections. Consider the composition of arrows $(hk)(gf)$. We have $(hk)(gf) = h\,(kg)f = hu_Df = hf = u_C$ from the associativity of the composition of arrows in $\mathbb{A}$. Thus, whenever f and g in $\mathbb{A}$ are sections, their composition fg is also an $\mathbb{A}$-section.

Proposition 4 If $f, g \in \mathbb{A}$ are two arrows and gf is an $\mathbb{A}$-section, then f is also an $\mathbb{A}$-section.

Proof If gf is a section, then there exists $m \in \mathbb{A}$ such that $m(gf) = u_C$ for some u_C. Using the associativity of the composition of arrows, we get $(mg)f = u_C$.

Definition 26 (retraction) An arrow f in $\mathbb{A}$ is called an $\mathbb{A}$-retraction provided there exists some arrow g in $\mathbb{A}$ such that $fg = u_B$ for some u_B. Arrow f is then said to have a right-inverse. Diagrammatically, the following commutes:

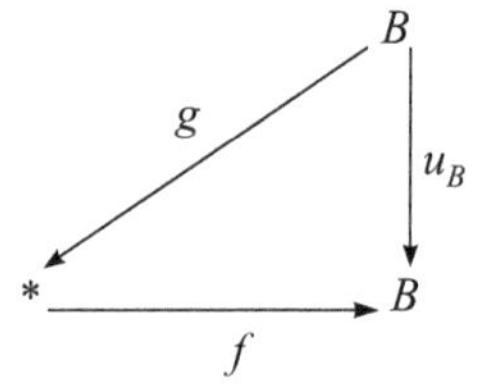

The right-inverse of a retraction need not be unique. (We invite the reader to provide an example.)

Proposition 5 A section is dual to a retraction.

Proof View the definition of a section as a statement $\mathcal{P}_{\mathbb{A}}$: an arrow f in category $\mathbb{A}$ has the property that there exists an arrow g in category $\mathbb{A}$ such that $g \circ f = u_A$ for some u_A.

Changing the occurrences of $\mathbb{A}$ to those of $\mathbb{A}^{\text{op}}$ we get the statement $\mathcal{P}_{\mathbb{A}^{\text{op}}}$. An arrow f in category $\mathbb{A}^{\text{op}}$ has the property that there exists an arrow g in category $\mathbb{A}^{\text{op}}$ such that $g \circ f = u_A$ for some u_A.

We convert this to its logically equivalent statement $\mathcal{P}_{\mathbb{A}}^{\text{op}}$: an arrow f in category $\mathbb{A}$ has the property that there exists an arrow g in category $\mathbb{A}$ such that $f \circ g = u_A$ for some u_A.

Clearly, $\mathcal{P}_{\mathbb{A}}^{\text{op}}$ is definition 26 of a retraction.

Observation 1 Because the left-inverse is not unique for a section and the right-inverse is not unique for a retraction, if f is an $\mathbb{A}$-section and g is an $\mathbb{A}$-retraction, the compositions $g \circ f$ and $f \circ g$, even when defined, need not be such that $g \circ f = \mathrm{id}_A$ and $f \circ g = \mathrm{id}_B$ for some objects $A, B \in \mathbb{A}$.

Having proved the results for a section, we need not prove the following two results for a retraction but can appeal to the useful principle of duality.

Henceforth, we will also mark the dual result, whenever stated, by a symbol "♣" at the end to indicate that its dual statement was considered earlier.

Proposition 6 If arrows f and g in a category $\mathbb{A}$ are $\mathbb{A}$-retractions, then so is fg an $\mathbb{A}$-retraction, whenever the composition is defined. ♣

Proposition 7 If $f, g \in \mathbb{A}$ are two arrows and gf is an $\mathbb{A}$-retraction, then g is also an $\mathbb{A}$-retraction. ♣

5.4 Isoarrow and Related Topics

In many fields of mathematical study, the conception of an isomorphism occurs and various results then naturally hold modulo or up to an isomorphism. Categorically, this is the concept of an isoarrow defined as follows.

Definition 27 (isoarrow) An arrow $A \xrightarrow{f} B$ in a category is called an isomorphism or an isoarrow provided there exists an arrow $B \xrightarrow{g} A$ such that $g \circ f = \mathrm{id}_A$ and $f \circ g = \mathrm{id}_B$. We often write $A \tilde{\to} B$.

Such an arrow g is called an inverse of f and is usually denoted by f^{-1}. Diagrammatically, the following commutes:

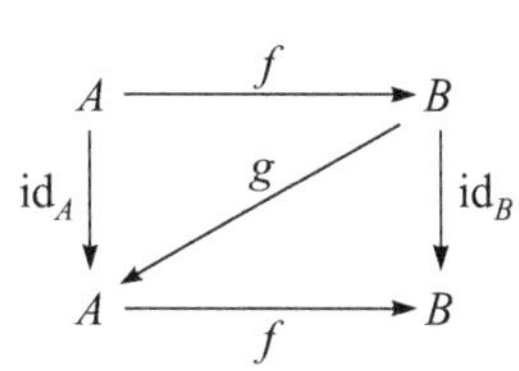

Proposition 8 The statement "An arrow f is an isoarrow in the category" is self-dual in any category.

Proposition 9 An inverse of an isoarrow f existing in any category is unique in its category.

Proof Let g be the inverse of an isoarrow f in a category $\mathbb{A}$. If there existed another inverse g' such that $fg' = \mathrm{id}_B$ *and* $g'f = \mathrm{id}_A$, then by composition, $gfg' = \mathrm{id}_A \circ g' = g \circ \mathrm{id}_B$, which implies $g = g'$.

Proposition 10 If f is an arrow in any category $\mathbb{C}$, then the following are equivalent (TFAE):

- f is an isoarrow.
- f has exactly one left-inverse, say h, and exactly one right-inverse, say k. Furthermore, they are equal ($h = k$).

Proof It follows from the definition of an isoarrow and from the properties of a section and a retraction in a category $\mathbb{C}$.

Proposition 11 If f is an isoarrow, then f^{-1} is also an isoarrow, and we have $f = (f^{-1})^{-1}$.

Proof That f^{-1} is an isoarrow is clear. Arrow $(f^{-1})^{-1}$ satisfies, by definition, $f \circ (f^{-1})^{-1} = \mathrm{id}_A$ and $(f^{-1})^{-1} \circ f^{-1} = \mathrm{id}_B$. But an inverse of an isoarrow is unique, and we also have $f^{-1} \circ f = \mathrm{id}_A$ and $f = (f^{-1})^{-1} f$.

Hence, we have:

Proposition 12 In any category, the composition of two isoarrows, whenever defined, is an isoarrow in that category.

Proof Let f and g be two isoarrows in category $\mathbb{A}$. Then by the definition of an isoarrow, there exist arrows $p, q \in \mathbb{C}$ such that $pf = \mathrm{id}_A$, $fp = \mathrm{id}_B$, $qg = \mathrm{id}_B$, and $gq = \mathrm{id}_C$. If gf is as defined, then $(pq)(gf) = p(qg)f = pf = \mathrm{id}_A$ and $(gf)(pq) = g(fp)q = gq = \mathrm{id}_C$. Hence, the composite gf of two isoarrows f and g is an isoarrow in $\mathbb{C}$.

Proposition 13 If f and g are isoarrows in a category $\mathbb{A}$, then $(f \circ g)^{-1} = g^{-1} \circ f^{-1}$ whenever the compositions are defined.

Proof Let $g : A \to B$ and $f : B \to C$ be two isoarrows in category $\mathbb{A}$. Then $f \circ g$ is an isoarrow in that category. There exists a unique inverse arrow m such that $m \circ (f \circ g) = \mathrm{id}_A$ and $(f \circ g) \circ m = \mathrm{id}_C$. But $(g^{-1} f^{-1})(f \circ g) = \mathrm{id}_A$ and $(f \circ g) \circ (g^{-1} \circ f^{-1}) = \mathrm{id}_C$.

Proposition 14 An isoarrow in any category is both a section and a retraction in it.

We invite the reader to check whether the converse of the above proposition, that any arrow that is, simultaneously, a section and a retraction in a category is also an isoarrow in it, holds.

Proposition 15 If f and g are arrows in a category $\mathbb{A}$ such that $g \circ f$ is an isoarrow of $\mathbb{A}$, then f is an $\mathbb{A}$-section and g is an $\mathbb{A}$-retraction.

Proof Let arrows f and g in a category $\mathbb{A}$ be such that gf is an isoarrow. Then by the definition of an isoarrow, there exists a unique inverse m of gf such that $mgf = \text{id}_A$ and $gfm = \text{id}_C$. Then f has a left-inverse mg in category $\mathbb{A}$ and, hence, is an $\mathbb{A}$-section. Also, g has a right-inverse fm in category $\mathbb{A}$ and, hence, is an $\mathbb{A}$-retraction.

For our earlier example of a group as a category having a single object G, every arrow (every group element) is an isoarrow. In $\mathbb{S}$et, an isoarrow is precisely the one-to-one (1-1) and onto map or the 1-1 correspondence. In the case of a monoid of natural numbers under addition, 0 is the only section, the only retraction, and the only isoarrow in its category.

5.4.1 Isomorphic Objects

Definition 28 (isomorphic objects) We shall call any two objects of an arbitrary category $\mathbb{C}$, connected by an isoarrow in that category, as isomorphic objects of that category $\mathbb{C}$.

For any category $\mathbb{A}$, "is isomorphic to" is an equivalence relation on the collection Ob($\mathbb{A}$).

Notice, however, that being isomorphic does not have the same sense as that of being equal. Two objects (e.g., two tables) can be isomorphic without being equal or the same. An equality is sameness, while an isomorphism is the "sameness of the resemblance" of objects.

5.4.2 Isomorphism-Dense Subcategory

Definition 29 (isomorphism-dense subcategory) A subcategory $\mathbb{A}$ of a category $\mathbb{B}$ is called an isomorphism-dense subcategory of category $\mathbb{B}$ provided that for each $\mathbb{B}$-object B, there exists some $\mathbb{A}$-object A such that A is $\mathbb{B}$-isomorphic to B.

The following is an example of an important subcategory of any category:

Definition: A skeleton of a category is a full, isomorphism-dense subcategory in which no two distinct objects are isomorphic.

Example 15 A full subcategory of all cardinal numbers is a skeleton for $\mathbb{S}$et.

Example 16 A full subcategory determined by the powers $\mathbb{R}^m$, where m runs through all cardinal numbers, is a skeleton for $\mathbb{V}$ec.

Proposition: Every category has a skeleton.

We shall return to this notion of a skeleton at a later stage, in Chapter 6, when we consider functors, and there we will prove the above proposition.

5.4.3 Isomorphism-Closed Subcategory

Definition 30 (isomorphism-closed subcategory) A subcategory $\mathbb{A}$ of a category $\mathbb{B}$ is called isomorphism-closed provided every $\mathbb{B}$-isomorphism with a domain in A belongs to $\mathbb{A}$.

(It may be observed that if an arrow belongs to a category $\mathbb{A}$, then so do its source and its target.)

Proposition 16 Let $\mathbb{A}$ be a subcategory of a category $\mathbb{B}$. Then any $\mathbb{A}$-section (or a retraction or an isoarrow) is also a $\mathbb{B}$-section (or a retraction or an isoarrow).

Proof Recall that $[A, B]_{\mathbb{A}} \subseteq [A, B]_{\mathbb{B}}$. Consider an $\mathbb{A}$-section $f \in [A, B]_{\mathbb{A}}$ having a left-inverse g such that $gf = \mathrm{id}_A$. Then $g \in [B, A]_{\mathbb{A}} \subseteq [B, A]_{\mathbb{B}}$. $\mathbb{B}$-identity of object A is also the $\mathbb{A}$-identity of A. Hence, f is also a $\mathbb{B}$-section. Similar is the case for other parts of the proposition.

Proposition 17 If $\mathbb{A}$ is a full subcategory of a category $\mathbb{B}$, then every $\mathbb{A}$-arrow that is a $\mathbb{B}$-section (or a retraction or an isoarrow) is necessarily an $\mathbb{A}$-section (or a retraction or an isoarrow).

Proof Recall that for a full subcategory, we have $[A, B]_{\mathbb{A}} = [A, B]_{\mathbb{B}}$. Consider a $\mathbb{B}$-section $f \in [A, B]_{\mathbb{B}}$ having a left-inverse g such that $gf = \mathrm{id}_A$. Then $g \in [B, A]_{\mathbb{B}} = [B, A]_{\mathbb{A}}$. Thus arrows f and g are in $\mathbb{A}$. Also, a $\mathbb{B}$-identity of object A is the $\mathbb{A}$-identity of A. Hence, f is also a $\mathbb{A}$-section. Similar is the case for other parts of the proposition.

Note that category $\mathbb{A}$ is required to be a full subcategory of category $\mathbb{B}$ in the last proposition.

For example, let $\mathbb{C}/\sim$ be the quotient category of a category $\mathbb{C}$. Then if f is a $\mathbb{C}$-section (or a retraction or an isoarrow) then the equivalence class $\tilde{f}$ is a $(\mathbb{C}/\sim)$-section (or a retraction or an isoarrow).

As an aside, we note that an arrow g in a category $\mathbb{C}$ is called a quasi inverse for another arrow f of $\mathbb{C}$ iff $fgf = f$. Then every $\mathbb{C}$-arrow that has a quasi inverse is itself a quasi inverse of some $\mathbb{C}$-arrow.

5.5 Monoarrow and Related Topics

The main motivation now is an injective (1-1) function of one set into another set. The statement "For a function $f : A \to B$, $f(x) = f(y)$ iff $x = y$ with $x, y \in A$, that is, function f is injective" is equivalent to the statement "For all functions a, b such that $fa = fb$, it follows that $a = b$."

Consider an injective f and an arbitrary set Y. If a, $b : Y \to A$, with $fa = fb$, are such that $a \neq b$, then there would exist some $z \in Y$ such that $a(z) \neq b(z)$ and $f(a(z)) \neq f(b(z)$ for a 1-1 function, which contradicts $fa = fb$. Functions a and b then agree on all elements with $a(z) = b(z)$ and, hence, are identical. To see the other part, let $f : A \to B$ be a function. Consider a singleton set $X = \{x\}$ and functions a, $b : X \to A$ such that $a(x) = p$ and $b(x) = q$ with $p, q \in A$. It is now given that $a = b$ follows whenever we have $fa = fb$. The images of a and b are then required to be identical on all the singletons. Thus, for function $f(a(x)) = f(b(x))$ only if $p = q$. Then f is injective.

An inclusion function $\rightarrowtail : A \to B$ of a set A into another set B is then an injective function from A to B.

We can write this above as follows: given the diagram $X \underset{b}{\overset{a}{\rightrightarrows}} A \xrightarrow{f} B$ such that $fa = fb$, we have $a = b$ for an injective function f. We then say that an injective function f is left-cancelable in the composition of functions.

5.5.1 Monoarrow

The above conception leads us to the following definition of a monomorphism, also known as a monic arrow or a monoarrow:

Definition 31 (monoarrow) An arrow $A \xrightarrow{f} B$ in a category $\mathbb{C}$ is called a monomorphism (or monic or monoarrow) iff for all $\mathbb{C}$-arrows g and h such that $f \circ g = f \circ h$, we have $g = h$, that is, if f is left-cancelable with respect to the composition of arrows.

Diagrammatically, in $X \underset{g}{\overset{h}{\rightrightarrows}} A \xrightarrow{f} B$, $fg = fh \Rightarrow g = h$ in category $\mathbb{C}$ for any object X, f being a monoarrow. For a monoarrow, the following notation is sometimes used:

$$A \rightarrowtail B$$

Definition 32 Two monoarrows $f : A \to B$ and $g : C \to B$ are called equivalent iff there exists an isoarrow $h : A \tilde{\to} C$ with $f = gh$, that is, the following triangle commutes:

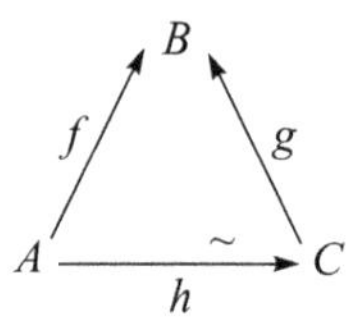

Proposition 18 If f and g are monoarrows in a category $\mathbb{C}$, then gf is a monoarrow in it whenever the composition is defined.

Proof Consider diagram $X \underset{k}{\overset{h}{\rightrightarrows}} A \xrightarrow{f} B \xrightarrow{g} C$ with f and g being monoarrows. Then if $(gf)h = (gf)k$, we have $g(fh) = g(fk)$. Since g is a monoarrow, we have $fh = fk$. As f is also a monoarrow, $h = k$.

Proposition 19 Every $\mathbb{C}$-section is also a $\mathbb{C}$-monoarrow for any category $\mathbb{C}$.

Proof Let $A \xrightarrow{f} B$ be a $\mathbb{C}$-section;, there then exists g such that $gf = \mathrm{id}_A$. Consider the diagram $X \underset{k}{\overset{h}{\rightrightarrows}} A \xrightarrow{f} B$ for which we have $fh = fk$. Consider the composition $g(fh) = g(fk)$.

Notice, however, that a $\mathbb{C}$-monoarrow f is not a $\mathbb{C}$-section necessarily, for there may not exist any arrow g in $\mathbb{C}$ with the property $gf = \mathrm{id}_A$ for some object A in $\mathbb{C}$.

Proposition 20 In the diagram $A \xrightarrow{f} B \xrightarrow{g} C$, if arrow gf is a monoarrow, then so is f a monoarrow in $\mathbb{C}$.

Proof Consider the diagram $X \underset{k}{\overset{h}{\rightrightarrows}} A \xrightarrow{f} B \xrightarrow{g} C$ and use the associativity property of the composition of arrows.

Proposition 21 In any category, TFAE:

- An arrow f is an isoarrow in it.
- An arrow f is a monoarrow as well as a retraction in it.

Proof If $f : A \to B$ is an isoarrow in a category, it is a retraction in that category as it has a right-inverse. It is a monoarrow, for in the diagram $X \underset{k}{\overset{h}{\rightrightarrows}} A \xrightarrow{f} B$, if $fh = fk$, then $f^{-1}(fh) = f^{-1}(fk)$ and $h = k$.

If $f : A \to B$ is any monoarrow that is also a retraction in the category, then it has a right-inverse $g : B \to A$ such that $fg = \mathrm{id}_B$. Consider $fgf = \mathrm{id}_B \circ f = f = f(gf)$. As f is a monoarrow, $f(gf) = f \circ \mathrm{id}_A$ implies $gf = \mathrm{id}_A$. Then f has a left-inverse and, hence, a retraction arrow f is also a section. Therefore, f is an isoarrow.

5.5.2 Subobject

Every inclusion function of a set into another set is a monoarrow in category $\mathbb{S}$et. Then whenever a monoarrow exists from an object A to object B in any category, we may think of an inclusion of A into B. This leads us to calling an object A as a subobject of object B.

However, category theory makes it clear that A being a subset of B is not only a property of set A but also a property of the naturally associated inclusion morphism from A into B. Notice that there can, in general, be more than one monoarrows from A to B. Consequently, the subobject of an object should be a monoarrow-dependent conception, as follows.

Definition 33 (subobject) Let $\mathcal{M}$ be a collection of monoarrows in any category $\mathbb{C}$. An $\mathcal{M}$-subobject A of an object B in any category $\mathbb{C}$ is then a pair (A, h) where $A \xrightarrow{h} B$ is a monoarrow. Furthermore, if h also happens to be a section in $\mathbb{C}$, then we will call the pair (A, h) as an $\mathcal{M}$-section of object B in category $\mathbb{C}$.

Because a subobject is an object-arrow pair, no notion of a subobject of a subobject exists to correspond to that of a subset of a subset. Nonetheless, in $\mathbb{S}$et, two subobjects (A, h) and (B, k) can be considered as being the same if $h[A] = k[B]$, that is, if the images of h and k are identical. A subobject (A, h) of C in $\mathbb{S}$et can also be considered to be smaller than the subobject (B, k) of C provided $h[A] \subseteq k[B]$.

There should then exist a preorder, that is, a reflective and transitive relation, on pairs (A, h) to correspond to these above notions in $\mathbb{S}$et. The following definitions are precisely such:

Definition 34 Let (A, h) and (B, k) be subobjects of C.

- Subobjects (A, h) and (B, k) are said to be isomorphic subobjects of C whenever there exists an isoarrow $f : A \to B$ with $h = k \circ f$, and we write $(A, h) \approx (B, k)$ for isomorphic subobjects.
- Subobject (A, h) is said to be smaller than the subobject (B, k), denoted by $(A, h) \leq (B, k)$, provided there exists an arrow $f : A \to B$ with $h = k \circ f$ such that the following triangle commutes:

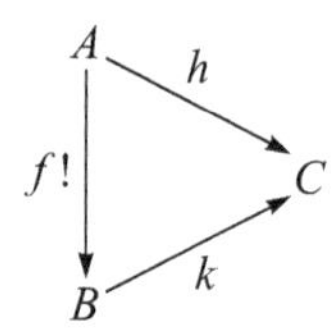

Notice that arrow f above must be a monoarrow. An exclamations mark is often used to indicate the existence and uniqueness of an arrow.

Proposition 22 For the collection of all subobjects of a given object:

- The relation of being isomorphic is an equivalence relation.
- The relation $\leq$ is a preorder, that is, it is reflexive and transitive. In general, it fails to be antisymmetric, but $(A, h) \leq (B, k)$ and $(B, k) \leq (A, h)$ imply that these two subobjects are isomorphic.

Proof The transitivity of the relation $(A, h) \approx (B, k)$ follows from the fact that the composition of isoarrows is an isoarrow, its symmetry follows from the fact that an inverse of an isoarrow is an isoarrow, and its reflexivity follows from the fact that an identity arrow is an isoarrow.

For the relation $\leq$, the transitivity follows from the fact that the composition of two monoarrows is a monoarrow and its reflexivity follows from the fact that an identity arrow is a monoarrow. The failure of the antisymmetry for the relation $\leq$ can be attributed to the fact that a composition of a section and a retraction is, in general, not an isoarrow.

Proposition 23 Two subobjects (A, a) and (B, b) of an object C in a category $\mathbb{C}$ are isomorphic subobjects of a $\mathbb{C}$-object C iff there exists a unique isoarrow $A \xrightarrow{h} C$ such that $b \circ h = a$.

In general, an object in a category need not have any subobjects, for there may not be any monoarrows to it from any other object in its category. (Think of an empty set in $\mathbb{S}$et.)

We also note here that the collection of all the subobjects of a given object is partially ordered by the relation $\leq$. This collection is a Heyting algebra that is discussed in Appendix B.

To note the often-used terminology in the literature, we note here the following definition:

Definition 35 ($\mathcal{M}$-well-powered and well-powered categories)

- A category $\mathbb{C}$ is called $\mathcal{M}$-well-powered provided no $\mathbb{C}$-object has a proper class of pairwise nonisomorphic $\mathcal{M}$-subobjects, "pairwise" here meaning that any pair consisting of distinct members is assumed to be nonisomorphic.
- In case $\mathcal{M}$ is the class of all monoarrows of $\mathbb{C}$, then $\mathcal{M}$-well-powered category $\mathbb{C}$ is called well-powered.

5.5.3 Epiarrow

Definition 36 (epiarrow) An arrow $A \xrightarrow{f} B$ in a category $\mathbb{C}$ is called an epiarrow or an epimorphism if for all $\mathbb{C}$-arrows g and h such that $g \circ f = h \circ f$ imply that $g = h$, that is, if f is right-cancelable with respect to the composition of arrows. Diagrammatically, $A \xrightarrow{f} B \underset{h}{\overset{g}{\rightrightarrows}} X \Rightarrow g = h$ in category $\mathbb{C}$ for any object X, f being an epiarrow.

It is easy to see that an epimorphism is an onto or surjective function for category $\mathbb{S}$et of sets and functions.

Definition 37 (thin category) If in a category, each arrow is an epiarrow, it is called a thin category.

Proposition 24 Given any objects A and B, the hom-collection $[A, B]_{\mathbb{C}}$ in a thin category $\mathbb{C}$ has at the most one member.

Proposition 25 Monoarrow and epiarrow are dual notions.

Thus, in $\mathbb{S}$et, a monoarrow is a 1-1 map and an epiarrow is an onto map, and these are the dual notions of each other. Then if we establish results for a 1-1 map, we obtain results for an onto map, without any need to prove the latter, by the principle of duality. Without the concepts of category theory, however, it is not immediately obvious that the concept of an onto function is dual to that of an injective or a 1-1 function!

Proposition 26 Every isoarrow in any category is a monoarrow as well as an epiarrow.

Proof Consider the diagram for an isoarrow $f: X \underset{k}{\overset{h}{\rightrightarrows}} A \xrightarrow{f} B$ with $fh = fk$. On composition with g, the inverse of f, we get $gfh = gfk$, or $\mathrm{id}_A \circ h = \mathrm{id}_A \circ k$, which implies that $h = k$.

Similarly, for the epiarrow part, consider the diagram for an isoarrow $A \xrightarrow{f} B \underset{k}{\overset{h}{\rightrightarrows}} X$ with $hf = kf$. On composition with g, the inverse of f, we get $hfg = kfg$, or $h \circ \mathrm{id}_B = k \circ \mathrm{id}_B$, which implies that $h = k$.

A converse, that an arrow that is a mono- as well as an epiarrow in a category is an isoarrow, of the above proposition is *not* true. We would not obtain an identity arrow on composition of f with g, as we have in the above proof, and would therefore not be able to conclude that $h = k$ for both the mono and the epi parts.

We then have the following propositions for an epiarrow:

Proposition 27 The composition of two epiarrows in any category is an epiarrow whenever the composition is defined. ♣

Proposition 28 If in any category, $g \circ f$ is an epiarrow, then g is an epiarrow whenever the composition is defined. ♣

Proposition 29 Every retraction arrow in any category is an epiarrow in that category. ♣

Proposition 30 In every category, TFAE:

- f is an isoarrow in a category.
- f is an epiarrow and a section in a category. ♣

Proposition 31 If an arrow f in category $\mathbb{C}$ is an epiarrow and its composition $g \circ f$ with any other arrow g in $\mathbb{C}$ is a section in $\mathbb{C}$, then g must also be a section in $\mathbb{C}$.

Proof As gf is a section, there exists $m \in \mathbb{C}$ such that $m(gf) = u_C$. Consider the composition $fmgf = f \circ u_C = f = u_B \circ f$. Now, given arrows $r, s \in \mathbb{C}$ such that $rf = sf$, we have $r = s$, for f is an epiarrow. Then (fm) $g = u_B$. Thus, g must be a section in category $\mathbb{C}$.

5.5.4 Quotient Object

The following is the categorical notion dual to that of a subobject of an object in a category. In conformity with the definition of a subobject related to the collection of monoarrows in a category $\mathbb{C}$, we define quotient objects in relation to the collection of epiarrows in $\mathbb{C}$ as follows.

Definition 38 (quotient object) Let ε be a collection of epiarrows in a category $\mathbb{C}$. An ε-quotient object of an object A in $\mathbb{C}$ is a pair (b, B), where $A \xrightarrow{b} B$ belongs to ε. In case ε consists of all the epiarrows in $\mathbb{C}$, ε-quotient objects are called simply quotient objects and (b, B) is called an ε-retract of A when arrow b is also a $\mathbb{C}$-retraction.

Further, in conformity with the definitions of isomorphic subobjects and of the smaller-than relation between the subobjects of an object, we have the following definitions of isomorphic quotient objects and of the larger-than relation between the quotient objects of an object:

Definition 39 Let (b, B) and (c, C) be quotient objects of A.

- (b, B) and (c, C) are called isomorphic, provided there exists an isomorphism $h : B \to C$ with $c = h \circ b$.
- (b, B) is said to be larger than (c, D), denoted by $(b, B) \geq (c, D)$, provided there exists some (necessarily unique) arrow $h : B \to C$ with $c = h \circ b$ such that the following triangle commutes:

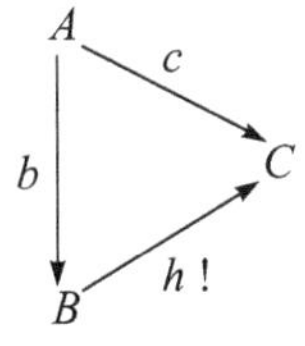

Notice that h is an epiarrow and that we have written for a quotient object as a pair (f, A) while we wrote for a subobject as a pair (A, f). This is only for our convenience.

In general, an object may not have any quotient objects. (Think of a singleton set in $\mathbb{S}$et.)

Note also the following:

Observation 2 In a category $\mathbb{C}$, it is possible that two subobjects (A, a) and (B, b) are nonisomorphic subobjects of an object C in $\mathbb{C}$, even when A and B are the isomorphic objects of $\mathbb{C}$.

By duality, the analogous result holds also for quotient objects of category $\mathbb{C}$.

On the basis of the above observation: Two subobjects (A, a) and (B, b), of an object C in a category $\mathbb{A}$, are isomorphic iff $a = bf$ with $f : A \to B$ being an isoarrow. Even when f is an isoarrow that is a monoarrow, we may not have a commuting triangle to yield $a = bf$.

5.5.5 Bidirectional Arrow

An arrow in a category can be a mono- and an epiarrow, both at the same time. This corresponds to bidirectionality of that arrow. Formally, we therefore define:

Definition 40 (bidirectional arrow) An arrow in a category $\mathbb{C}$ is called a bimorphism or bidirectional arrow provided it is both a monoarrow and an epiarrow in $\mathbb{C}$.

Proposition 32 The composition of two bidirectional arrows in any category is also a bidirectional arrow in that category whenever the composition is defined in that category.

Proof The composition of two monoarrows is a monoarrow and the composition of two epiarrows is an epiarrow whenever the compositions are defined in any category.

Proposition 33 If in any category $g \circ f$ is a bidirectional arrow, then f is a monoarrow and g is an epiarrow in that category.

Proof Given $g \circ f$ is a bidirectional arrow in the category. Then for any arrows r and s, $(gf)r = (gf)s$ implies $r = s$ as gf is a monoarrow. Then if we consider that $fr = fs$ and compose it with g we have $r = s$ as well. Hence, f is a monoarrow. Similarly, given arrows a and b, we have $a(gf) = b(gf)$, implying $a = b$. Then consider $ag = bg$. Composing it with f, we have $a = b$ as well. Hence, g is an epiarrow.

We invite the reader to check whether the converse of this proposition holds. Also, notice the following simple-to-prove results:

Observation 3 If a category $\mathbb{A}$ is a, full or not, subcategory of another category $\mathbb{B}$:

- Then any monoarrow (epiarrow or bidirectional arrow) in $\mathbb{A}$ is not necessarily a monoarrow (epiarrow or bidirectional arrow) in $\mathbb{B}$.
- But every arrow in $\mathbb{A}$ that is a monoarrow (epiarrow or bidirectional arrow) in category $\mathbb{B}$ is also a monoarrow (epiarrow or bidirectional arrow) in category $\mathbb{A}$.

Observation 4 If an arrow (h, k) in the arrow category $\mathbb{A}^2$ is a monoarrow, then arrow h is a monoarrow in category $\mathbb{A}$.

Observation 5 If $\mathbb{C}/\sim$ is the quotient category of category $\mathbb{C}$ and f is any monoarrow (epiarrow or bidirectional arrow) in $\mathbb{C}$, then:

- Then arrow $\tilde{f}$ is not necessarily a monoarrow (epiarrow or bidirectional arrow) in quotient category $\mathbb{C}/\sim$.
- But if $\tilde{f}$ is a monoarrow (epiarrow or bidirectional arrow) in quotient category $\mathbb{C}/\sim$, then arrow f is necessarily a monoarrow (epiarrow or bidirectional arrow) in category $\mathbb{C}$.

Definition 41 (balanced category) A category is called balanced if every arrow that is a monoarrow as well as an epiarrow in it is also an isoarrow in it. A category is balanced only if every bidirectional arrow in it is also its isoarrow.

The category of a group is an example of a balanced category. However, it may be noted that there exist many categories that are not balanced categories.

Proposition 34 If a category $\mathbb{C}$ is such that every epiarrow in it is also a retraction arrow in it, then $\mathbb{C}$ is balanced.

Henceforth, we avoid motivational details for definitions to follow. The reader may obtain the essence of any definition from corresponding considerations of category $\mathbb{S}$et of sets and functions or of any other category.

5.6 Intersection and Co-intersection

Definition 42 Consider a category $\mathbb{C}$. If B is a $\mathbb{C}$-object and (A_i, m_i) is any family of subobjects of B, then a pair (D, d) is called an intersection in $\mathbb{C}$ of $(A_i, m_i)_I$ provided:

- $d : D \to B$ is an arrow in $\mathbb{C}$.
- For each I $\in I$, there exists a $\mathbb{C}$-arrow $d_i : D \to A_i$ with the property that $m_i \circ d_i = d$.
- If $g : C \to B$ and for each $i \in I$, $g_i : C \to A_i$ such that $m_i \circ g_i = g$, then there exists a unique $\mathbb{C}$-arrow $f : C \to D$ such that $g = d \circ f$. That is to say, the following diagram commutes for every $i \in I$:

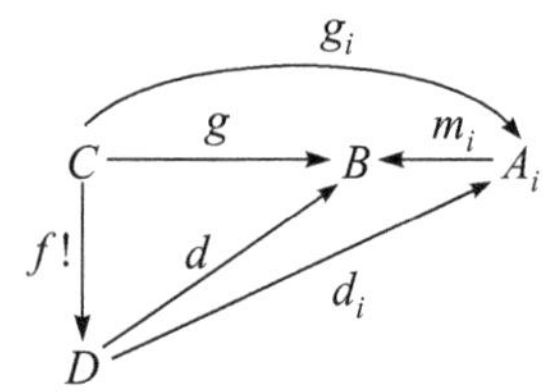

Proposition 35 Every intersection (D, d) of a family of subobjects $(A_i, m_i)_I$ of an object B is itself a subobject of B, that is, arrow d is necessarily a monoarrow, and that (D, d) is, up to an isoarrow, the largest subobject that is smaller than each of the subobjects (A_i, m_i).

Corollary 1 Any two intersections of a family of subobjects of an object A are isomorphic subobjects of A.

Definition 43 (has intersections/finite intersections) Any category $\mathbb{C}$ is said to possess intersections/finite intersections provided every set-indexed family of subobjects of each $\mathbb{C}$-object has an intersection.

Dually, we could then define a co-intersection of a family of quotient objects of any given object in a category. Results dual to those mentioned above then hold for a co-intersection.

5.7 Factorization

Sometimes, we want to consider all the arrows f in a category $\mathbb{C}$ that can be expressed as $f = me$ where m is a monoarrow and e is an epiarrow. We then have factored arrow f in terms of mono- and epiarrows. This factorization is useful to prove various results. This is what is formalized as the definition below. Of course, we can consider factorization of an arrow in terms of other properties. Hence, it is not specified in the definition below that m is a monoarrow and e is an epiarrow.

Consider now any two collections $\mathcal{E}$ and $\mathcal{M}$ of arrows in a category $\mathbb{C}$.

Definition 44 ((ε, $\mathcal{M}$)-factorization) A pair of arrows (e, m) is called an (ε, $\mathcal{M}$)-factorization of an arrow f in $\mathbb{C}$ when we have:

- $f = me$, where $e \in \varepsilon$ and $m \in \mathcal{M}$.

We use the notation that $f = me$ is an (ε, $\mathcal{M}$)-factorization of arrow f.

Definition 45 A category $\mathbb{C}$ is called an (ε, $\mathcal{M}$)-factorizable category when every $\mathbb{C}$-arrow is (ε, $\mathcal{M}$)-factorizable in it.

5.8 Separator, Separating Set, and Coseparator

Definition 46 (separator) An object S of category $\mathbb{A}$ is called a separator provided whenever $A \underset{g}{\overset{f}{\rightrightarrows}} B$ are distinct arrows, there exists in category $\mathbb{A}$ an arrow $S \xrightarrow{h} A$ such that $S \xrightarrow{h} A \xrightarrow{f} B \neq S \xrightarrow{h} A \xrightarrow{g} B$.

Remark 1 The existence of a separator in a category $\mathbb{A}$ often serves as a useful "smallness" condition for $\mathbb{A}$ that guarantees that there are not too many $\mathbb{A}$-objects. A slightly weaker condition that serves the same purpose is the existence of a separating set.

Definition 47 (separating set) A set $\mathcal{T}$ of objects is called a separating set provided that for any pair $A \underset{g}{\overset{f}{\rightrightarrows}} B$ of distinct arrows, there exists an arrow $S \xrightarrow{h} A$, with domain S being a member of $\mathcal{T}$ such that $f \circ h \neq g \circ h$.

Definition 48 (coseparator) An object C is called a coseparator provided that whenever $B \underset{g}{\overset{f}{\rightrightarrows}} A$ are distinct arrows, there exists an arrow $A \xrightarrow{h} C$ such that $B \xrightarrow{f} A \xrightarrow{h} C \neq B \xrightarrow{g} A \xrightarrow{h} C$.

5.9 Equalizer, Has Equalizer, and Coequalizer

Let $A \underset{g}{\overset{f}{\rightrightarrows}} B$ be a pair of arrows with an identical source and an identical target in category $\mathbb{C}$.

5.9.1 Equalizer Arrow

Definition 49 (equalizer) An object-arrow pair (E, e), $E \xrightarrow{e} A$ in $\mathbb{C}$ is called an equalizer of f and g provided $fe = ge$ and for any other arrow $K \xrightarrow{k} A$ with $fk = gk$, there exists a unique arrow $K \xrightarrow{h} E$, making the following commuting triangle, that is, $k = eh$ or k factors through E:

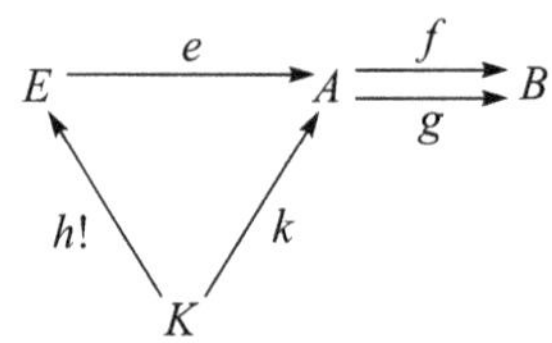

Proposition 36 If an object-arrow pair (E, e) is an equalizer of arrows $A \underset{g}{\overset{f}{\rightrightarrows}} B$, then (E, e) is a subobject of object A in that category. That is to say, an equalizing arrow e of arrows f and g is a monoarrow.

Proof Consider the diagram $X \underset{b}{\overset{a}{\rightrightarrows}} E \xrightarrow{e} A \underset{g}{\overset{f}{\rightrightarrows}} B$. Assuming $ea = eb$, we need to show that $a = b$. We are also given $fe = ge$. Then $f(ea) = g(ea)$ and $f(eb) = g(eb)$. From the definition of an equalizer, we have a unique arrow h such that $eh = ea$. Similarly, $eh = eb$. Since h is unique, $a = b$. Hence, the equalizing arrow e is a monoarrow.

Proposition 37 Equalizers are essentially unique; that is, given $X \underset{s}{\overset{r}{\rightrightarrows}} K \xrightarrow{k} A$ in a category, the following hold:

- If each of $E \xrightarrow{e} A$ and $K \xrightarrow{k} A$ is an equalizer of f and g, then there is an isoarrow $h : K \to E$ with $k = e \circ h$.
- If $E \xrightarrow{e} A$ is an equalizer of f and g and if $K \xrightarrow{h} E$ is an isoarrow, then $K \xrightarrow{e \circ h} A$ is also an equalizer of f and g.

Proof We are given $E \xrightarrow{e} A \underset{g}{\overset{f}{\rightrightarrows}} B$ and $K \xrightarrow{k} A \underset{g}{\overset{f}{\rightrightarrows}} B$, with (E, e) and (K, k) as two equalizers of f and g.

- Since $fk = gk$, there exists, by definition, a unique h such that $k = eh$. Similarly, as $fe = ge$, there exists a unique $\hat{h}$ such that

$e = k\hat{h}$. Then $e = e(h\hat{h}) = e \circ \mathrm{id}_E$. Hence, $h\hat{h} = \mathrm{id}_E$ for e is a monoarrow. Similarly, $\hat{h}h = \mathrm{id}_K$. Therefore, arrow h is an isoarrow.

- Whenever E and K are connected by an isoarrow h, there exists a unique h^{-1} to satisfy the conditions in the definition of an equalizer and we have $k = eh$ also as an equalizer of f and g.

Proposition 38 Any two equalizers of the pair of arrows $A \underset{g}{\overset{f}{\rightrightarrows}} B$ in any category are isomorphic subobjects of object A in that category.

Because of the above, we can think of an essentially generic equalizer of a pair of arrows f and g and denote it by Eq(f, g). Then when we talk about a particular subobject (A, a) of an object B as an equalizer of arrows $B \underset{g}{\overset{f}{\rightrightarrows}} C$, we write $(A, a) \approx \mathrm{Eq}(f, g)$.

Proposition 39 Let m be a monoarrow, and let mf and mg be defined in a category. Then TFAE statements:

- $(A, a) \approx \mathrm{Eq}(f, g)$.
- $(A, a) \approx \mathrm{Eq}(mg, mg)$.

Proposition 40 If $E \xrightarrow{e} A$ is an equalizer of $A \underset{g}{\overset{f}{\rightrightarrows}} B$, then TFAE statements:

- We have $f = g$.
- Arrow e is an epiarrow.
- Arrow e is an isoarrow.
- The subobject (A, id_A) is an equalizer of f and g.

Proof Consider an equalizer diagram with $fe = ge$.

- Now, if $f = g$, then there exists an arrow s such that $fes = ges$ and es = id. Then e has a right-inverse s, and it is a retraction arrow. But being an equalizer, it is also a monoarrow. It is therefore an isoarrow.
- An isoarrow is an epiarrow.
- If e is an epiarrow, then $fe = ge$ implies, by definition, $f = g$.

Proposition 41 If $A \xrightarrow{g} B \xrightarrow{f} A = A \xrightarrow{\mathrm{id}_A} A$, then g is an equalizer of $g \circ f$ and id_B.

Proof For g to be an equalizer of $g \circ f$ and id_B, we require $(g \circ f) \circ g = \mathrm{id}_B \circ g = g$. But given $f \circ g = \mathrm{id}_A$, $g \circ (f \circ g) = g \circ \mathrm{id}_A = g$.

Now, consider $X \underset{s}{\overset{r}{\rightrightarrows}} A \xrightarrow{g} B$ with $gr = gs$. As $fg = \mathrm{id}_A$, $f(gr) = f(gs)$ or $(fg)r = (fg)s$ implies that $r = s$. Hence, g is also a monoarrow.

Definition 50 (has equalizers) A category $\mathbb{C}$ is said to have equalizers when every pair of arrows, with a common source and a common target, in that category $\mathbb{C}$ has an equalizer in $\mathbb{C}$.

Proposition 42 Let a category $\mathbb{C}$ have equalizers. Then any arrow f in $\mathbb{C}$ is an epiarrow iff whenever $f = mg$ and m is an equalizer of some pair of arrows in $\mathbb{C}$, then m is an isoarrow.

5.9.2 Multiple Equalizers

Definition 51 If $(f_i)_{i \in I}$ is any nonempty indexed family of arrows in a category $\mathbb{C}$ with a common source and a common target, that is, for each $i \in I$, $f_i \in [A, B]$ for some objects A and B in $\mathbb{C}$, then (E, e) is called a multiple equalizer of $(f_i)_I$ and we write $(E, e) \approx \mathrm{Eq}((f_i)_{i \in I})$ iff $e : E \to A$ such that for all $i, j \in I$, $f_i \circ e = f_j \circ e$ and , the uniqueness in the equalizer diagram, if (E', e') is another pair for which $f_i \circ e' = f_j \circ e'$ for all $i, j \in I$, then there exists a unique arrow $k : E' \to E$ such that $e' = ek$.

Definition 52 (category has multiple equalizers) If a category $\mathbb{C}$ is such that each nonempty indexed family of its arrows with a common source and a common target has a multiple equalizer in it, we say that category $\mathbb{C}$ has multiple equalizers.

Proposition 43 Each multiple equalizer is a subobject.

Proposition 44 Any two multiple equalizers of the same family of arrows having a common source and a common target are isomorphic subobjects.

5.9.3 Regular Monoarrow

Every monoarrow need not be an equalizer arrow in a category. This is because any $f : A \to B$, given as a monoarrow, is an equalizer of g, $h : B \to C$ only when $[B, C] \neq \emptyset$. This need not be the case with every category. Think of a poset with a terminal object.

Nonetheless, we would like to deal with a monoarrow that also is an equalizer arrow of some pair of arrows in a category. This is formalized as the definition below.

Definition 53 (regular monoarrow) Any arrow $E \xrightarrow{e} A$ is called a regular monoarrow provided it is an equalizer of some pair of arrows. Inaccurately, we also say that (E, e) is a regular subobject of A.

Definition 54 (category: regular well powered) If every object in a category $\mathbb{C}$ has a representative collection of regular subobjects, then we say that category $\mathbb{C}$ is regular well powered.

Observation 6 Notice the following:

- The uniqueness requirement in the definition of an equalizer implies that a regular monoarrow must be a monoarrow.
- An equalizer is defined relative to a pair of arrows, whereas a regular monoarrow is an absolute notion.

Proposition 45 Every section is a regular monoarrow.

Proof Let $E \xrightarrow{e} A$ be a section. There then exists an arrow $m : A \to E$ such that $me = \mathrm{id}_E$. Consider the following diagram with $e \circ m : A \to A$:

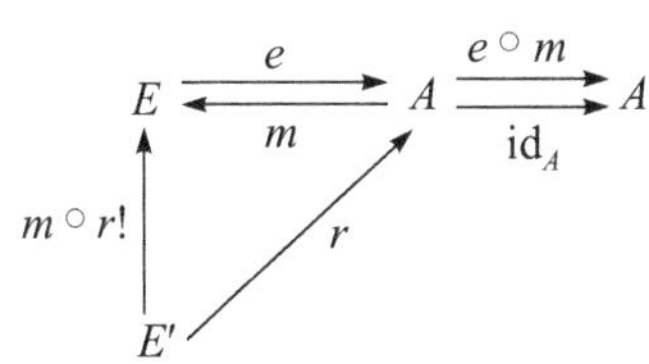

Then $eme = e \circ \mathrm{id}_E = e = \mathrm{id}_A \circ e$. Now, if there existed $r : E' \to A$ with $\mathrm{id}_A \circ r = emr$, then $r = e(mr)$ and mr is the unique arrow from $E' \to E$: for e is a section, and if there existed another arrow k with $e = ek$, then $mr = mek = \mathrm{id}_E \circ k = k$. Then any section arrow e always equalizes em and id_A and, therefore, it is a regular monoarrow.

Observation 7 The converse of the above proposition fails.

Proposition 46 In any category, TFAE statements:

- f is an isoarrow.
- f is a regular monoarrow and an epiarrow.
- f is a regular epiarrow and a monoarrow.

5.9.4 Extremal Monoarrow

A monoarrow $a : A \to B$ signifies that object A is a subobject of object B. In general, a monoarrow can factor through another object C. That is to say, in the following triangle $a = fe$ or that it commutes:

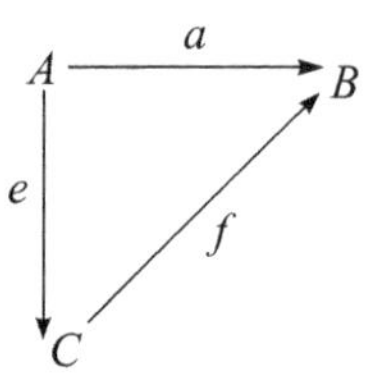

We now want to characterize monoarrows that factor through objects that are isomorphic to their source. If $e : A \to C$ in the above triangle is an isoarrow, we know that it is also an epiarrow. Then we can impose the condition that if $m = fe$ with e being an epiarrow, then e must be an isoarrow, to characterize monoarrows that factor through objects isomorphic to their source A. This situation is formalized below.

Definition 55 (extremal monoarrow) A monoarrow m is called extremal provided it satisfies the following extremal condition:

If $m = f \circ e$, where e is an epiarrow, then e must be an isoarrow.

Proposition 47 Let $A \xrightarrow{f} B$ and $B \xrightarrow{g} C$ be any arrows of category $\mathbb{C}$.

- If f is an extremal monoarrow and g is a regular monoarrow, then $g \circ f$ is an extremal monoarrow.
- If $g \circ f$ is an extremal monoarrow, then f is an extremal monoarrow.
- If $g \circ f$ is a regular monoarrow and g is a monoarrow, then f is a regular monoarrow.

Corollary 2 Every regular monoarrow is extremal.

Observation 8 A composition of two extremal monoarrows may fail to be extremal. Still, if all the extremal monoarrows in a category are regular, then in that category the class of extremal (=regular) monoarrows is closed under composition and is cancelable from the left, that is, its compositional behavior is similar to that o+f the class of all monoarrows.

Proposition 48 For any arrow f in a category, TFAE:

- f is an isoarrow.
- f is an extremal monoarrow and an epiarrow.

Proposition 49 For any category $\mathbb{A}$, TFAE:

- Category $\mathbb{A}$ is balanced.
- Each monoarrow is extremal in category $\mathbb{A}$.

5.9.5 Strict Monoarrow

Definition 56 (strict monoarrow) An arrow $A \xrightarrow{f} B$ is called a strict monoarrow provided that whenever $A' \xrightarrow{f'} B$ is an arrow with the property that for all arrows $B \overset{r}{\underset{s}{\rightrightarrows}} C$, $r \circ f = s \circ f$ implies that $r \circ f' = s \circ f'$, then there exists a unique arrow $A' \xrightarrow{k} A$ with $f' = f \circ k$. Diagrammatically, we have:

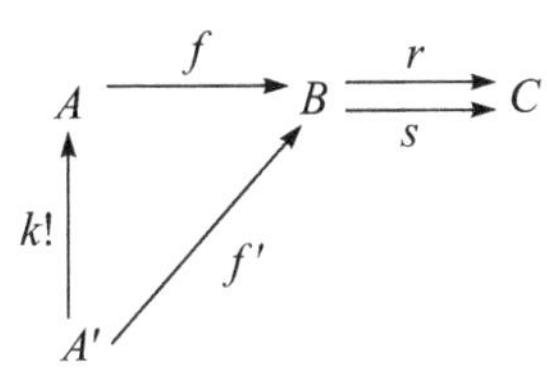

Notice that a strict monoarrow is a monoarrow by the uniqueness in the diagram for a strict monoarrow: for any arrows $p, q : X \to A$, we then have arrows $fp, fq : X \to B$ for which the unique arrow k in this diagram is equal to p and q both because of the strictness condition. Hence, $p = q$ and then the strict monoarrow f is a monoarrow.

Proposition 50 The following hold in any category:

- Every regular monoarrow is a strict monoarrow, but not vice versa.
- Every strict monoarrow is an extremal monoarrow, but not vice versa.
- If $A \xrightarrow{f} B$ is an extremal monoarrow and $B \xrightarrow{g} C$ is a strict monoarrow, then $A \xrightarrow{f} B \xrightarrow{g} C$ is an extremal monoarrow.
- If $A \xrightarrow{f} B$ is a strict monoarrow and $B \xrightarrow{g} C$ is a section, then $A \xrightarrow{f} B \xrightarrow{g} C$ is a strict monoarrow.

- The composite of two strict monoarrows is an extremal but not necessarily strict monoarrow.
- If $A \xrightarrow{f} B \xrightarrow{g} C$ is a strict monoarrow and $B \xrightarrow{g} C$ is a monoarrow, then $A \xrightarrow{f} B$ is a strict monoarrow.

5.9.6 Coequalizer

Recall that monoarrows deal with subobject structures while epiarrows deal with quotient object structures. For many purposes, mono- and epiarrows are weak notions, for these, respectively, correspond only to 1-1 maps and onto maps for category $\mathbb{S}$et. That is why we needed stronger notions, and that is what we achieved in the form of a regular monoarrow and an extremal monoarrow. Even these need not suffice for all the purposes of analyzing mathematical structures forming a category.

Given a pair of arrows $A \underset{g}{\overset{f}{\rightrightarrows}} B$, the dual of the notion of an equalizer is the following:

Definition 57 (coequalizer) An object-arrow pair (B, c), $B \xrightarrow{c} C$, or simply arrow c, in $\mathbb{C}$ is called a coequalizer of f and g provided $cf = cg$ and for any other arrow $B \xrightarrow{h} D$ with $hf = g$, there exists a unique arrow $C \xrightarrow{k} D$ making the following commuting triangle:

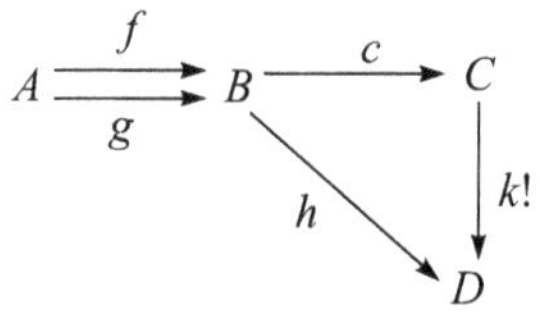

Definition 58 (regular epiarrow) An arrow $B \xrightarrow{c} C$ is called a regular epiarrow provided it is a coequalizer of some pair of arrows. ♣

Definition 59 (extremal epiarrow) An epiarrow e is called extremal provided it satisfies the following extremal condition:

If $e = m \circ f$, where m is a monoarrow, then m must be an isoarrow. ♣

We need not state and prove any results for coequalizers because of the duality. Still, we may note:

Proposition 51 The following hold in any category:

- Every retraction arrow is a regular epiarrow.
- Every regular epiarrow is an extremal epiarrow. ♣

5.10 Constant and Co-constant Arrows

Definition 60 (constant arrow) A constant arrow in category $\mathbb{C}$ is an arrow $A \xrightarrow{f} B$ such that for each object C and for all arrows $r, s \in [C, A]_{\mathbb{C}}$, we have $f \circ r = f \circ s$. That is, in the diagram $C \underset{s}{\overset{r}{\rightrightarrows}} A \xrightarrow{f} B$, we have $fr = fs$ for any object C and any two arrows r and s as shown.

Definition 61 (co-constant arrow) A co-constant arrow in $\mathbb{C}$ is then an arrow f that is a constant arrow in $\mathbb{C}^{\text{op}}$.

Notice that we used the opposite category $\mathbb{C}^{\text{op}}$ in the above definition. This convenient practice is often followed in the literature.

5.11 Initial and Terminal Objects

Definition 62 (terminal object) An object T is called a null or terminal object for category $\mathbb{C}$ if collection $[A, T]$ has precisely one element for each object A in category $\mathbb{C}$.

Definition 63 (initial object) An object I is called a conull or initial object for category $\mathbb{C}$ if collection $[I, A]$ has precisely one element for each object A in category $\mathbb{C}$.

Proposition 52 If T and T' (I and I') are terminal (initial) objects in any category $\mathbb{C}$, then T and T' (I and I') are connected by an isoarrow in $\mathbb{C}$.

Proof Let T' be another terminal object in a category that already has a terminal object T. Then there exists one and only one arrow $f : T \to T'$ and one and only one arrow $g : T' \to T$ in category $\mathbb{C}$. This is possible only if $gf = \text{id}_T$ and $fg = \text{id}_{T'}$, simultaneously. But then T and T' are connected by an isoarrow of category $\mathbb{C}$.

Similarly, any two initial objects in the same category $\mathbb{C}$ are connected by an isoarrow in $\mathbb{C}$.

Proposition 53 If T (I) is a terminal (initial) object in a category $\mathbb{C}$, then any other object that is isomorphic to T (I) is also a terminal (initial) object of that category $\mathbb{C}$.

The above two propositions amount to saying that a terminal (initial) object is essentially unique in a category.

5.12 Zero Object

Definition 64 (zero object) An object Z in category $\mathbb{C}$ is a zero object if it is simultaneously an initial and a terminal object.

Notice that a zero object is unique up to an isoarrow.

5.13 Zero Arrow

Definition 65 (zero arrow) Any arrow in a category $\mathbb{C}$ is called a zero arrow iff it is, simultaneously, a constant and a co-constant arrow.

Sometimes, a zero arrow is also defined using a zero object as follows.

Definition 66 (alternative definition of a zero arrow) An arrow $0 : A \to B$ in a category $\mathbb{C}$ with a zero object Z is called a zero arrow if it factors through Z, that is, if the triangle commutes:

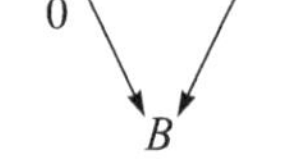

(Arrows $A \to Z$ and $Z \to B$ are the only such existing arrows as Z is a null (terminal) and a conull (initial) object, both.)

Proposition 54 A zero arrow of definition 66 is both a constant and a co-constant arrow in $\mathbb{C}$.

Proof Notice that a zero arrow factors through a zero object Z that, by definition, has only one arrow from it to any other object and only one arrow to it from any other object. In the diagram

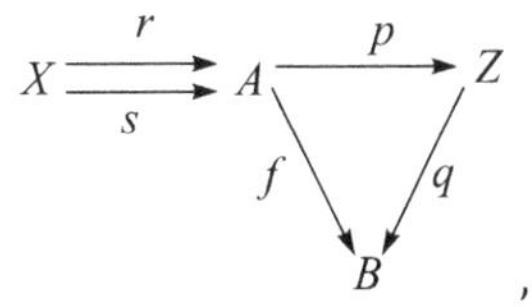

with f being a zero arrow, we have $ps = pr$ since there is only one arrow from X to Z. Thus, for the commuting triangle, we obtain $fr = qpr = qps = fs$ for all arrows r, s. Notice that p and q are unique arrows.

Similar is the case for the co-constant part.

Proposition 55 Each collection of arrows $[A, B]$ in category $\mathbb{C}$ can have precisely one zero arrow.

Proof Assume that $h, k : A \to B$ are both zero arrows. Then both arrows h and k factor through the zero object:

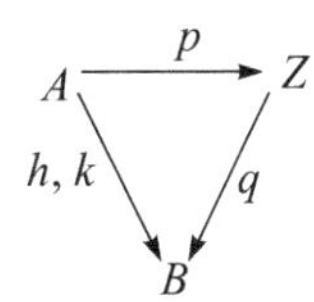

But as p and q are unique, we have $h = qp = k$.

Proposition 56 The composition, whenever it is defined, of a (constant, co-constant) zero arrow with any other arrow in $\mathbb{C}$ is also a (constant, co-constant) zero arrow.

Proof Let $A \xrightarrow{f} B$ be a zero arrow and $B \xrightarrow{g} C$ be another arrow such that gf is defined. Let the unique arrows involving the zero object be $A \xrightarrow{p} Z$ and $Z \xrightarrow{h} C$. Then forgetting the middle part of the diagram involving object B, we have $gf = hp$ for the triangle

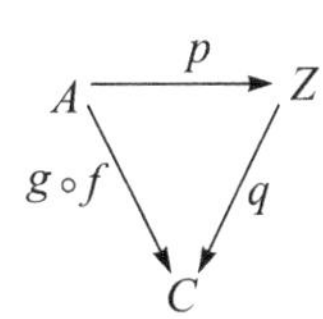

and the composite gf of f and g also factors through Z.

Now, for a constant arrow $A \xrightarrow{f} B$, we have the diagram

$$X \underset{k}{\overset{h}{\rightrightarrows}} A \xrightarrow{f} B \xrightarrow{g} C$$

and $g(fh) = (gf)h = g(fk) = (gf)k$.

Similarly, for the co-constant part of the proposition

$$X \underset{k}{\overset{h}{\leftleftarrows}} A \xleftarrow{f} B \xleftarrow{g} C$$

and $h(fg) = (hf)g = k\,(fg) = (kf)g$.

Proposition 57 Let $A \xrightarrow{f} B \xrightarrow{g} C$ be any arrows in a category $\mathbb{C}$. Then if g is a monoarrow and gf is a constant arrow in $\mathbb{C}$, then f is a constant arrow in category $\mathbb{C}$.

Proposition 58 Let $A \xrightarrow{f} B$ be any arrow in category $\mathbb{C}$, and let T be the terminal object of $\mathbb{C}$. Then when f can be factored through T, f is necessarily a constant arrow. If, furthermore, $[T, A] \neq \emptyset$, then any constant arrow f can be factored through T.

Proof Let $A \xrightarrow{f} B = A \xrightarrow{h} T \xrightarrow{k} B$ with a unique arrow h, and consider $X \underset{s}{\overset{r}{\rightrightarrows}} A \xrightarrow{h} T \xrightarrow{k} B$. As there must be only one arrow from X to T, we have $khr = khs$ or $fr = fs$. Hence, f is a constant arrow.

Now, let $[T, A] \neq \emptyset$ and $X \underset{s}{\overset{r}{\rightrightarrows}} A$. Consider for a constant arrow f the diagram $T \underset{s}{\overset{r}{\rightrightarrows}} A \xrightarrow{f} B$ with $fr = fs$. Then there is only one arrow to B from T (call it g) and $g = fr = fs$. Now, if $A \xrightarrow{h} T$, then $f = gh$ and, hence, the constant arrow f factors through T.

Proposition 59 If $A \xrightarrow{f} B$ is any arrow in category $\mathbb{C}$ and Z is a zero object of $\mathbb{C}$, then TFAE:

- Arrow f is a zero arrow.
- Arrow f is a constant arrow.
- Arrow f is a co-constant arrow.
- Arrow f can be factored through Z.

Corollary 3 Let $f, g : A \to B$ be such that f is a constant arrow and g is a co-constant arrow in a category $\mathbb{C}$. Then, provided $[B, A]_{\mathbb{C}} \neq \emptyset$, we always have $f = g$.

Proposition 60 In any category, TFAE:

- The hom-collection $[A, B]$ contains a zero arrow for all objects A, B in it.

- The hom-collection $[A, B]$ contains exactly one constant arrow for all objects A, B in that category.
- The hom-collection $[A, B]$ contains exactly one co-constant arrow for all objects A, B in that category.
- The hom-collection $[A, B]$ contains at least one constant arrow and at least one co-constant arrow for all A, B in it.
- There exists a choice function for selecting exactly one arrow from each of the hom-collections $[A, B]$ in the category such that the composition, whenever defined, of the selected arrow with any other arrow in that category is again one of the selected arrows.

Proof We have already shown that there can be only one zero arrow in any hom-collection. We have also shown that if the hom-collection $[B, A] \neq \emptyset$, then any constant arrow in $[A, B]$ is also co-constant. Thus, we will only prove here the last two statements of this proposition.

Let $A \xrightarrow{f} B$ be the selected arrow from $[A, B]$, and let $T \overset{h}{\underset{k}{\rightrightarrows}} A$ be any other arrows in the category. Then it is given that fh and fk are also the two arrows selected from the hom-collection $[X, B]$. But the selection is unique and, hence, $fh = fk$. Therefore, arrow f is a constant arrow in the hom-collection $[A, B]$. By duality, it also follows that arrow f is a co-constant arrow in the hom-collection $[A, B]$.

Notice then that if in any category $\mathbb{C}$, there exists in every collection $[A, B]$ of arrows from object A to object B, a distinguished arrow such that its composition with any other arrow is again a distinguished arrow, that is, any arrow $f \in [A, B]$ such that fg (with $g \in \mathbb{C}$) has the distinguishing property possessed by f, then there can, at the most, be one such collection of distinguished arrows in category $\mathbb{C}$ and a zero object can be concatenated to $\mathbb{C}$ so that the distinguished arrows become zero arrows without essentially changing category $\mathbb{C}$.

Notice also that the choice function of proposition 60 is uniquely determined and that it selects exactly the zero arrows of the category under consideration.

Proposition 61 The following hold in any connected category $\mathbb{C}$:

- In category $\mathbb{C}$, TFAE:

- There exists a constant monoarrow with the source as object A in category $\mathbb{C}$.
- Every arrow with the source as object A is a constant monoarrow in category $\mathbb{C}$.
- Object A is a terminal object of category $\mathbb{C}$.

- Consider the diagram $A \xrightarrow{h} B \underset{g}{\overset{f}{\rightrightarrows}} C$. If f and g are constant arrows such that $f \neq g$, then $fh = gh$.
- If $A \xrightarrow{f} B$ is a constant arrow, then there exists a unique constant arrow $B \xrightarrow{g} B$ such that $g \circ f = f$.
- There exists a 1-1 correspondence between the collection of constant arrows in $[A, B]$ and the collection of constant arrows in $[C, B]$ for any objects $A, B, C \in \mathbb{C}$.

Proof Let $\mathbb{C}$ be connected and A be any of its objects.

- A constant monoarrow with source A has, for $X \underset{s}{\overset{r}{\rightrightarrows}} A \xrightarrow{f} B$, $fr = fs$ and $r = s$ for all objects X in $\mathbb{C}$. Then $a) \Rightarrow b$. And, of course, $b) \Rightarrow a)$. Now, as there is only one arrow to A from every object X in $\mathbb{C}$, A is a terminal object of $\mathbb{C}$. Clearly, $c) \Rightarrow a)$.
- The composition of any arrow with two known distinct arrows gives distinct arrows, always.
- Let f be a constant arrow in $[A, B]$ and g be a constant arrow in $[C, B]$. As $\mathbb{C}$ is connected, $[A, C] \neq \emptyset$ and $[C, A] \neq \emptyset$. Consider commuting triangles involving objects A, B, and C and constant arrows f and g.
- This is left as an exercise for the reader.

5.13.1 Pointed Category

Definition 67 (pointed category) A category $\mathbb{C}$ will be called a pointed category provided each of its hom-collections $[A, B]$ contains a zero arrow for any objects A, B in $\mathbb{C}$.

Now, we should observe:

Proposition 62 For a pointed category $\mathbb{C}$, TFAE:

- An object Z in it is a zero object.
- The hom-collection $[Z, Z] = \{\mathrm{id}_Z\}$.

Proof For the first part, consider a zero arrow $f: Z \to Z$, apart from id_Z. Use corollary 3 to show that $[Z, Z] = \{\mathrm{id}_Z\}$.

For the second part, we can use the fact that the identity arrow id_Z is the only zero arrow in $[Z, Z]$ and that the composition of any arrow with a zero arrow is also a zero arrow.

From the definition of a pointed category, we can prove the following:

Proposition 63 The statement "A category $\mathbb{C}$ is a pointed category" is a self-dual statement.

Proposition 64 The following then hold:

- Every category having a zero object is pointed.
- Every full subcategory of a pointed category is also pointed.

Proposition 65 If $A \xrightarrow{f} B \xrightarrow{g}$ are any arrows in a pointed category $\mathbb{C}$ and if g is a monoarrow and gf is a zero arrow in $\mathbb{C}$, then f is also a zero arrow in category $\mathbb{C}$.

5.14 Kernel Arrow

Let $A \xrightarrow{f} B$ be any arrow in a category with a zero object.

Definition 68 We call an arrow $K \xrightarrow{k} A$ the kernel of f if $fk = 0$ and if for every arrow $K' \xrightarrow{k'} A$ such that $fk' = 0'$, there exists a unique arrow $K' \xrightarrow{g} K$ with $kg = k'$. Here $K \xrightarrow{0} B$ and $K' \xrightarrow{0'} B$ denote zero arrows. Object K is usually denoted by $\ker(f)$. Diagrammatically,

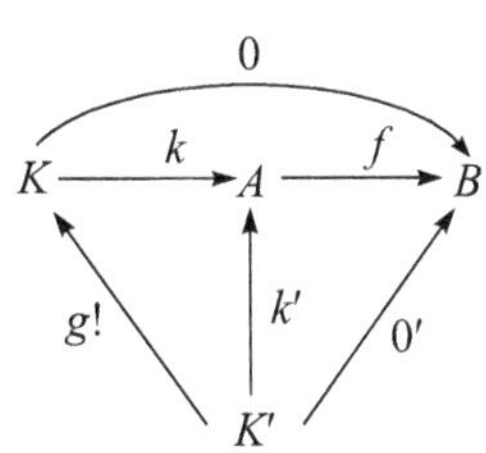

where, we recall, an exclamation mark after the arrow (as in $g!$) indicates its existence and uniqueness.

A kernel is a special case of an equalizer.

Proposition 66 In a pointed category, an equalizer of $A \xrightarrow{f} B$ and a zero arrow $0_{AB}: A \to B$ is a kernel of f.

Proof Analyze the diagram

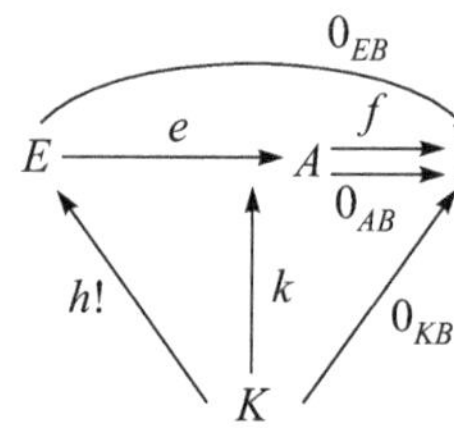

keeping in mind that the composition of any arrow with a zero arrow is also a zero arrow.

Definition 69 (category has kernels) Whenever in a category with a zero object, every arrow has a kernel, we say that the category has kernels.

Proposition 67 Every nonempty pointed category that has kernels also has a zero object.

Proposition 68 In a pointed category $\mathbb{C}$, a subobject (K, k) of a $\mathbb{C}$-object A is in the kernel of a monoarrow $f : A \to B$ iff K is a zero object of $\mathbb{C}$. In other words, in a pointed category, only an arrow from a zero object can be the kernel of any of its monoarrows.

Proof Let (K, k) be a subobject of an object A in a pointed category $\mathbb{C}$ and $f : A \to B$ be a monoarrow in $\mathbb{C}$.

For the "if" part, let K be the zero object of $\mathbb{C}$. Then k is the one and only one arrow $k : K \to A$. We have $fk = 0_{KB}$ since k is a zero arrow as it factors through K via its identity. Now, given any K' with $k' : K' \to A$ with $fk' = 0_{K'B}$, there exists one and only one and, hence, unique arrow $g : K' \to K$ such that $k' = kg$ for K is a zero object. The (only such existing) arrow $k : K \to A$ is also a monoarrow, for if $X \overset{r}{\underset{s}{\rightrightarrows}} K \xrightarrow{k} A$, then $r = s$ as there can be only one arrow to a zero object. Thus, (K, k) is a subobject in the kernel of monoarrow f.

Now, for the "only if" part, if the subobject (K, k) is in the kernel of monoarrow f, then we show that K is a zero object of $\mathbb{C}$. Clearly, (K, k) is also a subobject of B for fk is a monoarrow by proposition 18. We now have $fk = 0_{KB}$ and, given K' with $k' : K' \to A$, we have $fk' = 0_{K'B}$ss and $k' = kg$, with g being unique in the kernel diagram. Now, let $K' = K$ and if K were not a zero object, then let k' be different than k. Then we have $fk = fk' = 0_{KB}$ and $k' = kg$. Thus, $k = k'$ and $g = \mathrm{id}_K$ for f is a monoarrow and is left-cancelable. Then the hom-collection $[K, K]$ has only one arrow—the identity id_K. By proposition 62, K is then a zero object of the pointed category $\mathbb{C}$.

The dual of a kernel arrow is a cokernel arrow.

5.14.1 Normal Monoarrow

A kernel arrow of any arrow is then a monoarrow. Every monoarrow need of course not be in the kernel of some arrow in the category. To characterize a monoarrow that is the kernel of some arrow of a category, we adopt the following nomenclature:

Definition 70 (normal monoarrow) Any arrow in a category is called a normal monoarrow provided it is the kernel of some arrow.

Proposition 69 Note the following

- Every normal monoarrow is regular.
- If $\mathbb{A}$ is a pointed category such that each of its arrows has a kernel and a cokernel, then the following hold:
  - An arrow f is a normal monoarrow iff it is a kernel of a cokernel of f.
  - For each object A the preordered class of all nonisomorphic normal subobjects of A is anti-isomorphic with the preordered class of all nonisomorphic normal quotient objects of A.

5.15 Reflection Arrow and Co-reflection Arrow

Let $\mathbb{A}$ be a subcategory of $\mathbb{B}$, and let B be a $\mathbb{B}$-object.

Definition 71 An $\mathbb{A}$-reflection (or $\mathbb{A}$-reflection arrow) for B is a $\mathbb{B}$-arrow $B \xrightarrow{r} A$ from B to an $\mathbb{A}$-object A with the following universal property: for any $\mathbb{B}$-morphism $B \xrightarrow{f} A'$ from B into some $\mathbb{A}$-object A', there exists a unique $\mathbb{A}$-arrow $f': A \to A'$ such that the following triangle commutes:

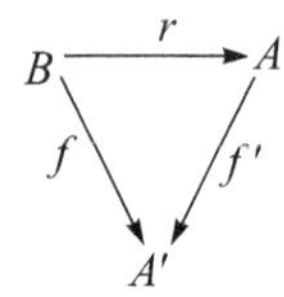

By an "abuse" of language an object A is called an $\mathbb{A}$-reflection for B whenever an $\mathbb{A}$-reflection $r : B \to A$ exists for B with codomain A.

Definition 72 A category $\mathbb{A}$ is called a reflective subcategory of category $\mathbb{B}$ provided each $\mathbb{B}$-object has an $\mathbb{A}$-reflection.

Proposition 70 Reflections are essentially unique, that is:

- If $B \xrightarrow{r} A$ and $B \xrightarrow{r'} A'$ are $\mathbb{A}$-reflections for $\mathbb{B}$, then there exists an $\mathbb{A}$-isomorphism $k : A \to A'$ such that the following triangle commutes:

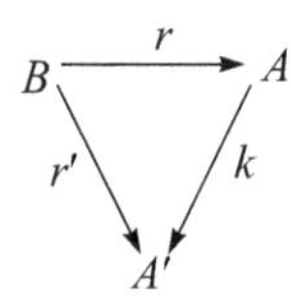

- If $B \xrightarrow{r} A$ is an $\mathbb{A}$-reflection for B and $A \xrightarrow{k} A'$ is an $\mathbb{A}$-isomorphism, then $B \xrightarrow{k \circ r} A'$ is an $\mathbb{A}$-reflection for $\mathbb{B}$.

Proposition 71 If $\mathbb{A}$ is a reflective subcategory of $\mathbb{B}$, then the following conditions are equivalent:

- $\mathbb{A}$ is a full subcategory of $\mathbb{B}$.
- For each $\mathbb{A}$-object A, $A \xrightarrow{\mathrm{id}_A} A$ is an $\mathbb{A}$-reflection.
- For each $\mathbb{A}$-object A, $\mathbb{A}$-reflection arrows $A \xrightarrow{r_A} A^*$ are $\mathbb{A}$-isomorphisms.
- For each $\mathbb{A}$-object A, $\mathbb{A}$-reflection arrows $A \xrightarrow{r_A} A^*$ are $\mathbb{A}$-morphisms.

Proof That (1) $\Rightarrow$ (2) $\Rightarrow$ (3) $\Rightarrow$ (4) is clear. To see that (4) $\Rightarrow$ (1), let $A \xrightarrow{f} A'$ be a $\mathbb{B}$-morphism between $\mathbb{A}$-objects. By the definition of reflection there is an $\mathbb{A}$-morphism $A^* \xrightarrow{\bar{f}} A'$ with $f = \bar{f} \circ r_A$. Thus f is the composite of $\mathbb{A}$-morphisms and so must be an $\mathbb{A}$-morphism.

Note 1 There exist nonfull reflective subcategories $\mathbb{A}$ of $\mathbb{B}$ such that every $\mathbb{A}$-reflection arrow is a $\mathbb{B}$-isomorphism. Note also that reflection arrows in $\mathbb{S}\mathrm{et}$ need not be surjective.

Proposition 72 An isomorphism-closed full reflective subcategory $\mathbb{A}$ of any category $\mathbb{B}$ contains with any object its retract in $\mathbb{B}$.

Proposition 73 If $\mathbb{A}$ is a full subcategory of $\mathbb{B}$, then an $\mathbb{A}$-reflection arrow is a section iff it is an isoarrow.

Now, several of the familiar mathematical constructions, such as certain completions, certain formations of quotients, and certain modifications of structures, can be regarded in a natural way as reflections. The following partial list is provided so that a reader may obtain an idea of the extent of the usefulness of the notion of a reflection arrow and also for introducing the standard notations for some more categories.

Example 17 Modifications of the structure

Making a relation symmetric: $\mathbb{B}$ = $\mathbb{R}$el, $\mathbb{A}$ = $\mathbb{S}$ym, the full subcategory of symmetric relations.

$$(X,\rho)\xrightarrow{\text{id}_X}(X,\rho\bigcup\rho^{-1})$$

is an $\mathbb{A}$-reflection for (X, ρ).

Example 18 Improving objects by forming quotients:

- Making a preordered set partially ordered: $\mathbb{B}$ = $\mathbb{P}$rost (preordered sets and order preserving maps), $\mathbb{A}$ = $\mathbb{P}$os. If $(X, \leq)$ is a preordered set, define an equivalence relation $\approx$ on X by $x \approx y \Leftrightarrow (x \leq y \text{ and } y \leq x)$. Let $p : X \rightarrow X/\approx$ be the canonical map. Then

  $$(X,\leq)\xrightarrow{p}(X/\approx,(p\times p)[\leq])$$

  is an $\mathbb{A}$-reflection for $(X, \leq)$.
- Making a group abelian
- Making a topological space T_0
- Making an abelian group torsion-free
- Making a reachable acceptor minimal: $\mathbb{B}$ = the full subcategory of Σ-Seq consisting of all reachable acceptors (i.e., those for which each state can be reached from the initial one by an input word), $\mathbb{A}$ = the full subcategory of $\mathbb{B}$ consisting of all minimal acceptors (i.e., those reachable acceptors with the property that no two different states are observably equivalent. The observability equivalence $\approx$ on a reachable acceptor B is given by $q \approx q'$ provided that whenever the initial

state of B is changed to q, the resulting acceptor recognizes the same language as it does when the initial state is changed to q'). Then the canonical map $\mathbb{B} \to \mathbb{B}/\approx$, where $\mathbb{B}/\approx$ is the induced acceptor on the set of all $\approx$-equivalence classes of states of B, is an $\mathbb{A}$-reflection for $\mathbb{B}$.

Example 19 Completions
$\mathbb{B} = \mathbb{M}\mathrm{et}_u$ (metric spaces and uniformly continuous maps) or $\mathbb{M}\mathrm{et}$ (metric spaces and nonexpansive maps), $\mathbb{A}$ = the full subcategory of complete metric spaces. In either case the metric completion $(X, d) \rightarrowtail (X^*, d^*)$ is an $\mathbb{A}$-reflection for (X, d).

The dual of the concept of a reflective subcategory is a co-reflective subcategory. That is, $\mathbb{A}$ is a co-reflective subcategory of $\mathbb{B}$ iff $\mathbb{A}^{\mathrm{op}}$ is a reflective subcategory of $\mathbb{B}^{\mathrm{op}}$. Although each of the above statements is an adequate definition, we provide a detailed dual formulation that does not involve the dual categories as an example of this construction.

Let $\mathbb{A}$ be a subcategory of $\mathbb{B}$ and let B be a $\mathbb{B}$-object.

Definition 73 An $\mathbb{A}$-co-reflection (or $\mathbb{A}$-co-reflection arrow) for B is a $\mathbb{B}$-arrow $A \xrightarrow{c} B$ from an $\mathbb{A}$-object A to B with the following universal property: for any $\mathbb{B}$-arrow $A' \xrightarrow{f} B$ from some $\mathbb{A}$-object A' to B there exists a unique $\mathbb{A}$-arrow $f' : A' \to A$ such that the following triangle commutes:

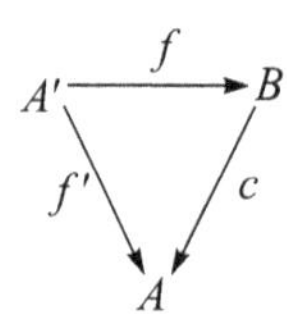

By an "abuse" of language an object A is called an $\mathbb{A}$-co-reflection for $\mathbb{B}$ provided an $\mathbb{A}$-co-reflection $A \xrightarrow{c} B$ exists for B with source A.

Definition 74 A category $\mathbb{A}$ is called a co-reflective subcategory of another category $\mathbb{A}$ provided each $\mathbb{B}$-object has an $\mathbb{A}$-co-reflection.

Examples Several mathematical constructions such as certain modifications of structures and certain selections of convenient subobjects can be regarded in a natural way as co-reflections. We list below some (only a few) of these co-reflection constructions.

Example 20 Modifications of the structure

- Making a relation symmetric: $\mathbb{B} = \mathbb{R}\mathrm{el}$, $\mathbb{A} = \mathbb{S}\mathrm{ym}$. Then
$(X,\rho\bigcap\rho^{-1})\xrightarrow{\mathrm{id}_X}(X,\rho)$
is an $\mathbb{A}$-co-reflection for (X, ρ). This example shows one of the rare instances where a subcategory is simultaneously a reflective and co-reflective category. ♣
- Making any topological space sequential: $\mathbb{B} = \mathbb{T}\mathrm{op}$, $\mathbb{A}$ = the full subcategory of sequential spaces (i.e., spaces in which every sequentially closed set is closed). If (X, τ) is a topological space, then
$\tau' = \{A \subseteq X \mid X\backslash A$ is sequentially closed in $(X, \tau)\}$ is a topology on X and $(X,\tau')\xrightarrow{\mathrm{id}_X}(X,\tau)$ is an $\mathbb{A}$-co-reflection for the topological space (X, τ). ♣

Example 21 Sorting out convenient subobjects

- Making an abelian group a torsion group: $\mathbb{B} = \mathbb{A}\mathrm{b}$, $\mathbb{A}$ = the full subcategory of abelian torsion groups. For any abelian group G the canonical embedding $TG \rightarrowtail G$ of the torsion subgroup TG of G into G is an $\mathbb{A}$-co-reflection for G. ♣
- Making a sequential acceptor reachable: $\mathbb{B} = \Sigma$-Seq, $\mathbb{A}$ = the full subcategory of reachable acceptors. For any acceptor A the canonical embedding $\mathbf{R}A \rightarrowtail A$ of the acceptor $\mathbf{R}A$, formed by removing from the state set of A all states that cannot be reached from the initial state, is an $\mathbb{A}$-co-reflection for A. Thus a minimalization of sequential acceptors is obtained in two steps: (i) the co-reflection, which yields a reachable acceptor, and (ii) the reflection, which gives the minimal quotient of the reachable part. ♣

The lists of examples of reflection and co-reflection arrows in some (really only very few) categories that have been provided above should suffice to convince the reader that these are naturally occurring concepts in many mathematical situations. We then see that reflection and co-reflection situations also abound in the mathematical theories.

Observation 9 We make the following observations:

- Every arrow with a terminal object as a target is a section.

- If $A \xrightarrow{f} B \xrightarrow{g} C$ is a section and $A \xrightarrow{f} B$ is an epiarrow, then $B \xrightarrow{g} C$ is a section.
- If $\mathbb{A}$ is a full subcategory of $\mathbb{B}$, then an $\mathbb{A}$-reflection arrow is a section iff it is an isoarrow.

5.16 Sources and Sinks

5.16.1 Sources

We have already noticed that objects and arrows of a category can be paired and treated together. This observation can now be put to use for the following important categorical concept. Needless to say, the following is then one of the basic concepts of category theory.

A categorical concept that treats objects and arrows together is that of sources and dually that of sinks.

Definition 75 (source) A source is a pair $(A, (f_i)_{i \in I})$ consisting of an object A and a family of arrows $f_i : A \to A_i$ with domain A, indexed by some collection I. A is called the domain of the source and the family $(A_i)_{i \in I}$ is called the codomain of the source. Diagrammatically, we have:

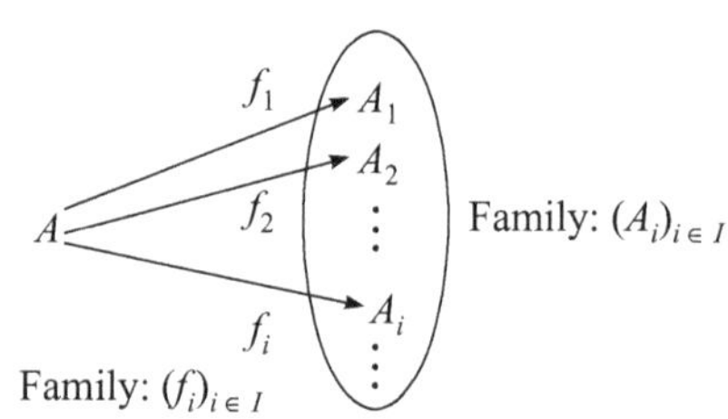

Remark 2 Note the following:

- Whenever convenient we use more concise notations, such as $(A, f_i)_I$, (A, f_i), or $(A \xrightarrow{f_i} A_i)_I$.
- The indexing collection I of a source $(A, f_i)_I$ may be any of the following: a conglomerate, a proper class, a nonempty set, or the empty set. In case $I = \emptyset$, the source is determined by A. In case $I \neq \emptyset$, the source is determined by the family $(f_i)_I$.
- Sources indexed by the empty set are called empty sources and are denoted by $(A, \emptyset)$. Objects are then empty sources.

- Sources that are indexed by a set are called set-indexed or small.
- Sources indexed by set $\{1, \ldots, n\}$ are called n-sources and are denoted by $(A, (f_1, \ldots, f_n))$. Arrows $A \xrightarrow{f} B$ are 1-sources (A, f).
- Some properties of sources depend on $(f_i)_I$ being a family, that is, an indexed collection (e.g., the property of being a product). Some other properties of sources (A, f_i) depend on the domain A and the associated class $\{f_i \mid i \in I\}$ only (e.g., the property of being a monosource). So, to avoid distinction between indexed and nonindexed sources, we may regard collections as families indexed by themselves. Hence, for any object A and any collection S of arrows with domain A, the pair (A, S) is a source. A particularly useful example is the total source (A, SA), where SA is the collection of all arrows with domain A.

Definition 76 (composition of sources) If $S = (A \xrightarrow{f_i} A_i)_I$ is a source and, for each $i \in I$, $S_i = (A_i \xrightarrow{g_{ij}} A_{ij})_{J_i}$ is a source, then the source

$$(S_i) \circ S = (A \xrightarrow{g_{ij} \circ f_i} A_{ij})_{i \in I, j \in J_i}$$

is called the composite of S and the family $(S_i)_I$.

Note 2 Notice now the following:

- For any source $S = (A \xrightarrow{f_i} A_i)_I$ and arrow $f : B \to A$ we use the notation $S \circ f = (B \xrightarrow{f_i \circ f} A_i)_I$.
- The composition of arrows can be regarded as a special case of the composition of sources.

5.16.2 Monosources

Definition 77 (monosource) A source $S = (A, f_i)_I$ is called a monosource provided it can be canceled from the left, that is, provided that for any pair $B \underset{s}{\overset{r}{\rightrightarrows}} A$ of arrows the equation $S \circ r = S \circ s$ (i.e., $f_i \circ r = f_i \circ s$ for each $i \in I$) implies $r = s$.

Example 22 An empty source $(A, \emptyset)$ is a monosource iff for each object B there is at most one arrow from B to A.

Example 23 A 1-source (A, f) is a monosource iff f is a monoarrow.

Example 24 An equalizer is a monosource.

Example 25 In $\mathbb{S}$et monosources are precisely the point-separating sources $(A, f_i)_I$, that is, sources $(A, f_i)_I$ such that for any two different elements a and b of A there exists some $i \in I$ with $f_i(a) \neq f_i(b)$.

Example 26 In every preordered class, considered as a category, every source is a monosource.

Notice here that this last property characterizes thin categories that are, alternatively, definable as categories in which every hom-collection has at the most one arrow for any two objects of the category.

Proposition 74 Let $T = (S_i) \circ S$ be a composite of sources.

- If S and all S_i are monosources, then so is T.
- If T is a monosource, then so is S.

Proposition 75 Let $(A, f_i)_I$ be a source.

- If $(A, f_j)_J$ is a monosource for some $J \subseteq I$, then so is $(A, f_i)_I$.
- If f_j is a monoarrow for some $j \in I$, then $(A, f_i)_I$ is a monosource.

Definition 78 (extremal monosource) A monosource S is called an extremal monosource provided that whenever $S = S \circ e$ for some epiarrow e, then e must be an isoarrow.

Example 27 A 1-source (A, f) is an extremal monosource iff f is an extremal monoarrow.

Example 28 In balanced categories (in $\mathbb{S}$et, $\mathbb{V}$ec, etc.) every monosource is extremal. Conversely, if every monosource in a category $\mathbb{C}$ is extremal, then that category $\mathbb{C}$ is balanced.

Example 29 A source $(A \xrightarrow{f_i} A_i)_I$ in $\mathbb{P}$os is an extremal monosource provided the following equivalence holds: $a \leq b \Leftrightarrow i \in I f_i(a) \leq f_i(b)$.

Example 30 A monosource $(A \xrightarrow{f_i} A_i)_I$ in $\mathbb{T}$op is extremal iff A carries the initial (= weak) topology with respect to (f_i).

Example 31 A monosource $(A \xrightarrow{f_i} A_i)_I$ in Σ-$\mathbb{S}$eq is extremal iff a state q of A is final whenever each state $f_i(q)$ is final in A_i.

Example 32 In a poset, considered as a category, a source $(A \to A_i)_I$ is an extremal monosource iff A is a maximal lower bound of $A_i / i \in I$.

Proposition 76 Note the following:

- If a composite source $(S_i) \circ S$ is an extremal monosource, then so is S.
- If $S \circ f$ is an extremal monosource, then f is an extremal monosource.

Note 3 However, if S and each S_i are extremal monosources, then $(S_i) \circ S$ need not be extremal.

Proposition 77 Let $(A, f_i)_I$ be a source.

- If $(A, f_j)_J$ is an extremal monosource for some $J \subseteq I$, then so is $(A, f_i)_I$.
- If f_j is an extremal monoarrow for some $j \in I$, then $(A, f_i)_I$ is an extremal monosource.

We have not provided proofs for the above propositions since these are easily obtainable from the corresponding properties of the involved arrows and objects. However, to become familiar with the use of sources and sinks, it is suggested that the reader supply proofs for such propositions.

Note 4 The concept of source allows a simple description of coseparators, that is, any object A is a coseparator iff for any object B, the source $(B, \hom(B, A))$ is a monosource.

This suggests the following definition:

Definition 79 (extremal coseparator) An object A is called an extremal coseparator provided that for any object B the source $(B, \hom(B, A))$ is an extremal monosource.

Example 33 In a balanced category every coseparator is extremal.

5.16.3 Sinks

A concept dual to that of a source is called a sink.

Definition 80 (sink) A sink is a pair $((f_i)_{i\in I}, A)$, sometimes denoted also by $(f_i, A)_I$ or $(A_i \xrightarrow{f_i} A)_I$, consisting of an object A, called the codomain of the sink, and a family of morphisms $f_i : A_i \to A$ indexed by some collection I, called the domain of the sink. Diagrammatically, we have:

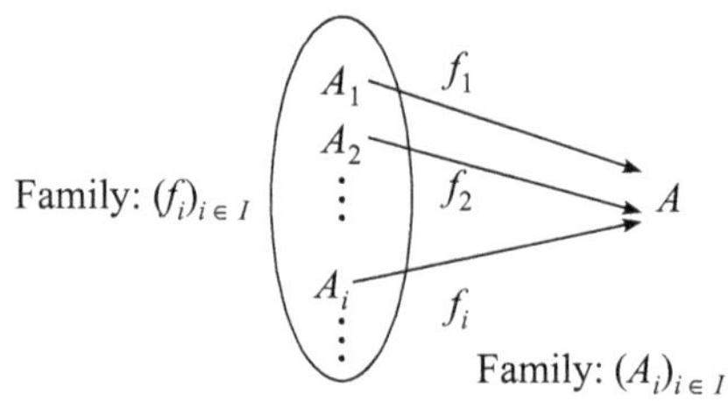

The composition of sinks is defined in the (obvious) way, which is dual to that of the composition of sources.

5.17 Product, Coproduct, and Biproduct

Now, the cartesian products of pairs of sets or, more generally, of families of sets (or direct products of families of vector spaces, topological products of families of topological spaces, etc.) can be regarded as objects together with families of (projection) arrows from them, that is, as sources defined earlier. Not as objects alone, but as a part of this specific collection of objects as well as arrows, these products can be characterized, up to isoarrows, by the following categorical property:

Definition 81 (product as a source) A source $P = (P \xrightarrow{\pi_i} A_i)_I$ is called a product provided that for every source $S = (S \xrightarrow{f_i} A_i)_I$ with the same codomain as P there exists a unique arrow $S \xrightarrow{f} P$ with $S = P \circ f$. A product with codomain $(A_i)_I$ is called a product of the family $(A_i)_I$. Diagrammatically, we have:

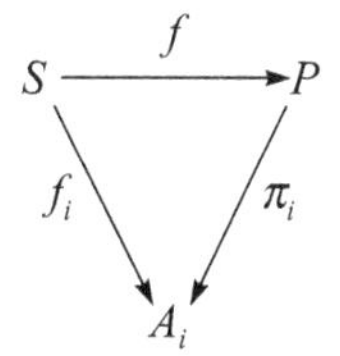

Using the arrows and objects explicitly, we could define a binary product also in the following manner:

Definition 82 (product as an object) In a category $\mathbb{C}$, a product of $\mathbb{C}$-objects A and B is a triple (π_A, π_B, P), where P is a $\mathbb{C}$-object and $\pi_A : P \to A$ and $\pi_B : P \to B$ are $\mathbb{C}$-arrows (called product projections) such that if D is any $\mathbb{C}$-object and $f: D \to A$ and $g : D \to B$ are arbitrary $\mathbb{C}$-arrows, then there exists a unique $\mathbb{C}$-arrow, usually denoted by $<f, g> : D \to P$, with $\pi_A \circ <f, g> = f$ and $\pi_B \circ <f, g> = g$. Diagrammatically, we have:

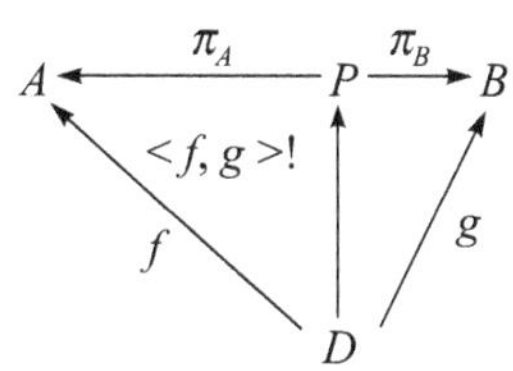

Often just object P is called the product: $A\prod B$. However, the product is not just an object but the triple (π_A, π_B, P).

This notion can be generalized to a family $(C_i)_{i\in I}$ of objects in category $\mathbb{C}$ by indexing them with collection I, as in the first definition. Note that the indexing collection I need not be a set.

Definition 83 In a category $\mathbb{C}$, a product of a family $(C_i)_{i\in I}$ of its objects is a pair $(\prod(C_i),(\pi_i))_{i\in I}$ such that:

- $\prod(C_i)_{i\in I}$ is an object of $\mathbb{C}$.
- For every $j \in I$, $\pi_j : \prod(C_i)_{i\in I} \to C_j$ is an arrow in $\mathbb{C}$— the projection arrow.
- For every pair $(C, (f_i)_{i\in I})$ with any object $C \in \mathbb{C}$ and for each $j \in I$, $f_j : C \to C_j$, there exists a unique arrow, denoted by $< f_i >: C \to \prod(C_i)_{i\in I}$ such that for every $j \in I$, we have $f_i = p_j \circ <f_i>$.

Example 34 In category $\mathbb{S}\text{et}$, let $(A_i)_I$ be a family of sets indexed by a set I and let $\prod_{i \in I} A_i$ be its cartesian product, that is, the set of all functions $g: I \to \bigcup_I A_i$ with the property that $g(i) \in A_i$. Then the family of projection functions $\pi_j : \prod_{i \in I} A_i \to A_j$, given by $g \mapsto g(j)$, is a product in $\mathbb{S}\text{et}$.

Example 35 In a partially ordered class, considered a category, a source $(P \xrightarrow{pi} A_i)_I$ is a product iff $P = \wedge_{i \in I} A_i$.

Example 36 An empty source $(T, \emptyset)$ is a product iff T is a terminal object. Therefore, a terminal object T is an empty product.

Example 37 A 1-source (P, p) is a product iff p is an isoarrow. Thus, regarding 1-sources as arrows, isoarrows are 1-products.

Proposition 78 Every product is an extremal monosource.

Proof Let $(\prod C_i, \pi_i)_{i \in I}$ be a product of $(C_i)_{i \in I}$ with projections π_i and let $A \overset{s}{\underset{t}{\rightrightarrows}} \Pi C_i$ be arrows with the property that for each $i \in I$, $\pi_i \circ s = \pi_i \circ t$. By the uniqueness of arrow $<f_i>$ in the definition of a product, the projection arrows π_I can, in conjunction, be canceled on the left. We then easily see that we have $s = <fi> = t$.

Proposition 79 For any family $(A_i)_I$ of objects, products of $(A_i)_I$ are essentially unique, that is, if $P = (P \xrightarrow{p_i} A_i)_I$ is a product of $(A_i)_I$, then the following hold:

- For each product $Q = (Q \xrightarrow{q_i} A_i)_I$, there exists an isoarrow $Q \xrightarrow{h} P$ such that $Q = P \circ h$.
- For each isoarrow $A \xrightarrow{h} P$ the source $P \circ h$ is a product of $(A_i)_I$.

Proof Let $(\prod (A_i), \pi_i)$ be one and $(\widehat{\prod} (A_i), \hat{\pi}_i)$ be another product of the same family $(A_i)_{i \in I}$ of objects in category $\mathbb{C}$. By the definition of a product, there then exist unique arrows h and k such that the following diagram commutes for each $j \in I$:

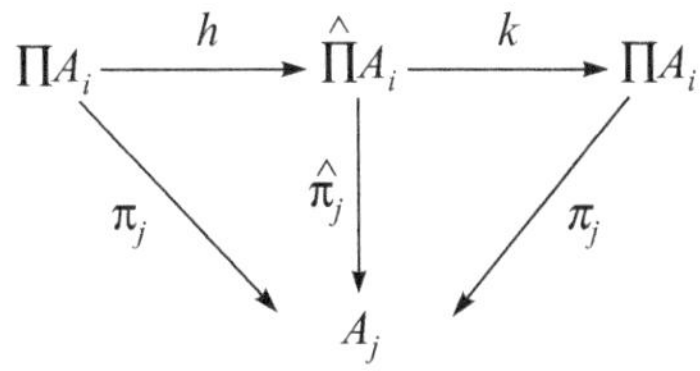

Then for every $j \in I$, we have $\pi_j \circ (k \circ h) = \pi_j = \pi_j \circ \mathrm{id}_{\prod A_i}$. But a product projection arrow can be canceled on the left. Hence, $k \circ h = \mathrm{id}_{\prod A_i}$. Similarly, we can show that $h \circ k = \mathrm{id}_{\widehat{\prod} A_i}$.

Proposition 79 also follows.

Notice that instead of using the concise notation of a source, we used the explicit notations for the product above. We sometimes do this for gaining clarity with the diagrams.

As a consequence of the above, we then have:

Corollary 4 Any two terminal objects of a category, when existing, are isomorphic objects of that category.

Note 5 The above uniqueness result allows us to introduce special notations for products (provided they exist):

- Products of $(A_i)_I$ is denoted by $(\prod_{i\in I} A_i \xrightarrow{\pi_j} A_j)_{j\in I}$, or, more simply, by $(\prod A_i \xrightarrow{\pi_j} A_j)_I$, and the arrows π_j are called projections.
- If $(\prod A_i \xrightarrow{\pi_j} A_j)_I$ is a product and $(A \xrightarrow{f_j} A_j)_I$ is a source with the same codomain, then the unique arrow $f : A \to \prod A_i$ with $f_j = \pi_j \circ f$ for each $j \in I$ will be denoted by $\langle f_i \rangle$:

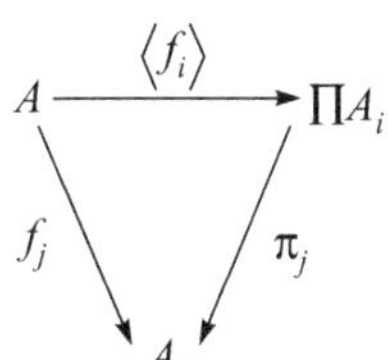

- In case $I = \{1, 2, \ldots, n\}$, the following notation often gets used instead of the one above:

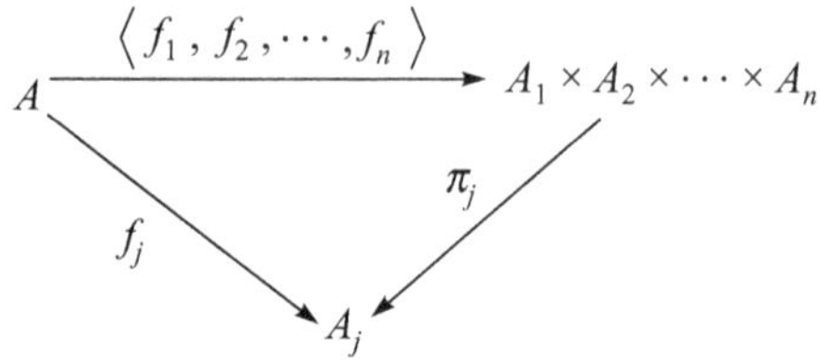

Remark 3 These above diagrams show, specifically, that the correspondence $(A, f_i) \mapsto \langle f_i \rangle$ provides a bijection from the collection of all sources with domain A and codomain $(A_i)_I$ to the set of all arrows with domain A and codomain $\prod_{i \in I} A_i$. Therefore, products allow us to treat sources as arrows. Essentially, a product is a bunching together of objects and arrows in a specific manner within a category.

The following propositions then illustrate this correspondence.

Proposition 80 Let $Q = (\mathcal{P}_i) \circ \mathcal{P}$ be a composite of sources.

- If $\mathcal{P}$ and all $\mathcal{P}_i$ are products, then so is Q.
- If Q is a product and all $\mathcal{P}_i$ are monosources, then $\mathcal{P}$ is a product.

Proposition 81 Consider this diagram:

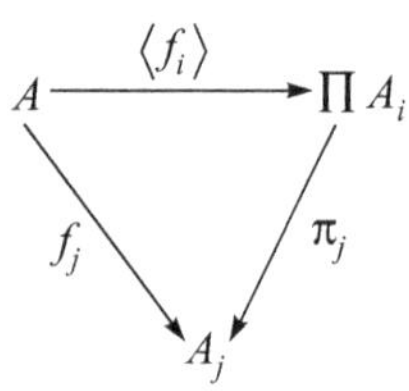

Then we have the following:

- $(A, f_i)_I$ is a monosource iff $\langle f_i \rangle$ is a monoarrow.
- $(A, f_i)_I$ is an extremal monosource iff $\langle f_i \rangle$ is an extremal monoarrow.
- $(A, f_i)_I$ is a product iff $\langle f_i \rangle$ is a product, that is, an isoarrow.

Remark 4 Whereas each product is an extremal monosource, the involved single or individual projections are usually retractions, as the following result shows. As can be noticed, this however is not always the case.

Proposition 82 If $(P \xrightarrow{\pi_i} A_i)_I$ is a product and if $i \in I$ is such that $\hom(A_{i'}, A_i) \neq \emptyset$ for each $i \in I$, then π_i is a retraction.

Proof Let π_i be a projection and let $f_i : A_i' \to A_i$, $i \in I$, be any arrows with $f_j : A_j \to A_j$ being the identities id_{A_j}, $j \in I$. Then by the definition of a product, there exists unique $< f_i >: A_i' \to \prod A_i$ with $\pi_j \circ < f_i > = \mathrm{id}_{A_j}$. Then π_i has a right-inverse.

Definition 84 We use the following terminology (has products):

- A category has products provided that for every set-indexed family $(A_i)_I$ of objects there exists a product $\left(\prod A_i \xrightarrow{\pi_j} A_j\right)_I$.
- A category has finite products provided that for every finite family $(A_i)_I$ of objects there exists a product $\left(\prod A_i \xrightarrow{\pi_j} A_j\right)_I$.

Proposition 83 A category has finite products iff it has a terminal object and products of pairs of objects.

We can of course consider products of products. To this end, let $(S_i)_{i \in I}$ be a family of pairwise disjoint sets. Suppose that for each $i \in I$, $(P^i, \pi_s^i)_{s \in S_i}$ is the product of the family $(A_s)_{s \in S_i}$ of objects of a category $\mathbb{C}$. Also, let (P, π_i) be the product of the family $(P^i)_{i \in I}$. Then:

Proposition 84 The pair $(P, (\pi_s^i \circ \pi_i)_{i \in I, s \in S_i})$ is the product of the families $(A_s)_{s \in \bigcup S_i}$ of $\mathbb{C}$-objects.

Proof The proof of this proposition involves keeping track of product diagrams without getting confused. We have therefore left this proof as an exercise for the reader.

As a consequence of the above, we have:

Proposition 85 If $\left(\prod A_i, \pi_i\right)$ and $\left(\prod B_i, \sigma_i\right)$ are the respective products of families (A_i) and (B_i) of objects in category $\mathbb{C}$, and if for each $i \in I$, there exists an arrow $f_i : A_i \to B_i$, then there exists a unique arrow that is denoted as $\prod f_i : \prod A_i \to \prod B_i$.

We use this in the form of a definition:

Definition 85 (product of a family of arrows) If $(A_i \xrightarrow{f_i} B_i)_I$ is a family of arrows and if $\left(\prod A_i \xrightarrow{\pi_j} A_j\right)_I$ and $\left(\prod B_i \xrightarrow{p_j} B_j\right)_I$ are products, then the unique arrow $\prod A_i \to \prod B_i$ that makes the following diagram commute for each $j \in I$

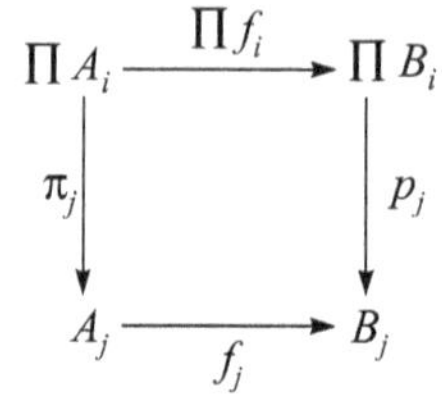

is denoted by $\prod f_i$ and is called the product of the family $(f_i)_I$. If I = $1, \ldots, n$ then $\prod f_i$ is usually written as $f_1 \times f_2 \times \ldots \times f_n$.

Proposition 86 Let $(f_i)_I$ be a set-indexed family of arrows with product $\prod f_i$. If each f_i has any of the properties of being a section, a retraction, an isoarrow, a monoarrow, a regular monoarrow (provided the category has products), or a constant arrow, then their product $\prod f_i$ also has that property.

Proof Let $\prod f_i$ be the product of the family $(f_i)_I$ of arrows in a category $\mathbb{C}$. Consider each of the above cases as follows:

- Let each f_i be a section, that is, let each f_i have a left-inverse. There then exists, for each $i \in I$, an arrow m_i in category $\mathbb{C}$ such that $m_i \circ f_i = \mathrm{id}_{A_i}$ for some object $A_i \in \mathbb{C}$. Pasting together the commuting squares in the definition of product of family of arrows

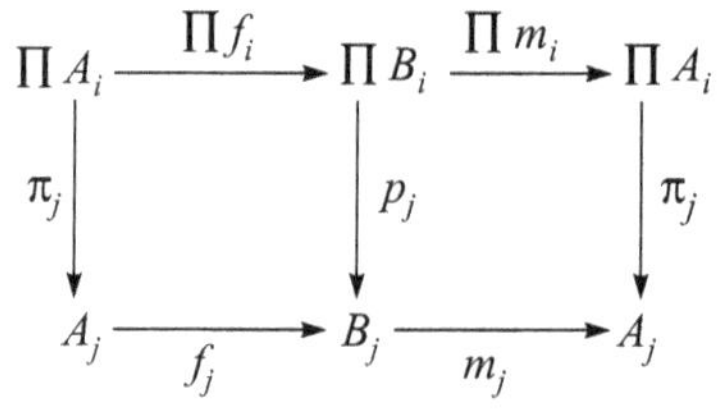

we have, for every $j \in I$,

$$\pi_j \circ \left(\prod m_i\right) \circ \left(\prod f_i\right) = m_j f_j \pi_j = \pi_j = \pi_j \circ \mathrm{id}_{\prod A_i}.$$

Then by the uniqueness in the definition of a product, we have

$$\left(\prod m_i\right) \circ \left(\prod f_i\right) = \mathrm{id}_{\prod B_i}$$

and then Πf_i has a left-inverse.

- It follows from considerations similar to the above.
- It follows from the above two considerations, since an isoarrow is a section and a retraction both.
- Let each f_i be a monoarrow, that is, a left-cancelable arrow. Then, given arrows h_j and k_j with $f_i \circ h_j = f_i \circ k_j$, we have $h_j = k_j$ for each $j \in I$. Consider next the following diagram:

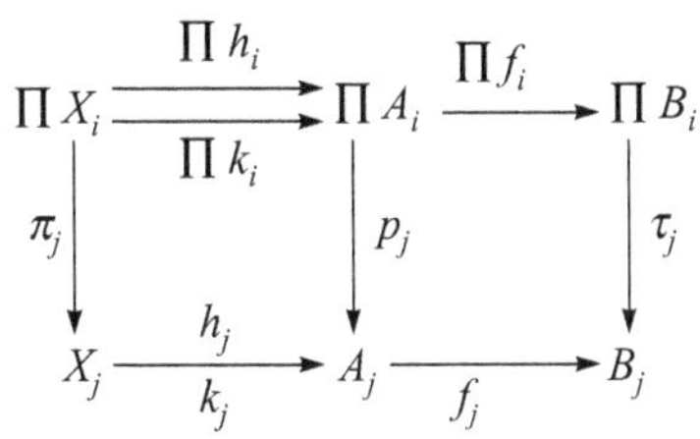

By the uniqueness in the outer product diagram, the bottom row being a diagram for a monoarrow for each $j \in I$, and the product projection arrows in conjunction being cancelable on the left, we have $\Pi f_i \circ \Pi h_i = \Pi f_i \circ \Pi k_i$ implying $\Pi h_i = \Pi k_i$. Thus, Πf_i is also a monoarrow.

We leave the remaining cases as an exercise for the reader.

Proposition 87 In a category with products, if I is a set and if $E_i \xrightarrow{e_i} A_i$ is an equalizer of $A_i \overset{f_i}{\underset{g_i}{\rightrightarrows}} B_i$ for each $i \in I$, then $\prod E_i \xrightarrow{\prod e_i} \prod A_i$ is an equalizer of $\prod A_i \overset{\prod f_i}{\underset{\prod g_i}{\rightrightarrows}} \prod B_i$. In alternative notation, let for every $i \in I$, $(E_i, e_i) \approx \mathrm{Eq}(f_i, g_i)$ and let the products Πe_i, Πf_i, and Πg_i exist. Then

$$\left(\prod E_i, \prod e_i\right) \approx \mathrm{Eq}\left(\prod f_i, \prod g_i\right).$$

In other words, a product of equalizers is an equalizer of the product.

Proof Consider the following diagram for every $j \in I$:

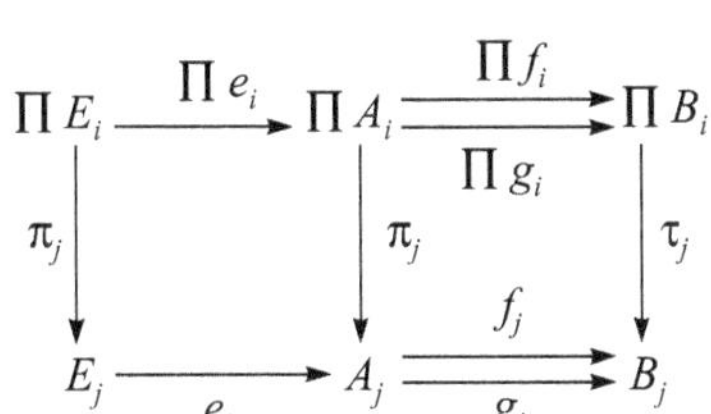

The product is a monosource, or by the uniqueness in the product diagram and since the bottom row is an equalizer diagram, we then see that

$$\prod f_i \circ \prod e_i = \prod g_i \circ \prod e_i .$$

Now, we want to show that given any other arrow $K \xrightarrow{k} \prod A_i$ such that $\prod f_i \circ k = \prod g_i \circ k$, and arrows h_j, for each $j \in I$, there exists arrow $<h_i>$ such that $k = \prod e_i \circ <h_i>$. Thus, consider the diagram:

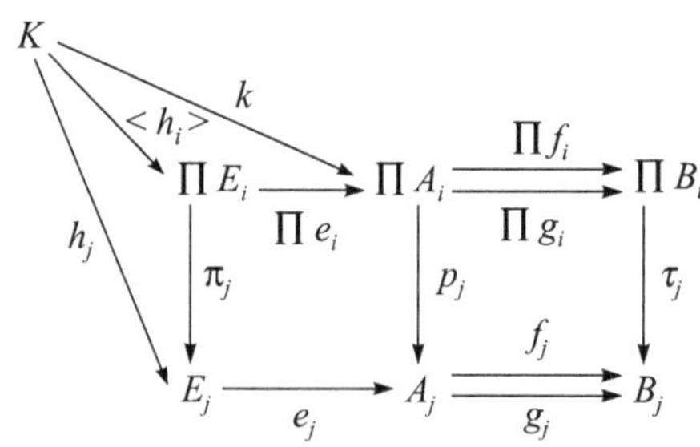

Then as the square on the left hand commutes and has the associated uniqueness of a product, we see that $k = \prod e_i \circ <h_i>$. Hence, a product of equalizers is an equalizer of the product.
Dual to a product is a coproduct.

Definition 86 (coproduct) A coproduct of $\mathbb{C}$-objects A and B is a triple (ϕ_A, ϕ_B, C), where C is a $\mathbb{C}$-object and $\phi_A : A \to C$ and $\phi_B : B \to C$ are $\mathbb{C}$-arrows (called coproduct injections) with the property that if D is any $\mathbb{C}$-object and $f : A \to D$, $g : B \to D$ are arbitrary $\mathbb{C}$-arrows, then there exists a unique $\mathbb{C}$-arrow, usually denoted by $[f, g] : C \to D$ such that $[f, g] \circ \phi_A = f$ and $[f, g] \circ \phi_B = g$. Diagrammatically, we have:

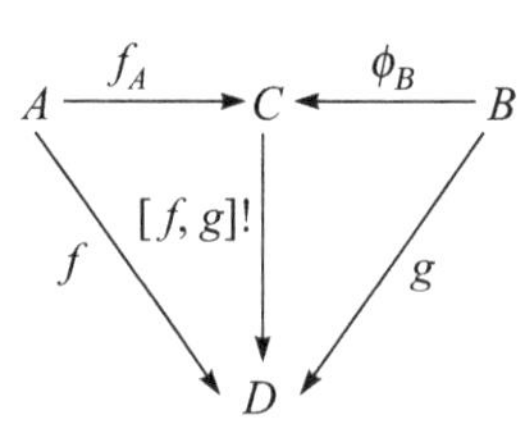

Often just object C is called the coproduct: $A \coprod B$.

The concept of coproduct in the case of the category of sets and functions is that of the disjoint union of sets: $\bigcup_{i\in I} X_i = \bigcup_{i\in I} (X_i \times \{i\})$. In this case, the coproduct injections are precisely the injective functions.

(Notice that measures are the set-valued functions defined on an appropriate collection, the σ-algebra of the subsets of a given set, of pairwise disjoint sets. Since a coproduct is a categorical generalization of disjointedness of sets, we may expect coproducts to play a central role in the definition of measures in the general categorical context. This will be the subject of our considerations in Chapter 6.)

Proposition 88 A category $\mathbb{A}$ that has a product of any family of its objects indexed by its collection Mor($\mathbb{A}$) of arrows is necessarily thin.

Proof Let a category $\mathbb{A}$ have a product of any family of its objects indexed by its collection Mor($\mathbb{A}$) of arrows, and let it also have the hom-collection hom (A, B) that has at least two arrows in it from its object A to another of its objects B. Now consider the family $(B_i)_{i\in I}$ with I = Mor($\mathbb{A}$) and $B_i = B$ for each $i \in I$. Since hom(A, B) has at least two members, there are at least as many distinct sources with domain A and codomain $(B_i)_I$ as there are subcollections of I. Hence, hom$(A, \Pi\ B_i)$ contains at least as many members as there are subcollections of Mor($\mathbb{A}$). However, this contradicts the fact that every hom-collection hom$(A, \Pi\ B_i)$ is contained in Mor($\mathbb{A}$). Then any hom-collection hom(A, B) has only one member, and category $\mathbb{A}$ is therefore seen to be necessarily thin.

Remark 5 Clearly, the existence of products for families of objects indexed by arbitrary collections forces a strong restriction on a category that it must be thin. It may then be noted that none of the familiar categories is thin, yet many of them do have products of all set-indexed families. In category theory, we can, if we so desire, distinguish then between sets and proper classes, as we have done in definition 84 by explicitly stating that we are using set-indexed families of objects.

In fact, this distinction already exists! In definition 3, Chapter 4, we have two collections: the first Mor($\mathbb{A}$) is the collection of all the arrows in a category, and the second hom(A, B) is the collection of all the arrows from an object A to an object B in the category. For internal consistency of this definition of a category, hom $(A, B) \subset \text{Mor}(\mathbb{A})$.

Consider a situation in which some sources exist within specific relations to each other; a product is one such. Existence of this situation for a family that can be indexed by $\text{Mor}(\mathbb{A})$ then means that all the sources in the category have this specific relationship. This is possible only when the category is thin, which is the essence of this proposition.

If we are given a nonthin category, then the permissible indexing of families of its objects is already appropriately restricted for forming their products within it. Thence, any nonthin category $\mathbb{C}$ certainly does not have a product of a family of its objects indexed by the collection $\text{Mor}(\mathbb{C})$ of its arrows. So, if $\text{Mor}(\mathbb{C})$ is a class, its nonthin category $\mathbb{C}$ can only have products for set-indexed families of its objects.

This issue should get clarified with the following corollary:

Corollary 5 A small category has all the set-indexed products iff it is equivalent to a complete lattice.

Proof Let $\mathbb{A}$ be a small category. Since $\mathbb{A}$ is small, $\text{Mor}(\mathbb{A})$ is a set so that if $\mathbb{A}$ has all the set-indexed products, we know from Proposition 88 that it is thin. Therefore, it is a preordered set that has intersections of all its subsets, and, hence, an empty set belongs to it. It then has elements 0 and 1 such that $0 \leq x \leq 1$ for all $x \in \mathbb{A}$. Hence, it is equivalent to a complete lattice. The converse of the above is evident.

(See Appendices A and B for details about a lattice.)

Remark 6 Notice therefore that, beginning with a generic category $\mathbb{C}$, we can construct a category of the families of its objects: category $\mathbb{F}\text{amily}(\mathbb{C})$, as was done in Section 4.9.6, Chapter 4. Then category $\mathbb{F}\text{amily}(\mathbb{C})$ is not a thin category, except when we allow families that can be indexed by $\text{Mor}(\mathbb{F}\text{amily}(\mathbb{C}))$. Notice that the underlying category of $\mathbb{F}\text{amily}(\mathbb{C})$ is, under these circumstances, not required to be thin.

Definition 87 (biproduct) A biproduct, denoted by $A \oplus B$, of objects A and B in category $\mathbb{C}$ is, simultaneously, a product and a coproduct.

Note also that an empty family of objects in a category $\mathbb{C}$ has a biproduct iff $\mathbb{C}$ has a zero object.

Now, as was done before, we may consider a family of the copies of the same object A, say $(C_i)_{i\in I} = A$ for every $i \in I$. Then:

Definition 88 (*I*th power of an object) If I is a collection and $\left(\prod A_i \xrightarrow{\pi_j} A_j\right)_I$ is a product with $A_i = A$ for each $i \in I$, then ΠA_i is denoted by A^I and called an Ith power of A.

We call the unique arrow $<(\mathrm{id}_A)_i> : A \to A^I$ as a *diagonal arrow* and denote it by ΔA or by Δ. In the case of a coproduct, we denote the coproduct $\coprod(C_i)$ by ${}^IA$ and call it the Ith copower of A and call the involved unique arrow a codiagonal arrow.

We may then restate proposition 88 as follows: any category $\mathbb{C}$ possessing the Mor($\mathbb{C}$)-th power of any of its objects is necessarily thin.

Proposition 89 In a category that has products, an object A is an (extremal) coseparator iff every object is a (extremal) subobject of some power A^I of A.

Definition 89 (Cartesian closed category) A category is called a Cartesian closed category if it has finite products, or equivalently, a terminal object and all the binary products.

5.18 Pullback and Push-Out

We will now introduce another important categorical concept, that of a pullback. As a comment on the history of category theory, we note that, unlike products and coproducts, the importance of the concept of a pullback was duly recognized only after the formulation of category theory. A push-out is the dual of the following notion of a pullback.

Definition 90

- The square

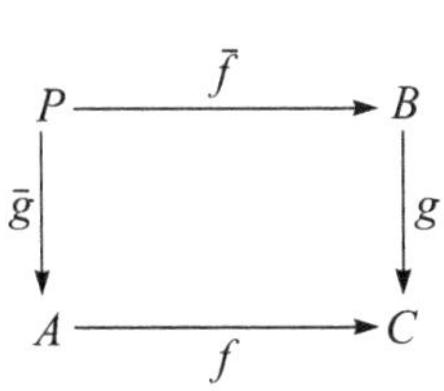

is called a pullback square provided it commutes and that for any other commuting square of the form

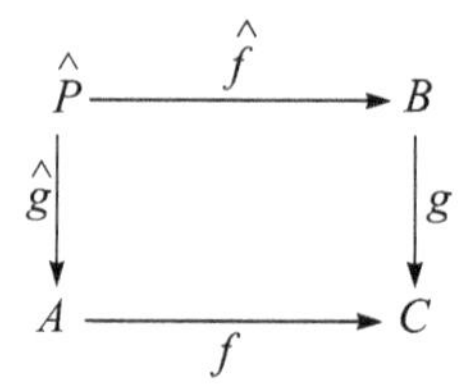

there exists a unique morphism $\hat{P}\xrightarrow{k}P$ for which the following diagram then commutes (uniqueness in the pullback diagram):

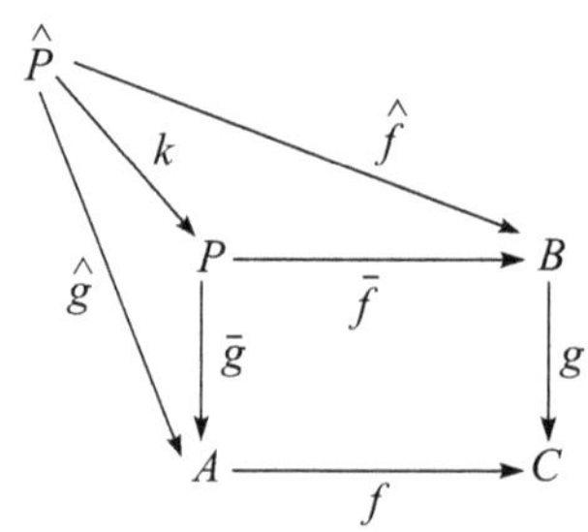

- If a pullback square is given, then the 2-source $A\xleftarrow{\bar{g}}P\xrightarrow{\bar{f}}B$ is called a pullback of the 2-sink $A\xrightarrow{f}C\xleftarrow{g}B$, $\bar{f}$ is called a pullback of arrow f along arrow g, and, symmetrically, $\bar{g}$ is called the pullback of the g along arrow f. In the case that g is a monoarrow, $\bar{f}$ is called an inverse image of g along f.
- We sometimes write for the pullback object P as $A \times_C B$ and say that P is a pullback of A and B over C. It is also called the fibred product of A and B relative to f and g. This is an inaccurate notation.
- Notice that the pullback square is characterized up to an isoarrow, as is usual with the categorical universality or uniqueness of diagrams.

Definition 91 (category has pullbacks) Whenever each diagram of the type

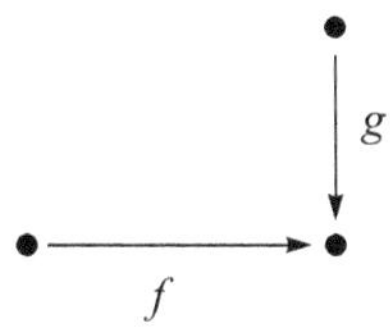

in any category $\mathbb{C}$ can be completed, with the resultant being a pullback square, we say that category $\mathbb{C}$ has pullbacks.
That is to say, whenever any two arrows f and g with a common target are given in a category, two more arrows can always be found in that category, with sources of f and g as their targets, making a commuting square and furthermore the categorical universality of the pullback square holds for this square and we say that the category has pullbacks.

(Category has inverse images): In the case that g in the above diagram is a monoarrow and every such diagram can be completed to a pullback square in category $\mathbb{C}$, we say that category $\mathbb{C}$ has inverse images.

Definition 92 (kernel pair of arrow) Consider now the pullback of an arrow f along itself as:

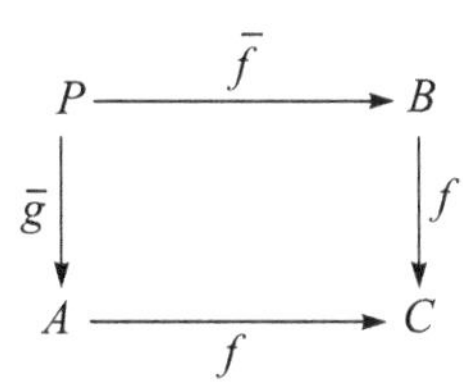

It then consists of a pair of arrows $\bar{f},\bar{g}:P\to A$. We call the pair of arrows $\bar{f},\bar{g}$ as the kernel pair of f.

Notice now the following characterization of a monoarrow:

Proposition 90 An arrow f is a monoarrow iff the pullback of f along itself is an identity arrow, that is, if the following square commutes:

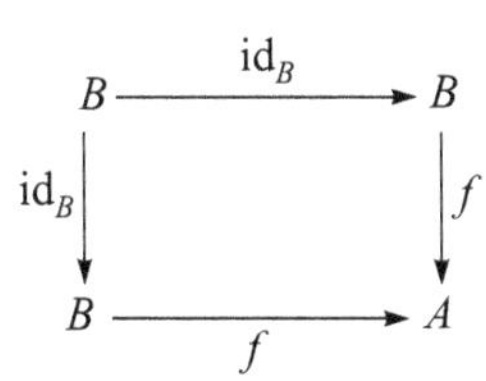

In other words, an arrow is a monoarrow iff identities form a kernel pair in its pullback along itself.

Proof Consider the following diagram:

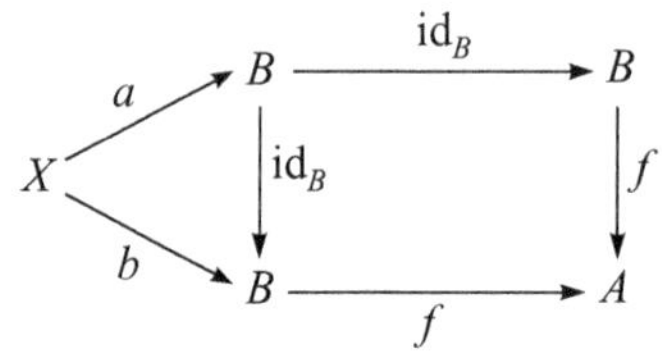

Then $f \circ \mathrm{id}_B \circ a = f \circ a = f \circ b$ and $b = \mathrm{id}_B \circ a = a$ in the triangle. Notice that an identity arrow in the above proposition can be replaced by an isoarrow. We leave it as an exercise for the reader to show this.

Proposition 91 If the square

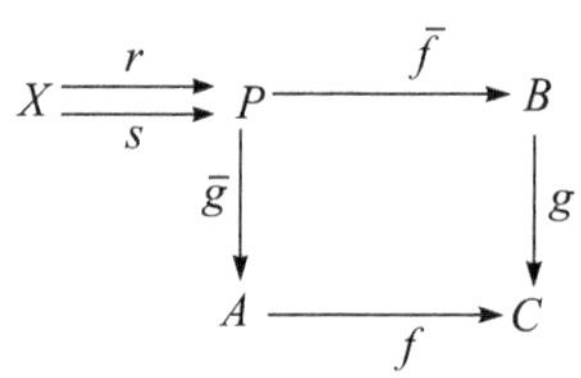

is a pullback, then $(P,(\bar{f},\bar{g}))$ is an extremal monosource.

Proof Let a pullback diagram be as given above. We then have $g\bar{f} = f\bar{g}$. Consider $X \overset{r}{\underset{s}{\rightrightarrows}} P$ as any pair of arrows, and let $S \circ r = S \circ s$ where we denote the source by $S = (P,(\bar{f},\bar{g}))$. Then we have

$$\bar{f}r = \bar{f}s, \quad \bar{g}r = \bar{g}s.$$

We then notice that X is also in the pullback with C, for $\bar{f}r : X \to B$, $\bar{g}s : X \to A$ and the square commutes. The uniqueness in the pullback diagram implies that there exists a unique arrow $k : X \to P$ and, hence, $k = r = s$. The source S is then a monosource.

Considering now any other pair of arrows $Y \overset{p}{\underset{q}{\rightrightarrows}} X$, we have $p = q$ whenever $kp = kq$, once again from the uniqueness in the pullback diagram involving Y in pullback with C. Thus, k is a monoarrow.

Now, let $k : X \to P$ be the unique arrow from the pullback diagram. Evidently, k is then an epiarrow, for we have $S = S \circ k$, that is,

whenever $g\bar{f}k = f\bar{g}k$ we have $g\bar{f} = f\bar{g}$.

Therefore, k is an isoarrow. Hence, S is an extremal monosource.

A pullback is thus a pair of a 2-source and a 2-sink forming a commuting square, with all the sources involving objects isomorphic to the object of the 2-source also forming a commuting square with that 2-sink.

Proposition 92 The pullback of a monoarrow along any arrow is a monoarrow.

Proof Consider the following pullback square:

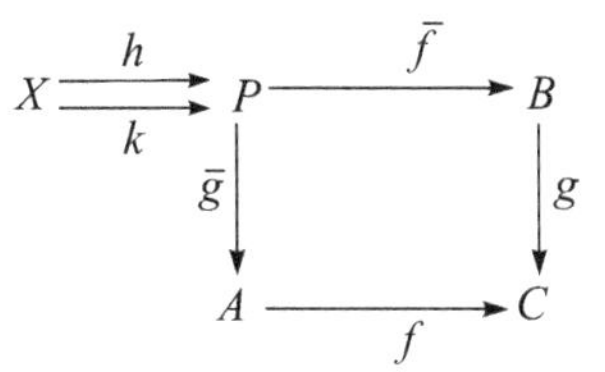

Here, g is a monoarrow and f is any arrow in the category of g. Then $g \circ \bar{f} = f \circ \bar{g}$. If h, $k : X \to P$ are such that $\bar{g}h = \bar{g}k$, then $f(\bar{g}h) = f(\bar{g}k) = g(\bar{f}h) = g(\bar{f}k)$. So, $\bar{f}h = \bar{f}k$ since g is a monoarrow. But $(P,(\bar{f},\bar{g}))$ is a (extremal) monosource. Hence, we can cancel $\bar{f}$ and $\bar{g}$ simultaneously from $\bar{g}h = \bar{g}k$ and $\bar{f}h = \bar{f}k$ so that $h = k$.

Lemma 1 (pasting lemma for pullbacks) Given the commuting diagram

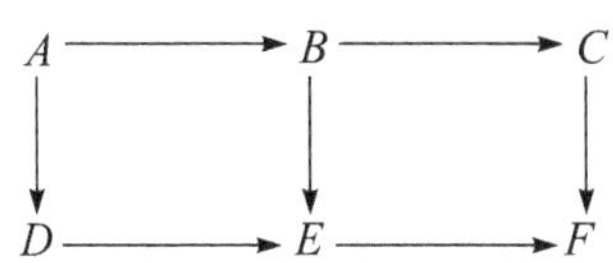

the following hold:

- The outer rectangle is a pullback diagram if both inner squares are pullback diagrams.
- If the outer rectangle as well as the right square are pullback diagrams, then the left square is a pullback diagram.

Proof (first part) Let the inner squares in the above diagram in a category be pullback squares. Notice that the arrows $A \to B \to C$, $A \to D \to E$, $C \to F$, and $E \to F$ form a commuting square. The uniqueness of the pullback diagram follows for this commuting square from the uniqueness of inner squares that are given to be pullbacks.

Proof for the second part is left as an exercise.

Proposition 93 Now, we note the following results:

- The pullback of a retraction arrow is also a retraction arrow.
- The pullback of a regular monoarrow is also a regular monoarrow.
- Consider the following commuting diagram:

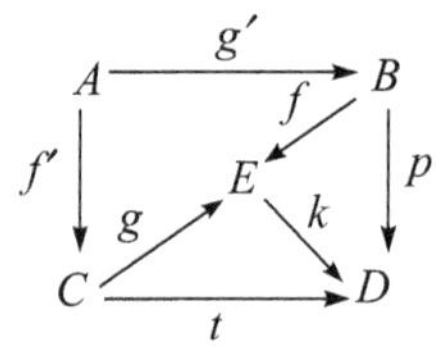

  If the inner square of this diagram is a pullback square and if arrow $k : E \to D$ is a monoarrow, then the outer square is also a pullback square.
- If

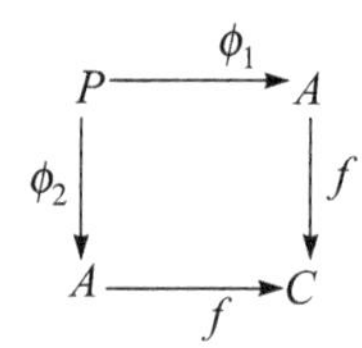

  is a pullback square, then arrow f is a regular epiarrow iff $(f, C) \approx \text{CoEq}(\phi_1, \phi_2)$.

Proof Let pullbacks exist in a category.

- Let $f : A \to C$ be a retraction arrow. Then by definition, there exists an arrow $m : C \to A$ such that $fm = \text{id}_C$. Consider now the following pullback diagram:

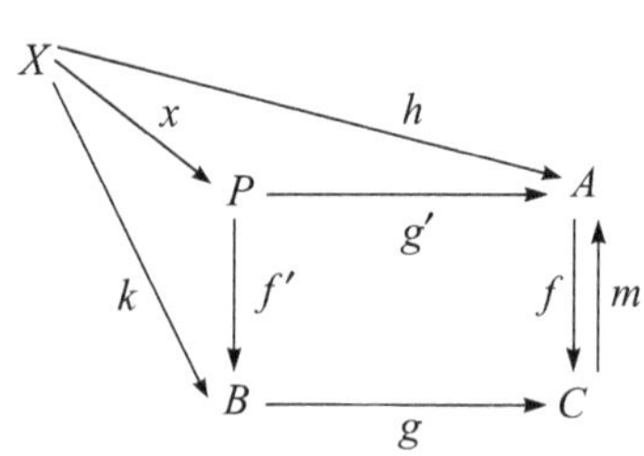

As the square commutes, $fg' = gf'$. By the uniqueness in the pullback diagram, we also have $k = f'x$ and $h = g'x$ with the unique arrow $x : X \to P$. If $fm = \mathrm{id}_C$, then $(fm)g = \mathrm{id}_C \circ g = g = g \circ \mathrm{id}_B = f(mg)$. By the overall uniqueness in the definition of a pullback, there then exists $n : B \to P$ such that $f'n = \mathrm{id}_B$.

- Consider the following diagram in a category:

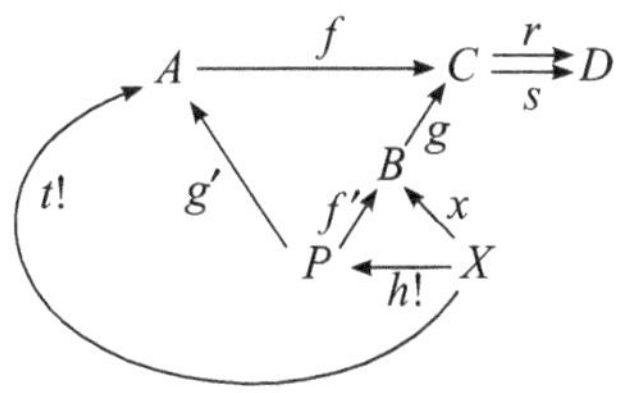

Given $f: A \to C$ is a regular monoarrow or that $(A, f) \approx \mathrm{Eq}(r, s)$, for some arrows $r, s : C \to D$, that is, $rf = sf$. From the given commuting square, we have $fg' = gf'$. Then $rfg' = sfg'$ or $(rg)f' = (sg)f'$. Given $x : X \to B$ with $(rg)x = (sg)x$, there then exists, by the definition of an equalizer arrow, a unique arrow t : $X \to A$ for which $ft = gx$. By the definition of a pullback, there then exists a unique arrow $h : X \to P$ with $x = f'h$. Arrow h is also unique with this property, as f' is a monoarrow. Thus, $(P, f') \approx \mathrm{Eq}(rg, sg)$.

- If the inner square is a pullback, then we have $fg' = gf'$ and $(fg')k = (gf')k$. The outer square then commutes with $p = kf$ and $t = kg$. The uniqueness of the outer square as a pullback follows from the uniqueness of the inner pullback diagram and the above.

Now, notice that if the outer square is a pullback diagram, then the inner square is also a pullback without k being a monoarrow.

- A regular epiarrow f is a coequalizer of some arrows in its category. Let h and k then be such that $fh = kf$. That is to say, let f be a coequalizer of h and k. Consider now the diagram

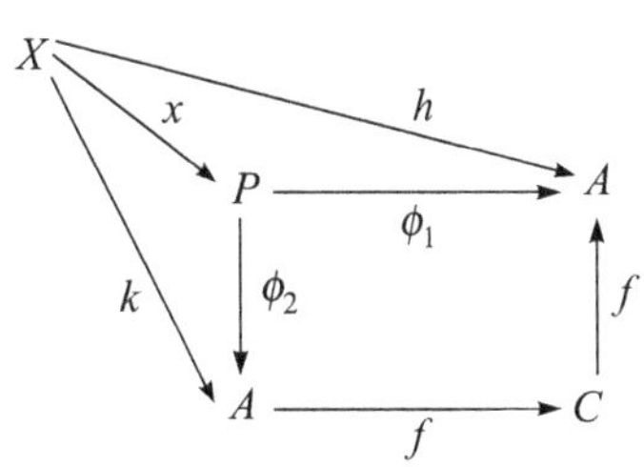

,

with $fh = fk$. Notice that X is in the pullback with C and there exists therefore a unique arrow $x : X \rightarrow P$ such that $k = f_2 \circ x$ and $h = f_1 \circ x$. We therefore have $(f \circ \phi_1) \circ x = (f \circ \phi_2) \circ x$ and $f \circ \phi_1 = f \circ \phi_2$. Then $(f, C) \approx \mathrm{CoEq}(\phi_1, \phi_2)$.

On the other hand, if $(f, C) \approx \mathrm{CoEq}(\phi_1, \phi_2)$, then there exist in the category arrows $h, k : X \rightarrow P$ and the same diagram then implies that $(f, A) \approx \mathrm{CoEq}(h, k)$. Thus, f is a regular epiarrow.

Definition 93 (pullback stability) A collection $\mathcal{M}$ of arrows in a category is called pullback stable or closed under the formations of pullbacks provided that for every arrow $f \in \mathcal{M}$, the pullback f' of f along any arrow g also belongs to $\mathcal{M}$. That is, in the pullback diagram

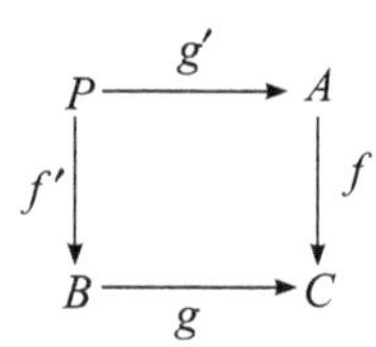

if $f \in \mathcal{M}$, then $f' \in \mathcal{M}$.

We then say that retraction arrows, monoarrows, and regular monoarrows are pullback stable.

The following provides the canonical construction for pullbacks. It begins with a 2-sink and involves a product and an equalizer.

Proposition 9 Let $f : A \rightarrow C$ and $g : B \rightarrow C$ be a pair of arrows in a category $\mathbb{C}$ or let $((f, g), C)$ be a 2-sink. Then provided that $(A \times B, \pi_A, \pi_B)$ is a product of A and B and if $(E, e) \approx \mathrm{Eq}(f \circ \pi_A, g \circ \pi_B)$, the outer square of the following diagram is a pullback square.

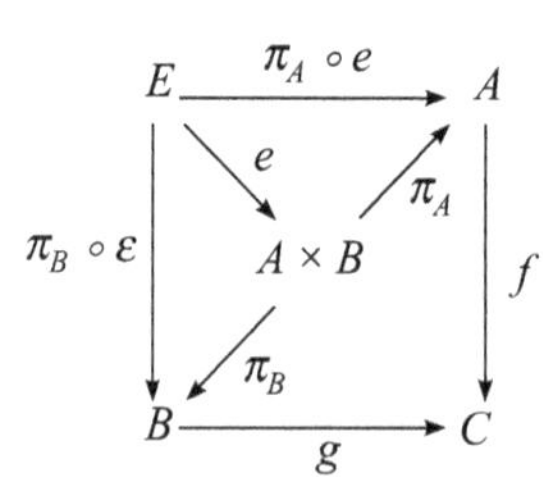

Proof The outer square commutes by its construction. If there exists arrows $p : X \rightarrow A$ and $q : X \rightarrow B$ such that $fp = gq$, then by the uniqueness in the product, there exists a unique arrow $h : X \rightarrow A \times B$

such that $\pi = \pi_A \circ h$ and $q = \pi_B \circ h$. Hence, we have $f \circ \pi_A \circ h = g \circ \pi_B \circ h$. But (E, e) is an equalizer of $f \circ \pi_A$ and $g \circ \pi_B$. There then exists a unique arrow $k : X \to E$ such that $h = e \circ k$. Furthermore, we then have $\pi_A \circ e \circ k = q$ and $\pi_B \circ e \circ k = q$. The uniqueness of k now follows from the fact that a product as well as an equalizer are monosources.

As a corollary to the above, we have:

Corollary 6 If a category $\mathbb{C}$ has finite products and equalizers, then it also has pullbacks.

Proposition 95 If T is a terminal object of a category $\mathbb{C}$ and P is in the pullback with T, then P is also a product.

Proof Let an object P of category $\mathbb{C}$ be in the pullback with the terminal object T of $\mathbb{C}$. Also, let $f : P \to B$ and $g : P \to A$ be the arrows of the following pullback diagram:

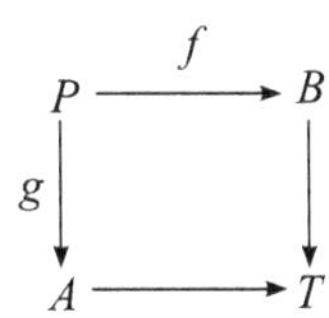

Then by the definition of the pullback square, there exists a unique arrow, $h : X \to P$, from any other object X, in the pullback with T, to object P, such that we have $f_B = f \circ h$ and $g_A = g \circ h$. Object P is then seen to be in the product with objects A and B.

(State and prove or disprove the converse of the above proposition.)

Corollary 7 If a category $\mathbb{C}$ has pullbacks and a terminal object, then it has finite products.

Now, if any object P is in the pullback with a product of two objects, then we can expect the uniqueness of the pullback diagram to impose restrictions on the arrows from P to the component objects of that product. In particular, object P could be a subobject of the component of the product and be in the intersection in relation to that product. This is the essential context of the following proposition:

Proposition 96 Let $A \overset{h}{\underset{k}{\rightrightarrows}} B$ be any arrows in a category $\mathbb{C}$. Let the product $(A \times B, \pi_A, \pi_B)$ also exist in $\mathbb{C}$. Also, let the following be a pullback diagram:

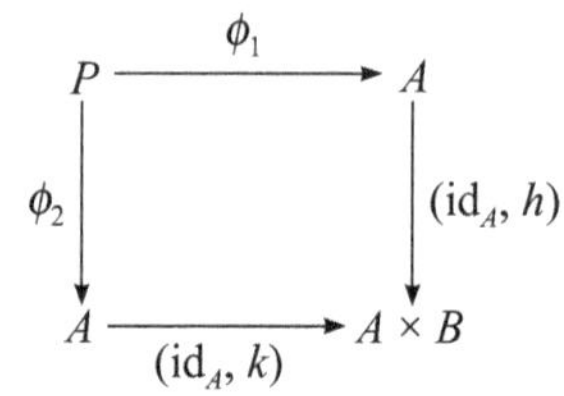

Then the following hold:

- $\phi_1 = \phi_2$.
- $(P, \phi_1) \approx \mathrm{Eq}(<\mathrm{id}_A, h>, <\mathrm{id}_A, k>)$.
- $(P, \phi_1) \approx \mathrm{Eq}(h, k)$.

Proof Let the conditions be as given.

- For the first part, we note that $<\mathrm{id}_A, h> \circ \phi_1 = <\mathrm{id}_A, k> \circ \phi_2$ as the square commutes. Then
  $f_1 = \pi_A \circ <\mathrm{id}_A, h> \circ f_1 = \pi_A \circ <\mathrm{id}_A, k> \circ \phi_2 = \phi_2$.
- The second part follows easily.
- For the third part, consider an arrow $x : X \to A$, with $hx = kx$. Then
  $\pi_B \circ <\mathrm{id}_A, h> \circ x = \pi_B \circ <\mathrm{id}_A, k> \circ x = \pi_A \circ <\mathrm{id}_A, k> \circ x$.
  But the product is a monosource and, hence, we have
  $<\mathrm{id}_A, h> \circ x = <\mathrm{id}_A, k> \circ x$.

But the given square is a pullback diagram. Therefore, we have a unique arrow $p : X \to P$ such that $p = \phi_1 \circ x$.

Corollary 8 If a category $\mathbb{C}$ has finite intersections and has finite products, then it also has equalizers.

Theorem 1 The following are equivalent (TFAE) for any category $\mathbb{C}$:

- $\mathbb{C}$ has finite products and equalizers.
- $\mathbb{C}$ has pullbacks and a terminal object.

5.18.1 Congruences and Pullbacks

Definition 94 (congruences and pullbacks) Given the following pullback square with $fp = fq$:

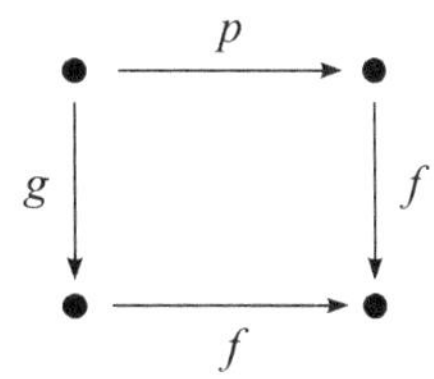

The pair (p, q) is called a congruence relation of arrow f.

Furthermore, any pair (p, q) of arrows p and q in a category $\mathbb{C}$ will be called a congruence relation if there exists in $\mathbb{C}$ an arrow f such that (p, q) is a congruence relation of f.

The motivation for the above definition is the congruence relation determined by a group homomorphism $f : A \to B$. Such a congruence relation is a subset S of $A \times A$ consisting of all the pairs (a, b) with $f(a) = f(b)$. The subset S can then be considered as a subgroup of $A \times A$ with the associated inclusion $m : S \rightarrowtail A \times A$ and the projections $A \times A \underset{\pi_2}{\overset{\pi_1}{\rightrightarrows}} A$. Then by the canonical construction of pullbacks, the following is a pullback diagram:

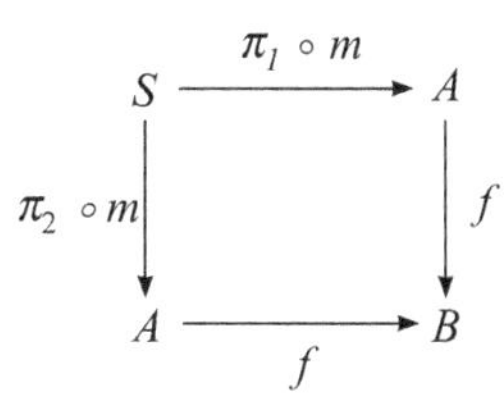

In general, we then exploit the aforementioned relationship of arrows in the above pullback diagram for the congruence relation (p, q) of f.

Proposition 97 Let (p, q) be a congruence relation of arrow f. Then the following hold:

- The pair (p, q) is a congruence relation also of arrow $m \circ f$ whenever the composition is defined and where m is a monoarrow.
- If $f = gh$ and $hp = hq$, then (p, q) is a congruence relation of h.
- If $c \approx \mathrm{CoEq}(p, q)$, then (p, q) is a congruence relation of c.

Proof This is left for the reader as an exercise.

5.18.2 Push-Out

The dual concept of a push-out can then be defined as follows.

Definition 95 (push-out) In the commuting diagram

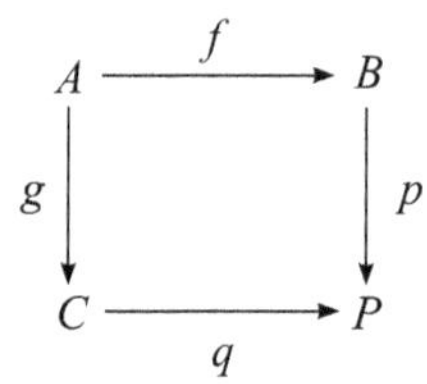

object P is called the push-out object and is denoted as $P = B\coprod_A C$, provided the dual of the categorical universality for pullbacks holds for it. Notice that any push-out diagram in category $\mathbb{C}$ is, clearly, a pullback diagram in $\mathbb{C}^{op}$.

A push-out is then a pair of a 2-source and a 2-sink forming a commuting square, with all the sinks involving objects isomorphic to the object of the 2-sink also forming a commuting square with that 2-source.

Push-out of f along g, direct image of f along g, has push-outs, and has direct images are notions dual to those that arose for pullbacks.

5.18.3 Pullation Square

Definition 96 (pullation or pull-push square) A commuting square that is simultaneously a pullback and a push-out square is called a pullation or pull-push square.

Proposition 98 For any square of the form

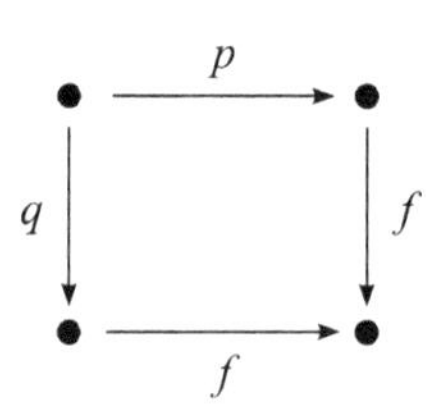

TFAE:

- The square is a pull-push or pullation square.
- (p, q) is a congruence relation of f and $f \approx \mathrm{CoEq}(p, q)$.
- (p, q) is a congruence relation of f and f is a regular epiarrow.

- (p, q) is a congruence relation and $f \approx \text{CoEq}(p, q)$.

Proof This is left for the reader as an exercise.

Proposition 99 In any category, TFAE statements:

- f is a monoarrow.
- (id, id) is a congruence relation of f.
- f has a congruence relation of the form (p, p).

We may now note the following:

- A group considered as a category has pullbacks.
- In any category $\mathbb{A}$, the square

$$\begin{array}{ccc} P & \xrightarrow{g'} & A \\ {\scriptstyle f'}\downarrow & & \downarrow{\scriptstyle f} \\ B & \xrightarrow[g]{} & C \end{array}$$

is a pullback square iff $P \xrightarrow{fg'=gf'} C$ together with f' and g' is a product of $A \xrightarrow{f} C$ and $B \xrightarrow{g} C$ in the comma category of $\mathbb{A}$ over C, that is, in category $\mathbb{A} \downarrow C$.

5.19 Subobject Classifier

We can now note that in category $\mathbb{S}$et, a subset $S \subset A$ is describable in two different ways:

- By a monoarrow $S \rightarrowtail A$ given by the inclusion function
- By a characteristic function of S as

$$\Phi_S(x) = \begin{cases} 0 & \text{for } x \in S \\ 1 & \text{for } x \notin S \end{cases}$$

The characteristic function Φ_S then takes values in a 2-element set $\{0, 1\}$, and we call these values truth values. Then our convention here is that 0 corresponds to "true" and 1 corresponds to "false." We can then notice that true can be regarded as the following monoarrow or injective function of category $\mathbb{S}$et:

$$\overrightarrow{true} : 1 = \{0\} \mapsto 2 = \{0,1\} \quad \text{with } 0 \mapsto 0$$

With this definition of a monoarrow $\overrightarrow{true}$, it is easy to see that we have the commuting square

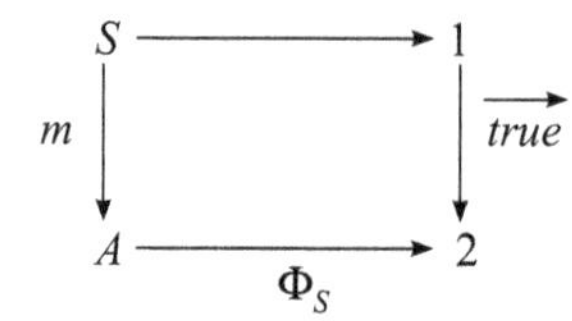

in which any monoarrow m is a pullback of arrow $\overrightarrow{true}$ along arrow Φ_S. Clearly, it determines m up to an isoarrow.

Notice here that $1 = \{0\}$, the singleton set, is a terminal object of $\mathbb{S}$et and $\overrightarrow{true}$ is a fixed monoarrow and, thus, given any monoarrow m, there exists a unique Φ_S, making the above a pullback diagram.

Observation 10 We now notice that for category $\mathbb{S}$et × $\mathbb{S}$et, there exist four truth values, and for category $\mathbb{S}\text{et}^n$, there exist 2^n truth values. Hence, the characteristic function does not, in general, take values in 2, but takes values in a suitable object, which we denote by Ω, of the category under considerations. We may call Ω the "truth value object."

Consequently, we can immediately arrive at the following generalization of a subobject classifier:

Definition 97 (subobject classifier) If T is a terminal object of a category $\mathbb{C}$ in which all the binary products exist, then a subobject classifier is a monoarrow $\overrightarrow{true}: T \to \Omega$, where Ω is an object of category $\mathbb{C}$ such that to every monoarrow m from object S to object X in $\mathbb{C}$, there exists a unique arrow Φ_S to correspond to that monoarrow and making the following a pullback diagram:

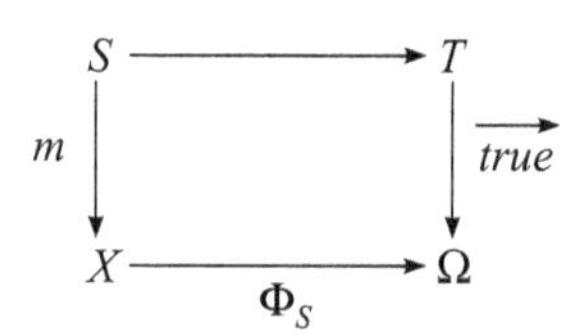

Then every subobject S of object X is a unique pullback of the universal monoarrow $\overrightarrow{true}: T \to \Omega$.

Observation 11 Notice, however, that the truth value object may not even exist in a category even when that category has a terminal object and also has all the binary products. There is then no subobject classifier in them. This, in fact, is the situation with many of the reasonable categories involving familiar mathematical structures.

For example, consider category $\mathbb{A}$b of (small) abelian groups. Then its terminal object is a zero group. All the binary products of (small) abelian groups also exist in $\mathbb{A}$b.

Now, a monoarrow from $T \to \Omega$ must send 0 to $0 \in \Omega$. Its pullback along any arrow $\Phi : X \to \Omega$ is then a subgroup $S = \ker\Phi = \Phi^{-1}(0)$ of group X. Then the truth value object Ω in category $\mathbb{A}$b is an abelian group such that it contains a copy of every quotient group X/S for every abelian group X! This is clearly an impossibility. Therefore, there is no object in $\mathbb{A}$b that can be used as a truth value object.

As was noted earlier, a collection of all the subobjects of a given object of a category is partially ordered by the relation $\leq$ on subobjects and this collection always has the structure of a Heyting algebra. The existence or nonexistence of a subobject classifier for a category has no role for this Heyting algebra structure that is always available.

5.20 Summary

To conclude this chapter, we note that the properties of arrows and objects that were considered here may, in general, not be available for many of the arrows and objects of a generic category. Nonetheless, these properties of arrows and objects are important to the analysis of the mathematical structure forming a category.

Of particular relevance is the property of being a left-cancelable arrow or a monoarrow and, dually, that of being a right-cancelable arrow or an epiarrow. We have extensively studied here situations in which monoarrows and epiarrows occur in diagrams in a category.

Now, functors as arrows connecting categories may also possess such properties. Furthermore, functors may preserve such properties of arrows of a category.

Chapter 6

Functors

6.1 Introduction

6.1.1 Arrows Connecting Categories

Consider two categories $\mathbb{A} = (\mathcal{A}_1, \circ_1)$ and $\mathbb{B} = (\mathcal{A}_2, \circ_2)$. Consider also a function $\mathfrak{F} : \mathcal{A}_1 \rightarrow \mathcal{A}_2$, that is, a function between the two collections of arrows $\mathcal{A}_1$ and $\mathcal{A}_2$ such that the identities in $\mathcal{A}_1$ are mapped to identities in $\mathcal{A}_1$: $\mathfrak{F}(u_C) = u_{\mathfrak{F}(C)}$ and that the composition of arrows in $\mathcal{A}_1$ is mapped under that function to the corresponding composition of arrows in $\mathcal{A}_2$, that is $\mathfrak{F}(f \circ_1 g) = \mathfrak{F}(f) \circ_2 \mathfrak{F}(g)$. We then say that $\mathfrak{F}$ is a function from category $\mathbb{A}$ to category $\mathbb{B}$ that preserves identities and the composition of arrows. This function is a partial binary algebra preserving map between $\mathbb{A} = (\mathcal{A}_1, \circ_1)$ and $\mathbb{B} = (\mathcal{A}_2, \circ_2)$. We call such a function a *functor* and we write $\mathfrak{F} : \mathbb{A} \rightarrow \mathbb{B}$ or $\mathbb{A} \xrightarrow{\mathfrak{F}} \mathbb{B}$. A formal definition of a functor is then:

Definition 98 (functor) Given categories $\mathbb{A}$ and $\mathbb{B}$, a functor $\mathfrak{F} : \mathbb{A} \rightarrow \mathbb{B}$ is defined to be a triple $(\mathbb{A}, \mathfrak{F}, \mathbb{B})$ with $\mathfrak{F}$ being a function from the collection of arrows in $\mathbb{A}$ to the collection of arrows in $\mathbb{B}$ such that:

- $\mathfrak{F}$ preserves identities, that is, if e is an identity arrow in $\mathbb{A}$, then $\mathfrak{F}(e)$ is an identity arrow in $\mathbb{B}$.

Subtlety in Relativity
Sanjay M. Wagh

ISBN 978-981-4774-44-4 (Hardcover), 978-1-351-23832-8 (eBook)
www.panstanford.com

- $\mathfrak{F}$ preserves compositions of arrows, that is, if f and g are arrows and $f \circ g$ is defined in $\mathbb{A}$, then $\mathfrak{F}(f \circ g) = \mathfrak{F}(f) \circ \mathfrak{F}(g)$ is defined in $\mathbb{B}$ and the equality holds under the compositions.

Notation for functors

To simplify notations, we omit suffixes on the operations of composition to write $\mathfrak{F}(f \circ g) = \mathfrak{F}(f) \circ \mathfrak{F}(g)$, and we write $\mathfrak{F}(A)$ as $\mathfrak{F}A$ and $\mathfrak{F}(f)$ as $\mathfrak{F}f$. It is then understood that the composition $\mathfrak{F}f\mathfrak{F}g$ relates to category $\mathbb{A}$ and that $\mathfrak{F}A$ is an object and $\mathfrak{F}f$ is an arrow of category $\mathbb{B}$. For the combined action of a functor, we will often write:

$$\mathfrak{F}(A \xrightarrow{f} B \xrightarrow{g} C) = \mathfrak{F}A \xrightarrow{\mathfrak{F}f} \mathfrak{F}B \xrightarrow{\mathfrak{F}g} \mathfrak{F}C$$

Many times, a functor whose source or domain is a small category is referred to as a small functor.

Functors being the functions of the collection of arrows of a category, their composition also preserves identities as well as the composition of arrows. Hence, functors can be composed to yield another functor.

Proposition 100 If $(\mathbb{A}, \mathfrak{F}, \mathbb{B})$ and $(\mathbb{B}, \mathfrak{G}, \mathbb{C})$ are functors, then $(\mathbb{A}, \mathfrak{G} \circ \mathfrak{F}, \mathbb{C})$ is a functor, denoted as the composition $\mathfrak{G} \circ \mathfrak{F}$ of functors $\mathfrak{F}$ and $\mathfrak{G}$.

Proposition 101 If $(\mathbb{A}, \mathfrak{F}, \mathbb{B})$, $(\mathbb{B}, \mathfrak{G}, \mathbb{C})$, and $(\mathbb{C}, \mathfrak{H}, \mathbb{D})$ are functors, then we have the associativity of their composition: $\mathfrak{H} \circ (\mathfrak{G} \circ \mathfrak{F}) = (\mathfrak{H} \circ \mathfrak{G}) \circ \mathfrak{F}$.

For convenience, we write, if $\mathbb{A} \xrightarrow{\mathfrak{F}} \mathbb{B}$ and $\mathbb{B} \xrightarrow{\mathfrak{G}} \mathbb{C}$, then $\mathbb{A} \xrightarrow{\mathfrak{G}\circ\mathfrak{F}} \mathbb{C}$. Notice that the notation is imprecise, for it gives the impression that a functor is a function from $\mathbb{A}$ to $\mathbb{C}$. Notice also the reversal, $\mathfrak{G} \circ \mathfrak{F}$, of the order in writing the composition of functors $\mathfrak{F}$ and $\mathfrak{G}$. This is in conformity with our convention about the composition of arrows.

Now the collection of the identity arrows in a category $\mathbb{A}$ is the same as the collection of the identity arrows in its dual or opposite category. Furthermore, we also have $\mathrm{Mor}(\mathbb{A}) = \mathrm{Mor}(\mathbb{A}^{\mathrm{op}})$, with the order of the composition of arrows in $\mathbb{A}^{\mathrm{op}}$ being the reverse of that in $\mathbb{A}$. Then functors $\mathfrak{F} : \mathbb{A} \to \mathbb{B}$ are generally different from functors $\mathfrak{b} : \mathbb{A}^{\mathrm{op}} \to \mathbb{B}$ and from functors $\mathfrak{H} : \mathbb{A} \to \mathbb{B}^{\mathrm{op}}$. This leads us to the following definitions:

Definition 99 (contravariant functor) Given any two categories $\mathbb{A}$ and $\mathbb{B}$, a functor $(\mathbb{A}^{op}, \mathfrak{G}, \mathbb{B})$ or a functor $(\mathbb{A}, \mathfrak{H}, \mathbb{B}^{op})$ is called a contravariant functor from category $\mathbb{A}$ to category $\mathbb{B}$.

Definition 100 (covariant functor) A functor $(\mathbb{A}, \mathfrak{G}, \mathbb{B})$ is called a covariant functor from category $\mathbb{A}$ to category $\mathbb{B}$.

Although we have the above distinct definitions for a contravariant and a covariant functor and a contravariant functor is not a functor from $\mathbb{A}$ to $\mathbb{B}$, it is often a practice in the literature to say that $\mathfrak{F} : \mathbb{A} \to \mathbb{B}$ is a contravariant functor. A reader is supposed to realize from the context that the functor under considerations is either $\mathfrak{G} : \mathbb{A}^{op} \to \mathbb{B}$ or $\mathfrak{H} : \mathbb{A} \to \mathbb{B}^{op}$. In this book, we will, however, maintain the above distinction of covariant and contravariant functors by explicitly saying, "Let us consider a (contravariant) functor $\mathfrak{F} : \mathbb{A} \to \mathbb{B}^{op}$ (or $\mathfrak{F} : \mathbb{A}^{op} \to \mathbb{B}$)."

Then with every functor $\mathfrak{F} : \mathbb{A} \to \mathbb{B}$, there are associated two contravariant functors that are denoted as $\mathfrak{F}^* : \mathbb{A}^{op} \to \mathbb{B}$ and $^*\mathfrak{F} : \mathbb{A} \to \mathbb{B}^{op}$.

6.1.2 Dual Role of a Functor

For a functor, function $\mathfrak{F}$ serves also as a map from the collection (of objects) indexing the units of $\mathbb{A}$ to the collection (of objects) indexing the units of $\mathbb{B}$, as it is a map from collection $\mathcal{A}_1$ of arrows of $\mathbb{A}$ to collection $\mathcal{A}_2$ of arrows of $\mathbb{B}$. Consequently, a functor $\mathfrak{F}$ plays a dual role: it assigns to an object C of $\mathbb{A}$ an object $\mathfrak{F}(C)$ in $\mathbb{B}$ and it also assigns to any arrow $A \xrightarrow{f} B$ in $\mathbb{A}$ an arrow $\mathfrak{F}(A) \xrightarrow{\mathfrak{F}(f)} \mathfrak{F}(B)$ in $\mathbb{B}$, while preserving the identities and composition of arrows as $\mathfrak{F}(f \circ_1 g) = \mathfrak{F}(f) \circ_2 \mathfrak{F}(g)$. This dual role surfaces particularly when we consider objects of categories and a functor connecting those categories. The object side of a functor is, nevertheless, completely determined by the arrows side for a functor, for identity arrows uniquely label objects of a category.

Notice therefore that a functor $\mathfrak{F} : \mathbb{A} \to \mathbb{B}$ induces a unique function (to be denoted by the same symbol $\mathfrak{F}$) from the collection of $\mathbb{A}$-objects to the collection of $\mathbb{B}$-objects in such a way that

$$\mathfrak{F}(\mathrm{id}_A) = \mathrm{id}_{\mathfrak{F}(A)}.$$

It should then be evident that

$$\mathfrak{F}([A, B])_{\mathbb{A}}) \{ [\mathfrak{F}(A),\mathfrak{F}(B)]_{\mathbb{B}}.$$

As a result of this, any functor $\mathfrak{F} : \mathbb{A} \to \mathbb{B}$ can be built using the object-function $\mathfrak{F} : \text{Ob}(\mathbb{A}) \to \text{Ob}(\mathbb{B})$ and all of the restrictions.

It is then a common practice to describe a functor using its object function and its aforementioned hom-collection restrictions.

6.2 Some Examples of Functors

Some examples of functors are the following:

- **Identity functor:** $\mathbf{1} : \mathbb{A} \to \mathbb{A}$. It is an assignment of $u_C \to u_C$ for every u_C and $f \circ g \to f \circ g$ for all f and g whenever the composition is defined. Then $\mathbf{1}(u_C) = u_{1(C)} = u_C$ and $\mathbf{1}(f \circ g) = \mathbf{1}(f) \circ \mathbf{1}(g) = f \circ g$. We will also denote this functor as $\text{id}_{\mathbb{A}}$.
- **Constant functor:** $\mathfrak{Const} : \mathbb{A} \to \mathbb{B}$. It is an assignment of all the arrows of $\mathbb{A}$ to the same unit u_C in $\mathbb{B}$. Then all the identities in $\mathbb{A}$ are mapped to an identity u_C in $\mathbb{B}$ and the composition $f \circ g$ in $\mathbb{A}$, whenever defined, is mapped to the composition $u_C \circ u_C$ $(=u_C)$ in $\mathbb{B}$.
- **Inclusion functor:** If $\mathbb{A} = (\mathcal{A}, \circ)$ is a subcategory of category $\mathbb{B} = (\mathcal{B}, \circ)$ and $\hookrightarrow : \mathcal{A} \to \mathcal{B}$ is the inclusion function, then $\hookrightarrow : \mathbb{A} \to \mathbb{B}$ is called an inclusion functor from $\mathbb{A}$ to $\mathbb{B}$.
- **Canonical or natural functor:** If $\mathbb{C}/\sim$ is a quotient category of category $\mathbb{C}$ and $Q : C \to C/\sim$ is a canonical function that assigns to each arrow in $\mathbb{C}$ its equivalence class under $\sim$, then $\mathbb{C} \xrightarrow{\mathfrak{Q}} \mathbb{C}/\sim$ is a functor, called the canonical or natural functor from $\mathbb{C}$ to $\mathbb{C}/\sim$.
- **Opposite functor:** If $(\mathbb{A}, \mathfrak{F}, \mathbb{B})$ is a functor, then $(\mathbb{A}^{\text{op}}, \mathfrak{F}^{\text{op}}, \mathbb{A}^{\text{op}})$ is also a functor, called the functor opposite to $\mathfrak{F}$. Notice here that $\mathfrak{F}$ and $\mathfrak{F}^{\text{op}}$ are identical as functions on the collection of arrows of the two categories, for $\text{Mor}(\mathbb{A}) = \text{Mor}(\mathbb{A}^{\text{op}})$. However, as functors, $\mathfrak{F}$ and $\mathfrak{F}^{\text{op}}$ are different, for they have different sources and targets.
- **Forgetful functor:** For any structured collections (set) as a category, there exists a map from that category to the underlying collection. Using this map, we can construct a functor, called a forgetful functor, that "forgets" the structure. For example, consider a functor from $\mathbb{T}$op to $\mathbb{S}$et that maps each object of $\mathbb{T}$op to its underlying set in $\mathbb{S}$op. A forgetful functor is also called an *underlying functor*.

- **Covariant hom-functor:** For any category $\mathbb{A}$ and a $\mathbb{T}$-object $\mathbb{A}$, we have a covariant hom-functor $[A]-] : \mathbb{A} \to \mathbb{C}\mathrm{ol}$, defined by its action on any $\mathbb{A}$-arrow $B \xrightarrow{f} C$ as
  $[A,-](B \xrightarrow{f} C) = [A,B] \xrightarrow{[A,-]f} [A,C]$,
  where $[A,-]f(g) = f \circ g$.
- **Contravariant hom-functor:** For any category $\mathbb{A}$ and an $\mathbb{A}$-object A, we have a contravariant hom-functor $[-, A] : \mathbb{A}^{\mathrm{op}}$ * $\mathbb{C}\mathrm{ol}$, defined by its action on any $\mathbb{A}^{\mathrm{op}}$-arrow $B \xrightarrow{f} C$ as
  $[-,A](B \xrightarrow{f} C) = [B,A] \xrightarrow{[-,A]f} [C,A]$,
  where $[-, A]f(g) = f \circ g$.

Occasionally, as in the next proposition, we will call a functor from the product of categories $\mathbb{A}^{\mathrm{op}} \times \mathbb{A}$ to category $\mathbb{C}\mathrm{ol}$ also as the hom-functor. This should, however, cause no confusion.

Proposition 102 If $\mathbb{A}$ is any category, then there exists a functor of the form

$$\mathrm{hom}_{\mathbb{A}} : \mathbb{A}^{\mathrm{op}} \times \mathrm{A} \to \mathbb{C}\mathrm{ol}$$

such that $\mathrm{hom}_{\mathbb{A}}(A, B)$ is the collection of $\mathbb{A}$-arrows from an $\mathbb{A}$-object A to another $\mathbb{A}$-object B and $\mathrm{hom}_{\mathbb{A}}(f, g)(h) = g \circ h \circ f$. Diagrammatically, we therefore have:

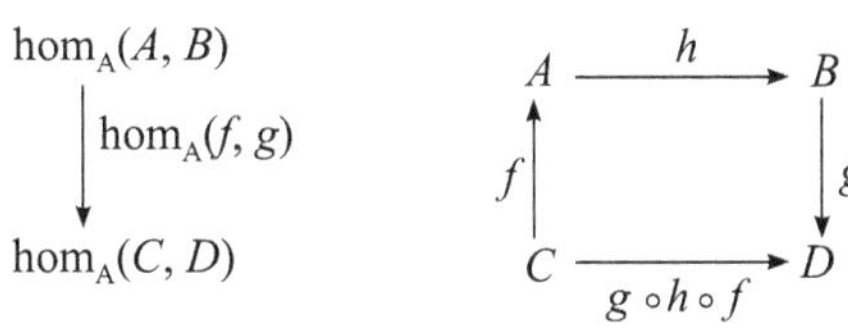

Proof We show that identities and compositions are preserved under the action of functor $\mathrm{hom}_{\mathbb{A}}$. (Remember that in (f, g), f is an arrow from $\mathbb{A}^{\mathrm{op}}$ and g is an arrow from $\mathbb{A}$.)

Firstly, we have $\mathrm{hom}_{\mathbb{A}}(\mathrm{id}_A, \mathrm{id}_B)(h) = \mathrm{id}_B \circ h \circ \mathrm{id}_A = h$ and, hence, $\mathrm{hom}_{\mathbb{A}}(\mathrm{id}_A, \mathrm{id}_B)$ is an identity on the collection $\mathrm{hom}_{\mathbb{A}}$ (A, B).

Next, for any other pair (a, b) in $\mathrm{hom}_{\mathbb{A}}(A, B)$, we have

$$\begin{aligned} \mathrm{hom}_{\mathbb{A}}(f, g)^*(a, b)(h) &= \mathrm{hom}_{\mathbb{A}}(af, gb)(h) \\ &= g * \mathrm{hom}_{\mathbb{A}}(a, b)(h) \circ f \\ &= \mathrm{hom}_{\mathbb{A}}(f, g)\,(\mathrm{hom}_{\mathbb{A}}(a, b)(h)) \\ &= \mathrm{hom}_{\mathbb{A}}(f, g) \circ \mathrm{hom}_{\mathbb{A}}(a, b)(h). \end{aligned}$$

Thus, function $\hom_{\mathbb{A}}$ preserves identities as well as composition and is therefore a functor.

With a little more effort, we also see:

Proposition 103 Given any category $\mathbb{A}$ and an $\mathbb{A}$-object A, we have

$$\hom_{\mathbb{A}}(-, A) = \hom_{\mathbb{A}^{op}}(A, -).$$

That is to say, the hom-functor $\hom_{\mathbb{A}}(-, A)$ is identical to the hom-functor $\hom_{\mathbb{A}^{op}}(A, -)$.

6.3 Important Nomenclature about Functors

Recall that the hom-collection plays the crucial role in definition 3 of a (generic) category, Chapter 4, in that, apart from collection A of arrows in a category, it is this hom-collection that determines the internal consistency of that definition of a category since it is also required to be free of any set theoretical paradoxes.

As we have seen now, functors are the arrows connecting the categories. Consequently, the action of a functor on a hom-collection of a category can be expected to play an important role in determining the properties of a functor. The following now illustrates the importance of hom-collections vis-à-vis actions of functors:

Definition 101 Let $\mathfrak{F}$ be a functor from $\mathbb{A}$ to $\mathbb{B}$.

- $\mathfrak{F}$ is called an embedding if $\mathfrak{F}$ is injective (one-to-one, or 1-1) on arrows.

  $\mathfrak{F}$ is called faithful if *all* the hom-set restrictions, that is, $\mathfrak{F} : \hom_{\mathbb{A}}(A, B) \to \hom_{\mathbb{B}}(\mathfrak{F}A, \mathfrak{F}B)$ are injective (1-1).
- $\mathfrak{F}$ is called full if *all* hom-set restrictions, that is, $\mathfrak{F} : \hom_{\mathbb{A}}(A, B) \to \hom_{\mathbb{B}}(\mathfrak{F}A, \mathfrak{F}B)$ of functor $\mathfrak{F}$ are surjective (onto).
- $\mathfrak{F}$ is called amnestic provided that an $\mathbb{A}$-isomorphism f is an identity whenever $\mathfrak{F}f$ is an identity.

Observation 12 Observe the following:

- Any inclusion functor from a subcategory to a category is an embedding.
- Any forgetful or underlying functor is faithful.
- Any canonical or natural functor from a category to its quotient category is necessarily full.
- Fullness and faithfulness are self-dual notions.

6.4 Properties of Functors

Since they relate categories, functors have a very special position within category theory. It would thus be helpful to first establish various properties for functors.

6.4.1 Preservation of Categorical Property by Functors

Definition 102 (categorical property being preserved by functors) A functor $\mathfrak{F} : \mathbb{A} \to \mathbb{B}$ is said to preserve a categorical property P provided that the image under $\mathfrak{F}$ of each arrow (or object) in a category $\mathbb{A}$ having property P has that property P in category $\mathbb{B}$.

As an example of the preservation of a categorical property by functors, consider the following proposition:

Proposition 104 All functors $\mathfrak{F} : \mathbb{A} \to \mathbb{B}$ preserve isoarrows, that is, whenever $A \xrightarrow{k} B$ is an $\mathbb{A}$-isoarrow, then $\mathfrak{F}$k is a $\mathbb{B}$-isoarrow.

Proof Fk $\circ\, \mathfrak{F}k^{-1} = \mathfrak{F}(k \circ k^{-1}) = \mathfrak{F}(\mathrm{id}_B g) = \mathrm{id}_{\mathfrak{F}B}$ and, similarly, we have $\mathfrak{F}k^{-1} \circ \mathfrak{F}k = \mathfrak{F}(k^{-1} \circ k) = \mathfrak{F}(\mathrm{id}_A) = \mathrm{id}_{\mathfrak{F}A}$.

This trivial proof can be profitably used to show that certain objects in a category $\mathbb{A}$ are not isomorphic by going to a useful category $\mathbb{B}$ where it is easier to establish that $\mathfrak{F}A$ and $\mathfrak{F}A'$ are not isomorphic.

As another such example of the preservation of a categorical property by functors, we have:

Proposition 105 Every functor preserves a commutative triangle.

Proof Let $h = gf$ for a commutative triangle. Since a functor preserves identities as well as compositions, we have $\mathfrak{F}h = \mathfrak{F}g\mathfrak{F}f$.

Proposition 106 Every functor preserves a section and a retraction.

Proof Let an arrow f in a category $\mathbb{A}$ be a section. Then there exists in $\mathbb{A}$ an arrow g such that $gf = \mathrm{id}_A$ for some object A and g is then a retraction. Now, let $\mathfrak{F} : \mathbb{A} \to \mathbb{B}$ be a functor. Then

$$\mathrm{id}_{\mathfrak{F}(A)} = \mathfrak{F}(\mathrm{id}_A) = \mathfrak{F}(gf) = \mathfrak{F}(g) \circ \mathfrak{F}(f).$$

Thus, $\mathfrak{F}f$ is a section and $\mathfrak{F}f$ is a retraction.

As a summary of various preservation properties of a functor that have been considered above, we have the following:

- Every functor preserves identities, composition of arrows, an isoarrow, a section, a retraction, and a commutative triangle.

Now, let us notice the following preservation property of a hom-functor.

Proposition 107 Each covariant hom-functor, hom$(A,-)$, preserves monoarrows.

Proof Let $A \xrightarrow{f} B$ be a monoarrow. Then we have

$$\text{hom}(A,-)(f)(h) = \text{hom}(A,-)(f)(k) \Rightarrow fh = fk.$$

But f is a monoarrow and, hence, $h = k$. Consequently, the covariant hom-functor is a monoarrow-preserving functor and, hence, is a 1-1 function on hom-collections.

Now, let us note that any functor, between connected categories and with its source having a terminal object, preserves constant arrows. *However, this preservation is a restriction on the category, since every category need not have a terminal object.*

Proposition 108 Whenever $\mathbb{A}$ and $\mathbb{B}$ are two connected categories such that $\mathbb{A}$ has a terminal object T, then for a functor $\mathfrak{F} : \mathbb{A} \to \mathbb{B}$ the following are equivalent (TFAE):

- $\mathfrak{F}$ preserves constant arrows.
- $\mathfrak{F}$ preserves the terminal object T.

Proof Let $A \xrightarrow{f} B$ be a constant arrow of the connected category $\mathbb{A}$ whose terminal object is T.

- Let $\mathfrak{F}$ preserve constant arrows. Then consider $\text{id}_T: T \to T$. It is a constant arrow. Therefore, $\mathfrak{F}(\text{id}_T) = \text{id}_{\mathfrak{F}(T)}$ must also be a $\mathbb{B}$-constant arrow. Then $X \underset{s}{\overset{r}{\rightrightarrows}} \mathfrak{F}(T) \xrightarrow{\text{id}_{\mathfrak{F}}(t)} \mathfrak{F}(T)$ implies $r = s$ for any arrows $r, s \in [X, \mathfrak{F}(\text{T})]$. Hence, there is only one arrow to $\mathfrak{F}(T)$ from any other object X in category $\mathbb{B}$. Thus, $\mathfrak{F}(T)$ is terminal in $\mathbb{B}$.
- Let functor $\mathfrak{F}$ preserve a terminal object. If arrow f is constant in $\mathbb{A}$, then it can be factored through T, the terminal object of $\mathbb{A}$. Let this factorization be $A \xrightarrow{f} B = A \xrightarrow{h} T \xrightarrow{k} B$. Under the action of functor $\mathfrak{F}$, we have $\mathfrak{F}A \xrightarrow{\mathfrak{F}f} \mathfrak{F}B = \mathfrak{F}A \xrightarrow{\mathfrak{F}h} \mathfrak{F}T \xrightarrow{\mathfrak{F}k} \mathfrak{F}B$.

But $\mathfrak{F}T$ is a terminal object of $\mathbb{B}$, and arrow $\mathfrak{F}f$ factors through it. Hence, $\mathfrak{F}f$ must be a constant arrow in category $\mathbb{B}$.

(Where and why is the connectedness of categories $\mathbb{A}$ and $\mathbb{B}$ required?) Then unless the source category of functors, in general, has a terminal object, a constant arrow is not preserved by functors.

A dual of the above proposition would then read as:

Proposition 109 Whenever $\mathbb{A}$ and $\mathbb{B}$ are two connected categories such that $\mathbb{A}$ has an initial object I, then for a functor $\mathfrak{F} : \mathbb{A} \to \mathbb{B}$ TFAE:

- $\mathfrak{F}$ preserves co-constant arrows.
- $\mathfrak{F}$ preserves the initial object I.

Thus, unless the source category of functors, in general, has an initial object, co-constant arrows are not preserved by functors.

Notice therefore that it follows from the duality principle, definition 12, Chapter 4, that if all the functors have a certain property, then all the functors also have the dual property.

6.4.2 Reflection of Categorical Property by Functors

Now, we make the following observations:

Observation 13 If $\mathfrak{F} : \mathbb{A} \to \mathbb{B}$ and $\mathfrak{F}k$ is a section, then k need *not* be a section in $\mathbb{A}$. By Duality, the same holds for a retraction.

Observation 14 If $\mathfrak{F} : \mathbb{A} \to \mathbb{B}$ and $\mathfrak{F}k$ is an isoarrow, then k need *not* be an isoarrow in $\mathbb{A}$.

On the basis of the above observations: As $\mathfrak{F} : \mathrm{Mor}(\mathbb{A}) \to \mathrm{Mor}(\mathbb{B})$ is a function, associate a nonisoarrow in $\mathbb{A}$ with an isoarrow in $\mathbb{B}$ by functor $\mathfrak{F}$

Consequently, we will talk about those properties of arrows in categories for which the following holds:

Definition 103 (categorical property being reflected by functors) A functor $\mathfrak{F} : \mathbb{A} \to \mathbb{B}$ is said to reflect a categorical property P provided that the preimage, under $\mathfrak{F}$, of each arrow (or object) in $\mathbb{B}$ having property P, has that property P *in* $\mathbb{A}$.

We therefore say that a functor, in general, need not reflect a section, a retraction, and isoarrows.

Proposition 110 Every faithful functor reflects monoarrows, epiarrows, bidirectional arrows, constant arrows, co-constant arrows, zero arrows, and commutative triangles.

Proof Let $\mathfrak{F}A \xrightarrow{\mathfrak{F}f} \mathfrak{F}B$ be a monoarrow in category $\mathbb{B}$, and consider two arrows $X \underset{k}{\overset{h}{\rightrightarrows}} A$ such that $fh = fk$. Then $\mathfrak{F}f\mathfrak{F}h = \mathfrak{F}f\mathfrak{F}k$. But $\mathfrak{F}f$ is a monoarrow, and we have $\mathfrak{F}f = \mathfrak{F}h$. Furthermore, $\mathfrak{F}$ is faithful (1-1 or injective) on hom-collection restrictions. Thus, $h = k$.

By duality, it then follows that a faithful functor reflects epiarrows and, hence, bidirectional arrows as well.

Now, let $\mathfrak{F}A \xrightarrow{\mathfrak{F}f} \mathfrak{F}B$ be a constant arrow in category $\mathbb{B}$ and let $X \underset{k}{\overset{h}{\rightrightarrows}} A$ be any two arrows. Then $\mathfrak{F}f\mathfrak{F}h = \mathfrak{F}f\mathfrak{F}k$. Any functor preserves the composition of an arrow and, therefore, $\mathfrak{F}(fh) = \mathfrak{F}(fk)$. But the given functor is faithful and, hence, $fh = fk$. Thus, f is a constant arrow in category $\mathbb{B}$.

By duality, it follows that a faithful functor reflects a co-constant arrow and, hence, also a zero arrow.

Now, consider the following commuting triangle:

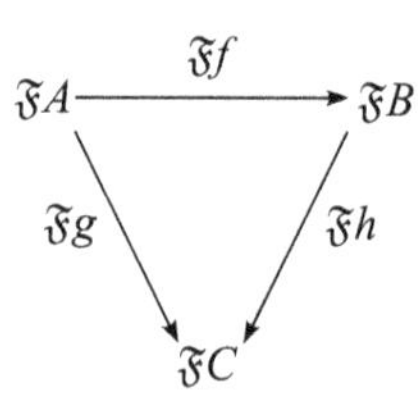

For this $\mathfrak{F}g = \mathfrak{F}h\mathfrak{F}f$. But functor $\mathfrak{F}$ is faithful, and, hence, we have $g = hf$. Therefore, the triangle in category $\mathbb{A}$ also commutes.

Proposition 111 Every full and faithful functor reflects a section and, hence, a retraction arrow.

Proof 64 Consider a full and faithful functor $\mathfrak{F} : \mathbb{A} \to \mathbb{B}$. Let $\mathfrak{F}f$ be a section and g be its left-inverse in category $\mathbb{B}$ so we have $g\mathfrak{F}f = \mathrm{id}_{\mathfrak{F}A}$. Since $\mathfrak{F}$ is full (onto), there exists some $h \in \mathbb{A}$ such that $\mathfrak{F}h = g$. Hence, we have $\mathfrak{F}h\mathfrak{F}f = \mathfrak{F}(hf) = \mathrm{id}_{\mathfrak{F}\mathrm{A}} = \mathfrak{F}(\mathrm{id}_A)$. But $\mathfrak{F}$ is also faithful (1-1 or injective). Therefore, $hf = \mathrm{id}_A$ and f is a section in category $\mathbb{A}$.

By duality, it follows that a retraction is also reflected by a full and faithful functor.

Proposition 112 If $\mathfrak{F} : \mathbb{A} \to \mathbb{B}$ is a full and faithful functor, then for every $\mathbb{B}$-arrow $f : \mathfrak{F}A \to \mathfrak{F}B$, there exists a unique $\mathbb{A}$-arrow $g : A \to B$

with $\mathfrak{F}g = f$. Furthermore, g is an $\mathbb{A}$-isoarrow if and only if (iff) f is a $\mathbb{B}$-isoarrow.

Proof Arrow g exists by fullness (onto) and is unique by faithfulness (1-1) of $\mathfrak{F}$. Since functors preserve isoarrows, f is an isoarrow if g is. If f is a $\mathbb{B}$-isoarrow, let $g' : B \to A$ be the unique $\mathbb{A}$-arrow with $\mathfrak{F}g' = f^{-1}$. Then $\mathfrak{F}(g' \circ g) = \mathfrak{F}g' \circ \mathfrak{F}g = f^{-1} \circ f = \mathrm{id}_{\mathfrak{F}A} = \mathfrak{F}(\mathrm{id}_A)$, so by faithfulness $g' \circ g = \mathrm{id}_A$. Likewise, $g \circ g' = \mathrm{id}_B$. Hence, g is an isoarrow.

Proposition 113 A functor is an embedding iff it is faithful as well as injective (1-1) on objects.

Observation 15 Full and faithful functors need not preserve monoarrows and, also, epiarrows, by duality.

We only need to note that a functor can map an object A with a monoarrow $A \xrightarrow{f} B$ in $\mathbb{A}$ to some object $\mathfrak{F}A$ in $\mathbb{B}$ that may not have any monoarrow originating from it. The fullness and the faithfulness of a functor are, respectively, just the one-one-ness and onto-ness of the functor as a map of hom-collections $\mathfrak{F} : [A, B]_{\mathbb{A}} \to [\mathfrak{F}A, \mathfrak{F}B]_{\mathbb{B}}$.

Apart from being full and faithful, a functor $\mathfrak{F}$ then needs to possess some other property for preserving mono- and epiarrows.

6.4.3 Isofunctor and Isomorphism of Categories

Definition 104 (isofunctor) A functor $\mathfrak{F} : \mathbb{A} \to \mathbb{B}$ is called as an isofunctor provided that there exists a functor $\mathfrak{G} : \mathbb{B} \to \mathbb{A}$ such that we have $\mathfrak{F} \circ \mathfrak{G} = \mathrm{id}_{\mathbb{B}}$ and $\mathfrak{G} \circ \mathfrak{F} = \mathrm{id}_{\mathbb{A}}$.

The unique functor $\mathfrak{G}$ will be denoted by $\mathfrak{F}^{-1}$.

Proposition 114 The composition of two isofunctors is an isofunctor.

Evidently, this is just the definition of an isoarrow applied to functors as arrows connecting categories. Analogous to isomorphic objects in a category, we then have the notion of isomorphic categories as follows.

Definition 105 (isomorphic categories) Two categories $\mathbb{A}$ and $\mathbb{B}$ are said to be isomorphic categories provided there exists an isofunctor or functor isomorphism between them.

Isomorphic categories are the same in all of their categorical properties, that is, they have all the categorical properties identical.

Proposition 115 A functor is an isoarrow iff it is full, faithful, as well as bijective (1-1 and onto) on objects.

Then an isomorphism of categories is seen to be a very restrictive concept about the relationship of the collections of arrows and objects in two categories connected by a functor.

6.4.4 Equivalence Functor

In this last context, what we desire is a certain notion of the essential sameness of categories rather than their exact sameness. This essential sameness should then concern the properties of arrows being identical in the two categories rather than their collections of objects being identical. In this case, we would not be able to distinguish the two categories by using any categorical properties.

No doubt, we expect essentially same categories to have similar collections of arrows. Thus, we would expect the essentially same categories to be connected by a full and, simultaneously, faithful functor. But it should suffice for us if any isomorphic objects of category $\mathbb{A}$ are also carried by that functor to the isomorphic objects of category $\mathbb{B}$. This leads us to defining isoarrow denseness for functors as follows.

Definition 106 (isoarrow dense functor) Given a functor $\mathfrak{F} : \mathbb{A} \to \mathbb{B}$, if for any $\mathbb{B}$-object B, there exists some $\mathbb{A}$-object A such that $\mathfrak{F}A$ is isomorphic to B, then $\mathfrak{F}$ is called an isoarrow dense functor. An isoarrow dense functor is also called a dense functor or a representative functor.

Proposition 116 Being isoarrow dense is a self-dual property.

Observation 16 Any canonical or natural functor from a category to its quotient category is necessarily full and isodense both.

Proposition 117 The composition of full (faithful, dense, embedding) functors is also full (faithful, dense, embedding).

Definition 107 (equivalence functor) A full, faithful, and isoarrow dense functor $\mathfrak{F} : \mathbb{A} \to \mathbb{B}$ is called an equivalence functor from $\mathbb{A}$ to $\mathbb{B}$.

6.5 Equivalence of Categories

Definition 108 (equivalent categories) Categories $\mathbb{A}$ and $\mathbb{B}$ are called equivalent provided that there is an equivalence functor from $\mathbb{A}$ to $\mathbb{B}$.

Proposition 118 Every isofunctor is an equivalence functor.

Proof 66 An isofunctor is full and faithful, and so is its inverse. We only need to show that it is isoarrow dense.

Proposition 119 If $\mathfrak{F} : \mathbb{A} \to \mathbb{B}$ is an equivalence, then there exists an equivalence functor $\mathfrak{G} : \mathbb{B} \to \mathbb{A}$.

Proof For each object B in $\mathbb{B}$, choose an $\mathbb{A}$-object $\mathfrak{G}(B)$, and a $\mathbb{B}$-isoarrow $\varepsilon_B = \mathfrak{F}(\mathfrak{G}(B)) \to B$. Since $\mathfrak{F}$ is full and faithful (onto and 1-1 on hom-collection restrictions $\mathfrak{F} : \hom_{\mathbb{A}} (A, A') \to \hom_{\mathbb{B}} (\mathfrak{F}A, \mathfrak{F}A')$), for each $\mathbb{B}$-arrow $g : B \to B'$, there exists a unique $\mathbb{A}$-arrow $\mathfrak{G}(g) : \mathfrak{G}B \to \mathfrak{G}B'$ with $\mathfrak{F}(\mathfrak{G}(g)) = \varepsilon_{B'}^{-1} \circ g \circ \varepsilon_B : \mathfrak{F}(\mathfrak{G}(B)) \to \mathfrak{F}(\mathfrak{G}(B'))$. Thus,$\mathfrak{G}(g)$ is the unique $\mathbb{A}$-arrow for which the corresponding square commutes:

$$
\begin{array}{ccc}
\mathfrak{F}(\mathfrak{G}(B)) & \xrightarrow{\mathfrak{F}(\mathfrak{G}(g))} & \mathfrak{F}(\mathfrak{G}(B')) \\
\downarrow{\scriptstyle \varepsilon_B} & & \downarrow{\scriptstyle \varepsilon_{B'}} \\
B & \xrightarrow[g]{} & B'
\end{array}
$$

Then the preservation of identities by $\mathfrak{G}$ follows from the uniqueness in the commuting square. That $\mathfrak{G}$ preserves composition follows from this uniqueness and the commutativity of two adjoining commuting squares

$$
\begin{array}{ccccc}
\mathfrak{F}(\mathfrak{G}(B)) & \xrightarrow{\mathfrak{F}(\mathfrak{G}(g))} & \mathfrak{F}(\mathfrak{G}(B')) & \xrightarrow{\mathfrak{F}(\mathfrak{G}(h))} & \mathfrak{F}(\mathfrak{G}(B'')) \\
\downarrow{\scriptstyle \varepsilon_B} & & \downarrow{\scriptstyle \varepsilon_{B'}} & & \downarrow{\scriptstyle \varepsilon_{B''}} \\
B & \xrightarrow[g]{} & B' & \xrightarrow[h]{} & B''
\end{array}
$$

and the fact that $\mathfrak{F}$ preserves composition. Therefore, $\mathfrak{G}$ is a functor. Now, $\mathfrak{G}$ is full (onto) since for each $\mathbb{A}$-arrow $f : \mathfrak{G}B \to \mathfrak{G}B'$, the $\mathbb{B}$-arrow $\varepsilon_{B'} \circ \mathfrak{F}f \circ \varepsilon_B^{-1} : B \to B'$, denoted by g, has the property

$g \circ \varepsilon_B = \varepsilon_{B'} \circ \mathfrak{F}f$ and by the uniqueness of the two adjoining commuting squares, $f = \mathfrak{G}B$.

Now, given $B \underset{g2}{\overset{g1}{\rightrightarrows}} B'$ with $\mathfrak{G}(g_1) = \mathfrak{G}(g_2) = f$, it follows from the two adjoining commuting squares that

$$g_1 = \varepsilon_{B'} \circ \mathfrak{F}(\mathfrak{G}(g_1)) \circ \varepsilon_B^{-1} = \varepsilon_{B'} \circ \mathfrak{F}(f) \circ \varepsilon_B^{-1}$$

$$= \varepsilon_{B'} \circ \mathfrak{F}(\mathfrak{G}(g_2)) \circ \varepsilon_B^{-1} = g_2 .$$

Thus, $\mathfrak{G}$ is faithful (1-1) as well.

Lastly, $\mathfrak{G}$ is isoarrow dense since for each $\mathbb{A}$-object A, the $\mathbb{B}$-isoarrow $\varepsilon_{\mathfrak{F}\,A}(\mathfrak{F}(\mathfrak{G}\ (\mathfrak{F}A) \to \mathfrak{F}A$ is the image of some $\mathbb{A}$-isoarrow $\mathfrak{G}\mathfrak{F}A \to A$, as $\mathfrak{G}$ is full (onto) as well as faithful (1-1).

Proposition 120 Being an equivalence is a self-dual property.

Proposition 121 If $\mathfrak{F} : \mathbb{A} \to \mathbb{B}$ and $\mathfrak{H} : \mathbb{B} \to \mathbb{C}$ are equivalences, then so is $\mathfrak{H} \circ \mathfrak{F} : \mathbb{A} \to \mathbb{C}$ an equivalence.

Proof 68 As both $\mathfrak{F}$ and $\mathfrak{H}$ are full and faithful, so is $\mathfrak{H} \circ \mathfrak{F}$ full and faithful. Now, given a $\mathbb{C}$-object C, the isoarrow denseness of $\mathfrak{F}$ and $\mathfrak{H}$ gives us a $\mathbb{B}$-object B and an isoarrow $h : \mathfrak{H}(B) \to C$ and a $\mathbb{B}$-object A and an isoarrow $k : \mathfrak{F}(A) \to B$. Thus, $h \circ \mathfrak{H}(k) : \mathfrak{H} \circ \mathfrak{F}(A) \to C$ is an isoarrow. So, $\mathfrak{H} \circ \mathfrak{F}$ is isoarrow dense.

The following proposition is then a consequence of various of the above propositions. (We have not mentioned the dual properties below, for it follows, by the duality principle, that the dual properties are also preserved and reflected by the functors that are full, faithful, and dense.)

Proposition 122 Every full, faithful, and isoarrow dense functor. That is to say, an equivalence functor, preserves as well as reflects monoarrows, bidirectional arrows, constant arrows, zero arrows, sections, isoarrows, and commutative triangles.

Proof Every functor preserves isoarrows, a section, and a commutative triangle. Hence, the equivalence functor also preserves them.

A faithful functor reflects monoarrows, bidirectional arrows, constant arrows, zero arrows, and commutative triangles. Hence, an equivalence functor also reflects them.

Every full and faithful functor reflects a section and an isoarrow. Hence, an equivalence functor also reflects them.

We need to show only that an equivalence functor preserves a monoarrow and a constant arrow.

Then let $A \xrightarrow{f} B$ be an $\mathbb{A}$-arrow and let $X \underset{s}{\overset{r}{\rightrightarrows}} \mathfrak{F}(T) \xrightarrow{\mathrm{id}_{\mathfrak{F}}(t)} \mathfrak{F}(T)$ be any $\mathbb{B}$-arrows. Consider now the diagrams:

$$C \underset{k}{\overset{h}{\rightrightarrows}} A \xrightarrow{f} B$$

$$\mathfrak{F}(C) \xrightarrow{r} X \underset{q}{\overset{p}{\rightrightarrows}} \mathfrak{F}(A) \xrightarrow{\mathfrak{F}f} \mathfrak{F}B$$

Since $\mathfrak{F}$ is an isoarrow dense functor, there exists an $\mathbb{A}$-object C and a $\mathbb{B}$-isoarrow $r : \mathfrak{F}C \to X$. Furthermore, as functor $\mathfrak{F}$ is full (onto), there exist $\mathbb{A}$-arrows $C \underset{k}{\overset{h}{\rightrightarrows}} A$ such that $\mathfrak{F}h = pr$ and $\mathfrak{F}k = qr$.

- Let the $\mathbb{A}$-arrow f be a monoarrow. Then $fh = fk \Rightarrow h = k$ in $\mathbb{A}$. Now, let $(\mathfrak{F}f)p = (\mathfrak{F}f)q$. We prove that $p = q$ so that $\mathfrak{F}f$ is a monoarrow in category $\mathbb{B}$. Then $(\mathfrak{F}f)pr = \mathfrak{F}f\mathfrak{F}h = \mathfrak{F}(fh) = (\mathfrak{F}f) qr = \mathfrak{F}f\mathfrak{F}k = \mathfrak{F}(fk)$. So, $\mathfrak{F}(fh) = \mathfrak{F}(fk)$. But $\mathfrak{F}$ is faithful (1-1) and, hence, $fh = fk$ in category $\mathbb{A}$, where we know that it implies $h = k$. Then $\mathfrak{F}h = \mathfrak{F}k \Rightarrow pr = qr$ in category $\mathbb{B}$ and, moreover, we have $p = q$ since r is an isoarrow that is also an epiarrow in category $\mathbb{B}$.
- Let f be a constant arrow in category $\mathbb{A}$. Then $fh = fk$ in $\mathbb{A}$. We now need to show that $(\mathfrak{F}f)p = (\mathfrak{F}f)q$ in $\mathbb{B}$ whenever $fh = fk$ in $\mathbb{A}$. Now, $\mathfrak{F}(fh) = \mathfrak{F}f\mathfrak{F}h = \mathfrak{F}f(pr) = \mathfrak{F}(fk) = \mathfrak{F}f\mathfrak{F}k = \mathfrak{F}f(qr)$. Then $\mathfrak{F}f \circ p \circ r = \mathfrak{F}f \circ q \circ r$. However, $\mathfrak{F}f \circ p = \mathfrak{F}f \circ q$ as r is an isoarrow and, hence, an epiarrow in category $\mathbb{B}$.

Notice now that essentially every categorical property is preserved and reflected by an equivalence functor. Therefore, equivalent categories cannot be distinguished using any of the categorical properties. We can then define a categorical property in the following way:

Definition 109 (categorical property) A property P in category theory is called a categorical property, provided that it is both preserved and reflected by an equivalence functor.

Observation 17 The cardinality of the collection of arrows of a category is not preserved by an equivalence functor.

This means that cardinality is not a categorical property. One could have guessed this observation from the fact that the basic structure underlying a category is that of a partial binary algebra and not that of a set or collection and the fact that a functor is only a partial binary algebra preserving function of the collection of arrows in a category, the cardinality of this collection being immaterial to this structure.

Proposition 123 A subcategory of any category is full (isoarrow dense) provided that its inclusion functor is full (isoarrow dense).

We may now establish that various definitions of a category provided in Chapter 4 are connected by equivalence functors. Be warned that this is a nontrivial task and we will refer the reader to an already existing literature for the proofs.

Definition 110 Categories $\mathbb{A}$ and $\mathbb{B}$ are called dually equivalent provided that $\mathbb{A}^{\mathrm{op}}$ and $\mathbb{B}$ are equivalent.

Definition 111 (self-dual category) A category A will be called a self-dual category provided that $\mathbb{A}$ and $\mathbb{A}^{\mathrm{op}}$ are equivalent.

Observation 18 Category $\mathbb{C}\mathrm{ol}$ or $\mathbb{S}\mathrm{et}$ is not self-dual.

Definition 112 Given functor $\mathfrak{F} : \mathbb{A} \to \mathbb{B}$ with action

$$\mathfrak{F}(A \xrightarrow{f} B) = \mathfrak{F}A \xrightarrow{\mathfrak{F}f} \mathfrak{F}B$$

the dual or opposite functor $\mathfrak{F}^{\mathrm{op}} : \mathbb{A}^{\mathrm{op}} \to \mathbb{B}^{\mathrm{op}}$ is defined by its action as

$$\mathfrak{F}^{\mathrm{op}}(B \xrightarrow{f} A) = \mathfrak{F}B \xrightarrow{\mathfrak{F}f} \mathfrak{F}A.$$

Proposition 124 We have $(\mathfrak{F}^{\mathrm{op}})^{\mathrm{op}} = \mathfrak{F}$.

Definition 113 (self-dual functor) If $\mathfrak{F} : \mathbb{A} \to \mathbb{B}$ is a functor such that $\mathfrak{F} = \mathfrak{F}^{\mathrm{op}}$, we call $\mathfrak{F}$ a self-dual functor.

Note 6 To form the dual of a categorical statement involving functors, make the same statement with each category and each functor replaced by its dual. Then translate this statement back into a statement about the original categories and functors.

Proposition 125 Being an isofunctor, embedding, faithful, full, isoarrow dense, and equivalence are the self-dual properties of a functor.

Proposition 126 Every isomorphism between categories reflects identities, but every equivalence of categories does not.

Proof We first show that isofunctors reflect identities.

- Let $\mathfrak{F} : \mathbb{A} \to \mathbb{B}$ be an isofunctor, and let id_B be an identity arrow of object B in category $\mathbb{B}$. By the fullness, faithfulness, and bijectiveness of an isofunctor $\mathfrak{F}$ on objects, there exists a unique object A and a unique arrow $A \xrightarrow{\text{id}_A} A$ in $\mathbb{A}$ to correspond to id_B.
- An equivalence functor is not bijective on objects.

Proposition 127 An equivalence of categories, that is, an equivalence functor is an embedding iff it reflects identities.

Proof Let $\mathfrak{F} : \mathbb{A} \to \mathbb{B}$ be an equivalence that reflects identities. Then if id_B is an identity in $\mathbb{B}$, there exists object A in $\mathbb{A}$ with identity id_A. But the identity arrow is unique. Hence, $\mathfrak{F}$ is 1-1 on arrows.

Observation 19 Category $\mathbb{C}\text{ol}$ can be embedded into $\mathbb{C}\text{ol}^{\text{op}}$, and $\mathbb{C}\text{ol}^{\text{op}}$ can be embedded into category $\mathbb{C}\text{ol}$. Note, however, that $\mathbb{C}\text{ol}$ and $\mathbb{C}\text{ol}^{\text{op}}$ are not equivalent categories.

Proposition 128 Equivalence functor preserves the connectedness and pointedness of a category.

Observation 20 We now observe that categories $\mathbb{A}$ and $\mathbb{B}$ are isomorphic iff there exists an equivalence $\mathfrak{C} : \mathbb{A} \to \mathbb{B}$ such that for each $\mathbb{A}$-object A, the number of isomorphic copies of A in $\mathbb{A}$ coincides with the number of isomorphic copies of $\mathfrak{C}(A)$ in $\mathbb{B}$.

Isomorphic categories are then seen to be equivalent categories with the same number of objects in each of them. As it is highly instructive to do so, we invite the reader to provide proof for the above observation that indeed can be considered to be a very natural notion, rather the basis, of the categorical conception of equivalence: equivalent categories differ by way of only the number of isomorphic copies of objects in them.

6.6 Functors and Subcategories

Now, notice the following:

Observation 21 Let $\mathfrak{F} : \mathbb{A} \to \mathbb{B}$ be a functor. Then:

- There exist functors $\mathfrak{G} : \mathbb{A} \to \mathbb{C}$ and $\mathfrak{H} : \mathbb{A} \to \mathbb{B}$ such that
  - $\mathfrak{F} = \mathfrak{H} \circ \mathfrak{G}$.
  - $\mathfrak{G}$ is full and bijective (1-1 and onto) on objects.
  - $\mathfrak{H}$ is faithful.
  - Whenever $\mathbb{A} \xrightarrow{\mathfrak{F}} \mathbb{B} = \mathbb{A} \xrightarrow{\mathfrak{G}'} \mathbb{C}' \xrightarrow{\mathfrak{H}'} \mathbb{B}$ and $\mathfrak{H}'$ is faithful, there exists a unique functor $\mathbb{C} \xrightarrow{\mathfrak{K}} \mathbb{C}'$ such that the square

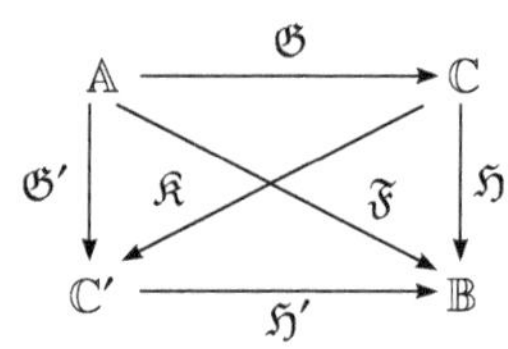

    commutes, with $\mathfrak{G}' = \mathfrak{K} \circ \mathfrak{B}$ and $\mathfrak{H} = \mathfrak{H}' \circ \mathfrak{K}$.
- There exist functors $\mathfrak{G} : \mathbb{A} \to \mathbb{C}$ and $\mathfrak{H} : \mathbb{C} \to \mathbb{B}$ with the following properties:
  - $\mathfrak{F} = \mathfrak{H} \circ \mathfrak{G}$.
  - $\mathfrak{G}$ is bijective (1-1 and onto) on objects.
  - $\mathfrak{H}$ is full and faithful.
  - Whenever $\mathbb{A} \xrightarrow{\mathfrak{F}} \mathbb{B} = \mathbb{A} \xrightarrow{\mathfrak{G}'} \mathbb{C}' \xrightarrow{\mathfrak{H}'} \mathbb{B}$ and $\mathfrak{H}'$ is full and faithful, there exists a unique functor $\mathbb{C} \xrightarrow{\mathfrak{K}} \mathbb{C}'$ such that the above square commutes with $\mathfrak{G}' = \mathfrak{K} \circ \mathfrak{G}$ and $\mathfrak{H} = \mathfrak{H}' \circ \mathfrak{K}$.

6.6.1 Inclusion of a Subcategory

Corollary 9 For every subcategory $\mathbb{A}$ of a category $\mathbb{B}$ there is a naturally associated inclusion functor $\mathfrak{I} : \mathbb{A} \hookrightarrow \mathbb{B}$. Moreover, each such inclusion is

- Always an embedding
- A full functor iff $\mathbb{A}$ is a full subcategory of $\mathbb{B}$

The inclusions of subcategories are (up to isofunctors) precisely the embedding functors and (up to equivalence) precisely the faithful functors.

Proposition 129 We have the following:

- A functor $\mathfrak{F} : \mathbb{A} \to \mathbb{B}$ is a (full) embedding iff there exists a (full) subcategory $\mathbb{C}$ of $\mathbb{B}$ with inclusion functor $\mathfrak{I} : \mathbb{C} \hookrightarrow \mathbb{B}$ and an isofunctor $\mathfrak{G} : \mathbb{A} \to \mathbb{C}$ with $\mathfrak{F} = \mathfrak{J} \circ \mathfrak{G}$. That is to say, the following triangle commutes:

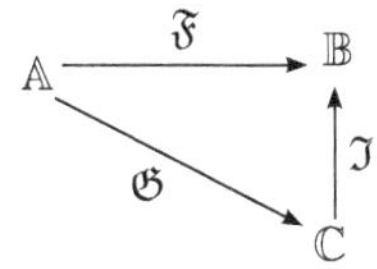

- A functor $\mathfrak{F} : \mathbb{A} \to \mathbb{B}$ is faithful iff there exist embedding functors $\mathfrak{E}_1 : \mathbb{D} \to \mathbb{B}$ and $\mathfrak{E}_2 : \mathbb{A} \to \mathbb{C}$ and an equivalence $\mathfrak{G} : \mathbb{C} \to \mathbb{D}$ such that the following square commutes:

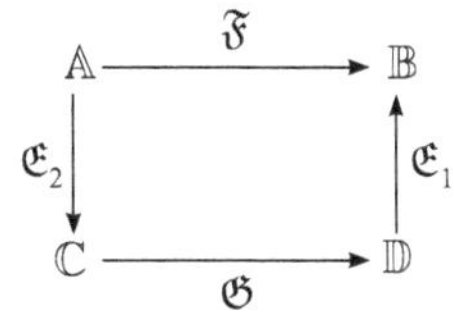

Proof We proceed as follows:

- The "If" part is immediate and the "Only if" part is a consequence of the fact that (full) embedding functors are closed under composition.
- The "If" part follows from the composition of faithful functors being faithful. To show, conversely, that every faithful functor can be decomposed as stated, let $\mathfrak{E}_1 : \mathbb{D} \to \mathbb{B}$ be the inclusion of the full subcategory $\mathbb{D}$ of $\mathbb{B}$ that has as objects all images (under $\mathfrak{F}$) of $\mathbb{A}$-objects. Let $\mathbb{C}$ be the category with $\mathrm{Ob}(\mathbb{C}) = \mathrm{Ob}(\mathbb{A})$, with $\hom_{\mathbb{C}}(A, A') = \hom_{\mathbb{B}}(FA, FA')$ and with identities and composition defined as in $\mathbb{B}$. $\mathbb{C}$ is then easily seen to be a category. Now define functors $\mathfrak{E}_2 : \mathbb{A} \to \mathbb{C}$ and $\mathfrak{G} : \mathbb{C} \to \mathbb{D}$ by $\mathfrak{E}_2(A \xrightarrow{f} A') = A \xrightarrow{\mathfrak{F}f} A'$ and $\mathfrak{G}(C \xrightarrow{g} C') = \mathfrak{F}C \xrightarrow{g} \mathfrak{F}C'$. Then $\mathfrak{E}_2$ is an embedding, $\mathfrak{G}$ is an equivalence, and $\mathfrak{F} = \mathfrak{E}_1 \circ \mathfrak{G} \circ \mathfrak{E}_2$.

6.6.2 Full Embedding of a Subcategory

Definition 114 A category $\mathbb{A}$ is said to be fully embeddable into $\mathbb{B}$ provided that there exists a full embedding $\mathbb{A} \to \mathbb{B}$, or, equivalently, provided that $\mathbb{A}$ is isomorphic to a full subcategory of $\mathbb{B}$.

Remark 7 Because *full subcategories* are determined by their object classes, they are often regarded as *properties of objects*. Since most of the interesting properties P satisfy the condition that whenever an object A has property P then every object isomorphic to A also has P, we often require that full subcategories have the property of being isoarrow closed.

Proposition 130 If $\mathbb{A}$ is a full subcategory of $\mathbb{B}$, then the following conditions are equivalent:

- $\mathbb{A}$ is an isoarrow dense subcategory of $\mathbb{B}$.
- The inclusion functor $\mathbb{A} \hookrightarrow \mathbb{B}$ is isoarrow dense.
- The inclusion functor $\mathbb{A} \hookrightarrow \mathbb{B}$ is an equivalence.

6.6.3 Subcategories of Subcategories

Note that every subcategory $\mathbb{A}$ of $\mathbb{B}$ can be embedded into a smallest isomorphism-closed subcategory $\mathbb{A}'$ of $\mathbb{B}$ that contains $\mathbb{A}$.

Proposition 131 Let $\mathbb{A}$ be a subcategory of $\mathbb{B}$ and $\mathbb{B}$ be a subcategory of $\mathbb{C}$. Then the following hold:

- $\mathbb{A}$ is a subcategory of $\mathbb{C}$.
- If $\mathbb{A}$ is reflective in $\mathbb{B}$ and $\mathbb{B}$ is reflective in $\mathbb{C}$, then $\mathbb{A}$ is reflective in $\mathbb{C}$.
- If $\mathbb{A}$ is reflective in $\mathbb{C}$ and $\mathbb{B}$ is a full subcategory of $\mathbb{C}$, then $\mathbb{A}$ is reflective in $\mathbb{B}$.
- If $\mathbb{A}$ is reflective in $\mathbb{C}$, then $\mathbb{A}$ need not be reflective in $\mathbb{B}$.

6.6.4 Skeleton of a Category

We may now note:

Definition 115 (skeleton of a category) A skeleton of a category is a full, isoarrow dense subcategory in which no two distinct objects are isomorphic.

Proposition 132 Now, we have the following:

- Every category has a skeleton.
- Any two skeletons of a category are isomorphic.
- Any skeleton of a category $\mathbb{C}$ is equivalent to $\mathbb{C}$.

Proof Let $A \cong B$ in a category $\mathbb{A}$ mean that there exists an isoarrow between its objects A and B. Clearly, $\cong$ is then an equivalence relation on the collection $\text{Ob}(\mathbb{A})$. Then by the axiom of choice for classes as conglomerates, there exists a function $C : \mathcal{U} \to \mathcal{U}$, where $\mathcal{U}$ is the collection of all sets, such that $C(X) \in X$ for any chosen set X and it represents the equivalence classes under the relation $\cong$. Let the collection of such equivalence classes be ε and the category generated by it be $\mathfrak{C}$. Then $\mathfrak{C}$ is a full subcategory of A and is not contained in any other full subcategory of $\mathbb{A}$. Furthermore, it can be seen to be a skeleton category.

Corollary 10 Two categories are equivalent iff they have isomorphic skeletons.

Proposition 133 Given two skeletons of a category $\mathbb{A}$, there exists an isofunctor $\mathbb{A} \to \mathbb{A}$ that is restricted to an isofunctor between the skeletons.

6.7 Some Relevant Conceptions

6.7.1 Graph of a Category

Definition 116 (graphs of categories) A graph is a quadruple (V, E, d, c) consisting of a set V of vertices, a set E of edges, and functions $d, c : E \to V$ (giving the source [domain] and target [codomain] of an edge).

A large graph is the same concept with V and E being classes.

The graph $G(\mathbb{A})$ of a category $\mathbb{A}$ is the obvious large graph with $V = \text{Ob}(\mathbb{A})$ and $E = \text{Mor}(\mathbb{A})$.

We have, in Section 4.10, Chapter 4, seen that some small categories can be conveniently represented by diagrams or graphs.

Observation 22 Nonisomorphic categories can possess the same graph.

Definition 117 The free category generated by a graph (V, E, d, c) is category $\mathbb{A}$ with $V = \text{Ob}(\mathbb{A}) = V$, $\text{Mor}(\mathbb{A})$ = all paths (= all finite sequences in E in which the source of each edge is the target of the preceding one), composition is the composition of paths, and the identity arrows are the empty paths.

One can easily verify that the above $\mathbb{A}$ is indeed a category.

Observation 23 Whenever $\mathbb{A} = (\mathcal{O}, \text{hom}, \text{id}, \circ)$ and $\mathbb{A}^* = (\mathcal{O}, \text{hom}, \text{id}^*, \circ)$ are categories, then $\mathbb{A} = \mathbb{A}^*$.

6.7.2 Internal and External Characterization

Consider the following definition of a projective object using the hom-functor. This is termed as the external characterization of a projective object.

Definition 118 (projective object) An object P in a category $\mathbb{C}$ is called $\mathbb{C}$-projective iff functor

$$\text{hom } \mathbb{C}(P, -) : \mathbb{C} \to \mathbb{C}\text{ol}$$

preserves epiarrows. The dual notion is that of a $\mathbb{C}$-injective object.

In terms of a commuting triangle, we then (internally, that is, without using functors) characterize a projective object within a category as:

Proposition 134 For a projective object P of a category $\mathbb{C}$, the following triangle commutes for each epiarrow $Q \xrightarrow{f} R$ and each arrow $P \xrightarrow{g} R$, there being a unique arrow $P \xrightarrow{q} Q$:

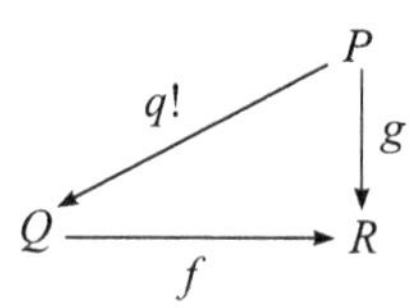

Characterization that is internal to the category is then easy to handle in many situations. Such characterizations of objects and arrows in a category are often very useful in particular categories.

We also note that many of the internally characterized concepts can also be defined externally using hom-functors. For example, a monoarrow can be externally characterized as:

Proposition 135 An arrow $A \xrightarrow{f} B$ in a category $\mathbb{C}$ is a monoarrow provided that the hom-functor action hom$(X, -)(f)$ is injective for every object X of category $\mathbb{C}$.

The internal and external characterizations of categorical properties are nonetheless the same. Considerations of both these characterizations often help us to grasp the essential content of a categorical property.

6.8 Category of All Categories?

It can now be noticed that a functor serves as an arrow, originating from one category and ending at another category. There also exists an identity functor associated with every category, which serves as an identity arrow to correspond to that category. It is also clear that the composition of functors is the usual composition of functions (between two partial binary algebras) that preserves the identities and composition of arrows of a category. This composition is also associative, and the identity functors serve also as the identity with respect to this composition.

This leads us to considering a category of all the categories. However, it is at this point that we have to be careful with the smallness condition in the definition (definition 3) of a category!

As seen earlier, in Chapter 4, if we are careful enough, even such mathematical structures are legitimate ones, for the basic structure of a generic category is that of its partial binary algebra.

However, in the development of category theory, it has been observed that such exceedingly large structures have not been needed to understand any of the existing mathematical theories.

6.9 Functors and Natural Transformations

Functors from any category $\mathbb{A}$ to any category $\mathbb{B}$ are arrows connecting their objects and arrows. Two such functors can then be related to each other on the basis of their actions on objects and arrows. This relationship is that of a natural transformation. This is formally defined as follows.

6.9.1 Natural Transformation and Natural Isomorphism

Definition 119 Let $\mathfrak{F}, \mathfrak{G} : \mathbb{A} \to \mathbb{B}$ be functors. A natural transformation τ from $\mathfrak{F}$ to $\mathfrak{G}$ (denoted by $\tau : \mathfrak{F} \to \mathfrak{G}$ or $\mathfrak{F} \xrightarrow{\tau} \mathfrak{G}$) is a function that assigns to each $\mathbb{A}$-object A a $\mathbb{B}$-arrow $\tau_A : \mathfrak{F}A \to \mathfrak{G}A$ in such a way that for each $\mathbb{A}$-arrow $A \xrightarrow{f} B$, the following square commutes:

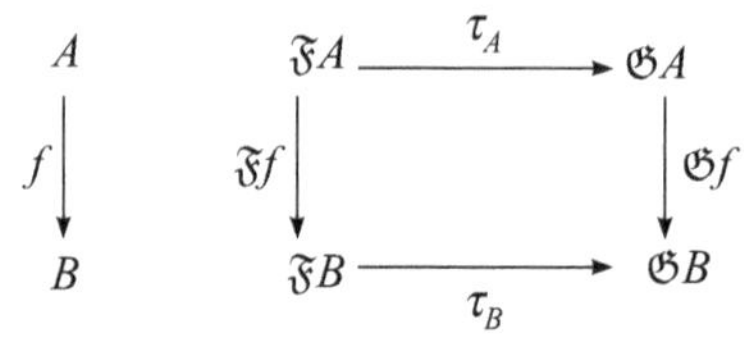

Natural transformations are also called functor morphisms. We also call τ_B the component of τ at B.

Notice that a natural transformation is then viewed as a triple $(\mathfrak{F}, \tau, \mathfrak{G})$ where $\tau : \mathrm{Ob}(\mathbb{A}) \to \mathrm{Mor}\ (\mathbb{B})$ is a function that assigns to each $\mathbb{A}$-object an arrow, denoted by τ_A or $\tau(A)$, from a $\mathbb{B}$-object $\mathfrak{F}A$ to another $\mathbb{B}$-object $\mathfrak{G}A$ such that the aforementioned square commutes. Because of this, the following definitions are then immediate:

Definition 120 (natural isomorphism) A natural transformation $(\mathfrak{F}, \tau, \mathfrak{G})$ is called a natural isomorphism when for each $\mathbb{A}$-object, τ_A is a $\mathbb{B}$-isoarrow.

Definition 121 (naturally isomorphic functors) Any two given functors $\mathfrak{F}, \mathfrak{G} : \mathbb{A} \to \mathbb{B}$ are said to be naturally isomorphic provided that there exists a natural isomorphism from $\mathfrak{F}$ to $\mathfrak{G}$. We write $\mathfrak{F} \cong \mathfrak{G}$ for naturally isomorphic functors $\mathfrak{F}$ and $\mathfrak{G}$.

Proposition 136 A functor $\mathbb{A} \xrightarrow{\mathfrak{F}} \mathbb{B}$ is an equivalence of categories $\mathbb{A}$ and$\mathbb{A}$ iff there exists a functor $\mathbb{B} \xrightarrow{\mathfrak{G}} \mathbb{A}$ such that $\mathrm{id}_{\mathbb{A}} \cong \mathfrak{G} \circ \mathfrak{F}$ and $\mathfrak{F} \circ \mathfrak{G} \cong \mathrm{id}_{\mathbb{B}}$.

6.9.2 Usual Product of Natural Transformations

If $\mathfrak{F}, \mathfrak{G}, \mathfrak{H} : \mathbb{A} \to \mathbb{B}$ are functors and $\mathfrak{F} \xrightarrow{\sigma} \mathfrak{G}$ and $\mathfrak{G} \xrightarrow{\tau} \mathfrak{H}$ are natural transformations, then the composition of natural transformations $\tau \circ \sigma : \mathfrak{F} \to \mathfrak{H}$ is the natural transformation that assigns to each $\mathbb{A}$-object A arrow $\tau_A \circ \sigma_A : \mathfrak{F}(A) \to \mathfrak{H}(A)$. Diagrammatically, the following commutes:

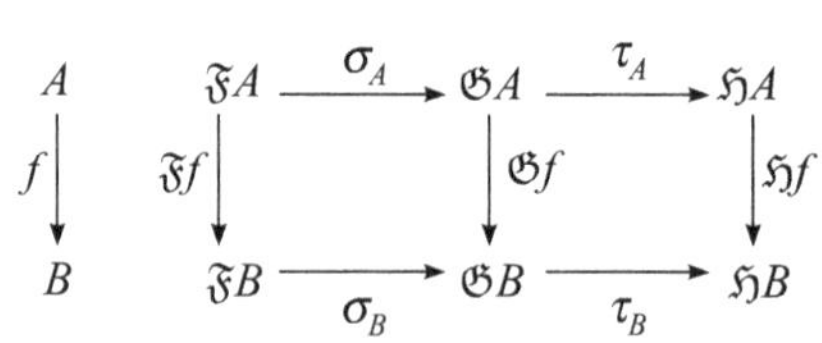

Associativity of the usual product

The composition of natural transformations can then be seen to be associative from the commutativity of the following diagram:

$$\begin{array}{ccccccccc} A & & \mathfrak{F}A & \xrightarrow{\sigma_A} & \mathfrak{G}A & \xrightarrow{\tau_A} & \mathfrak{H}A & \xrightarrow{\rho_A} & \mathfrak{K}A \\ f\big\downarrow & & \mathfrak{F}f\big\downarrow & & \big\downarrow\mathfrak{G}f & & \big\downarrow\mathfrak{H}f & & \big\downarrow\mathfrak{K}f \\ B & & \mathfrak{F}B & \xrightarrow[\sigma_B]{} & \mathfrak{G}B & \xrightarrow[\tau_B]{} & \mathfrak{H}B & \xrightarrow[\rho_B]{} & \mathfrak{K}B \end{array}$$

Proposition 137 Given functors $\mathfrak{F}$, $\mathfrak{G}$: $\mathbb{A} \to \mathbb{B}$, a natural transformation$\tau : \mathfrak{H}\mathfrak{F} \to \mathfrak{G}$ is a natural isomorphism iff there exists another natural transformation $\sigma : \mathfrak{G} \to \mathfrak{F}$ such that $\tau \circ \sigma = \mathrm{id}_{\mathfrak{G}}$ and $\sigma \circ \tau = \mathrm{id}_{\mathfrak{F}}$.

Proof Whenever τ is given as a natural isomorphism from a functor $\mathfrak{F}$ to a functor $\mathfrak{G}$, we have a function $\sigma : \mathrm{Ob}(\mathbb{A}) \to \mathrm{Mor}(\mathbb{B})$ as $\sigma(A) = \tau_A^{-1}$. Then σ is a natural transformation, and $\tau \circ \sigma = \mathrm{id}_{\mathfrak{G}}$ and $\tau \circ \sigma \circ \tau = \mathrm{id}_{\mathfrak{F}}$.

On the other hand, if we are given that $\tau \circ \sigma = \mathrm{id}_{\mathfrak{G}}$ and $\sigma \circ \tau = \mathrm{id}_{\mathfrak{F}}$, then for every $\mathbb{A}$-object A, we have $\tau_A \circ \sigma_A = (\tau \circ \sigma)_A = \mathrm{id}_{\mathfrak{G}A}$ and $\sigma_a \circ \tau_A = (\sigma \circ \tau)_A = \mathrm{id}_{\mathfrak{F}A}$. Then each $\mathbb{B}$-arrow τ_A is an isoarrow in $\mathbb{B}$. Hence, τ is a natural isomorphism.

The identity natural transformations can also be seen to act as the units or identities with respect to this composition.

6.9.3 Star Product of Natural Transformations

Now, as was mentioned in the Chapter 4, if $\mathbb{A}$ is any category, then we may construct another category whose objects are the same as those of $\mathbb{A}$ but whose arrows are different from those of $\mathbb{A}$. We now illustrate that the natural transformations connecting functors can be composed in a different manner than the composition as defined earlier. This new way of composing natural transformations is called their star product.

Let $\mathbb{A} \underset{\mathfrak{G}}{\overset{\mathfrak{F}}{\rightrightarrows}} \mathbb{B}$ and $\mathbb{B} \underset{\mathfrak{K}}{\overset{\mathfrak{H}}{\rightrightarrows}} \mathbb{C}$ be functors, and let $\tau : \mathfrak{F} \to \mathfrak{G}$ and $\sigma : \mathfrak{H} \to \mathfrak{K}$ be natural transformations.

Proposition 138 Given the functors and natural transformations as above, the following square commutes for each $\mathbb{A}$-object A:

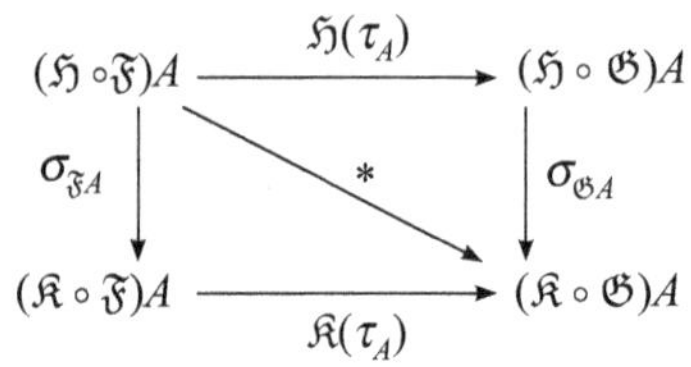

Furthermore, there exists a function, denoted by $* : \mathrm{Ob}(\mathbb{A}) \to \mathrm{Mor}(\mathbb{C})$ that sends an $\mathbb{A}$-object A along the shown diagonal of the above square to an object $(\mathfrak{H} \circ \mathfrak{G})A$ of category $\mathbb{C}$ such that

$$*_A = \sigma_{\mathfrak{G}A} \circ \mathfrak{H}(\tau_A) = \mathfrak{K}(\tau_A) \circ \sigma_{\mathfrak{G}A}.$$

The triple $(\mathfrak{H} \circ \mathfrak{F}, *, \mathfrak{H} \circ \mathfrak{G})$ is then a natural transformation called the star product of σ and τ and is denoted as $\sigma * \tau$.

Proof The given square commutes because τ_A is an arrow from $\mathfrak{F}A$ to $\mathfrak{G}A$ in category $\mathbb{B}$. Its top row depicts the action of functor $\mathfrak{H}$ on arrow τ_A in category $\mathbb{B}$, while the bottom row depicts the action of functor $\mathfrak{K}$ on the same arrow τ_A, and σ is a natural transformation from $\mathfrak{H}$ to $\mathfrak{K}$.

To show that $*$ is a natural transformation, consider an arrow $A \xrightarrow{f} B$ of category $\mathbb{A}$ and the following commuting square for the given natural transformation τ:

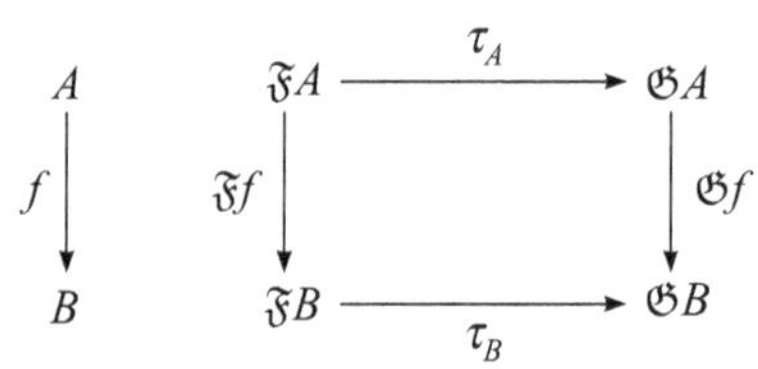

In addition, consider the action of functor $\mathfrak{H}$ on this commuting square of category $\mathbb{B}$. It follows that the left-hand square of the following diagram commutes:

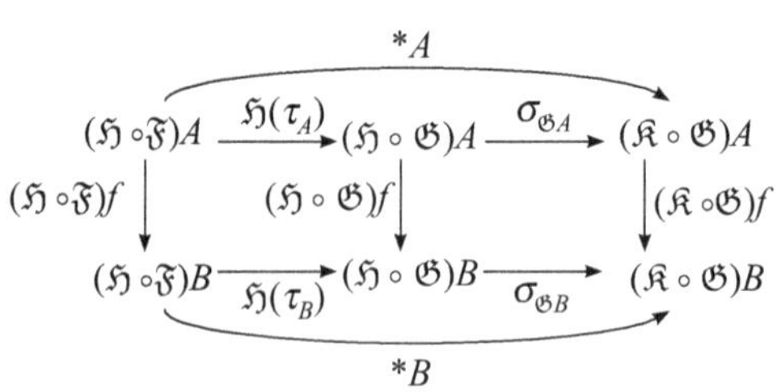

Now, as $\mathfrak{G}(f) : \mathfrak{G}A \to \mathfrak{G}B$ is a $\mathbb{B}$-arrow and σ is a natural transformation from $\mathfrak{H}$ to $\mathfrak{K}$, the right-hand square also commutes in the above diagram, thereby yielding a natural transformation * from $\mathfrak{H} \circ \mathfrak{F}$ to $\mathfrak{H} \circ \mathfrak{G}$, which is a composite of σ and τ.

Associativity of the star product

We then have the following proposition:

Proposition 139 The aforementioned star product of natural transformations is associative. That is to say, if σ, τ, and λ are natural transformations, then $(\sigma * \tau)*\lambda = \sigma * (\tau * \lambda)$.

Proof Now, either side of $(\sigma * \tau)*\lambda = \sigma * (\tau * \lambda)$ is defined iff the following diagram exists:

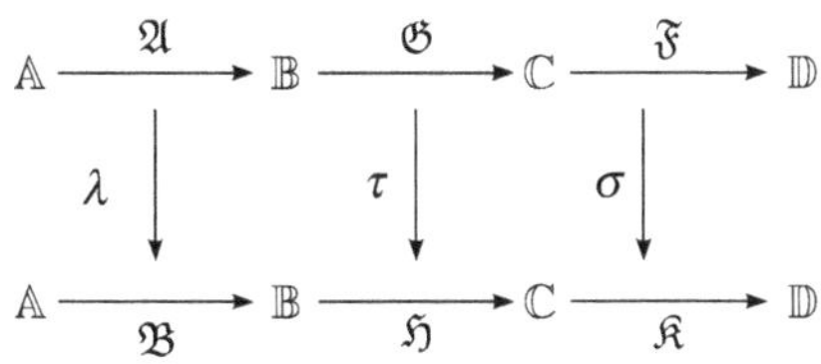

Using the star product, we then have $((\sigma * \tau)*\lambda)_A = (\mathfrak{K}\mathfrak{H}(\lambda_A) \circ (\sigma * \tau)_{\mathfrak{A}A} = \mathfrak{K}(\mathfrak{H}(\lambda_A)) \circ (\mathfrak{K}(\tau_{\mathfrak{A}A}) \circ \sigma_{(\mathfrak{G}\mathfrak{A})A}) = \mathfrak{K}(\mathfrak{H}(\lambda_A) \circ \tau_{\mathfrak{A}A}) \circ \sigma_{(\mathfrak{G}\mathfrak{A})A} = \sigma(\tau * \lambda))_A$ for any object A of category $\mathbb{A}$.

6.9.4 Law for the Interchange of the Usual Product and the Star Product of Natural Transformations

Proposition 140 Given any natural transformations λ, τ, σ, and ε, we have

$$(\lambda \circ \tau)*(\sigma \circ \varepsilon) = (\lambda * \tau) \circ (\sigma * \varepsilon)$$

whenever the compositions involved are defined.

Proof The left-hand side $(\lambda \circ \tau)*(\sigma \circ \varepsilon)$ is defined iff there exists the following diagram:

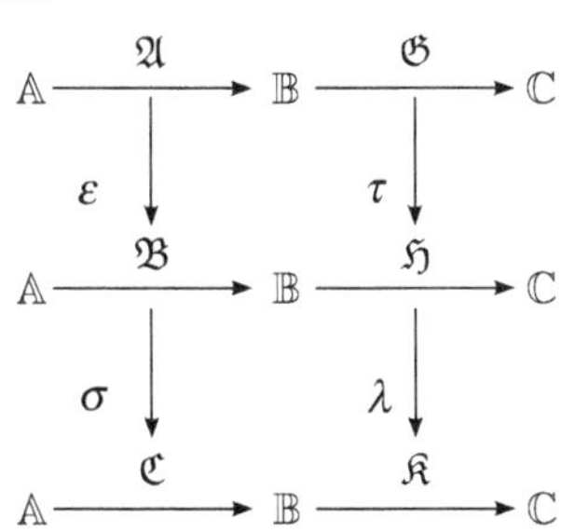

And then the right-hand side $(\lambda * \tau) \circ (\sigma * \varepsilon)$ is also defined.

Consider now any object A of category $\mathbb{A}$. From the definitions of the usual product and the star product of natural transformations, we then have

$$(\lambda \circ \tau)^*(\sigma \circ \varepsilon))_A = (\lambda \circ \tau)_{\mathfrak{C}A} \circ \mathfrak{G}((\sigma \circ \varepsilon)_A)$$
$$= (\lambda_{\mathfrak{C}A} \circ \lambda_{\mathfrak{C}A} \circ \mathfrak{G}(\sigma_A) \circ \mathfrak{G}(\varepsilon_A).$$

Now, τ is a natural transformation and $\mathfrak{B}A \xrightarrow{\sigma_A} \mathfrak{C}A$ is an arrow of category $\mathbb{B}$. We therefore have the commuting square:

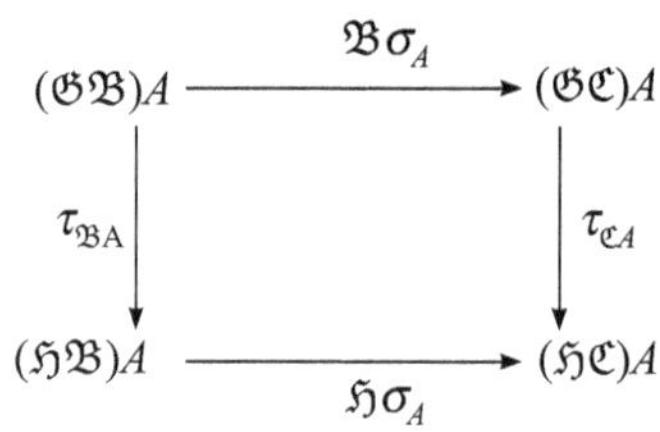

Then we can forget the middle to obtain

$$\lambda_{\mathfrak{G}A} \circ \mathfrak{H}(\sigma_A) \circ \tau_{\mathfrak{B}A} \circ \mathfrak{G}(\varepsilon_A) = (\lambda^* \tau) \circ (\sigma^* \varepsilon))_A,$$

where the right-hand side of this last equality follows from the definition of the star product.

6.9.5 Godement's Rules

As a result of the earlier propositions, we have the following rules, called Godement's rules, involving functors, the star, and the usual product of the natural transformations.

Proposition 141 Given the functors as well as the natural transformations of the diagram

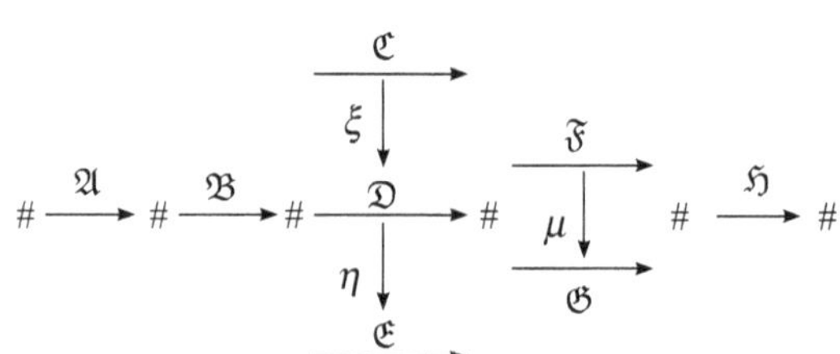

the following equalities, called Godement's rules, hold for any categories in the places of the symbols # in the above diagram:

- $(\mathfrak{H} \circ \mathfrak{F}) * \xi = \mathfrak{H} * (\mathfrak{F} * \xi)$.
- $\xi * (\mathfrak{B} \circ \mathfrak{A}) = \mathfrak{H} * (\xi * \mathfrak{B}) * \mathfrak{A}$.
- $\mathrm{id}_{\mathfrak{C}} * \mathrm{id}_{\mathfrak{B}} = \mathrm{id}_{\mathfrak{C} \circ \mathfrak{B}}$.
- $\mathrm{id}_{\mathfrak{F}} * \mathrm{id}_{\mathfrak{C}} = \mathrm{id}_{\mathfrak{F} \circ \mathfrak{C}}$.
- $\mathfrak{F} * (\xi * \mathfrak{B}) = (\mathfrak{F} * \xi) * \mathfrak{B}$.
- The following square commutes:

$$\begin{array}{ccc} \mathfrak{F} \circ \mathfrak{C} & \xrightarrow{\mathfrak{F} * \xi} & \mathfrak{F} \circ \mathfrak{D} \\ {\scriptstyle \mu * \mathfrak{C}} \downarrow & & \downarrow {\scriptstyle \mu * \mathfrak{D}} \\ \mathfrak{G} \circ \mathfrak{C} & \xrightarrow[\mathfrak{G} * \xi]{} & \mathfrak{G} \circ \mathfrak{D} \end{array}$$

- $\mathfrak{F} * (\eta * \xi) * \mathfrak{B} = (\mathfrak{F} * \eta * \mathfrak{B}) \circ (\mathfrak{F} * \xi * \mathfrak{B})$.

6.10 Functor Categories

Natural transformations therefore connect functors having the same source and target. The aforementioned properties of a natural transformation mean that natural transformations form a partial binary algebra. Consequently, we can now treat functors having the same source and target as being objects of a category and the natural transformations connecting these functors can be treated as being the arrows connecting such objects. Thus, it is easy to see that the following categories are now obtainable:

Observation 24 The quintuple ($\mathcal{F}$, $\mathcal{N}$, dom, cod, $\circ$), where $\mathcal{F}$ is a collection of all functors, $\mathcal{N}$ is a collection of all natural transformations, dom and cod are functions that, respectively, assign to each natural transformation its source (domain) and target (codomain), and $\circ$ is the usual product of natural transformations, is a (generic) category. It is, in general, a quasi category.

This quasi category is often denoted as $\mathfrak{FUNC}$ or as $\mathcal{FUNC}$. In the case that the underlying categories $\mathbb{A}$ and $\mathbb{B}$ are small categories, the same quasi category is also denoted as $\mathfrak{F}unc$ or *Func*.

Observation 25 The quintuple $\mathcal{C}$, $\mathcal{N}$, dom, cod, *), where $\mathcal{C}$ is a collection of all categories, $\mathcal{N}$ is a collection of all natural transformations, dom and cod are functions that, respectively, assign to each natural transformation its source (domain) and target (codomain), and * is the star product of natural transformations, is a (generic) category. It is, in general, a quasi category with these collections as conglomerates.

Therefore, the triple ($\mathcal{N}$, $\circ$, *) is often called a double quasi category of natural transformations.

Definition 122 For categories $\mathbb{A}$ and $\mathbb{B}$, the functor category $\mathbb{B}^{\mathbb{A}}$ has as objects all the functors from $\mathbb{A}$ to $\mathbb{B}$, as arrows from its object $\mathfrak{F}$ to another object $\mathfrak{G}$ all natural transformations from functor $\mathfrak{F}$ to functor $\mathfrak{G}$, as identities the identity natural transformations, and as composition the composition of natural transformations, the usual or the star product.

The following is then an equivalent definition of a functor category $\mathbb{B}^{\mathbb{A}}$:

Definition 123 (alternative definition of a functor category) Let $\mathbb{A}$ and $\mathbb{B}$ be categories. Then a full subcategory, denoted by [$\mathbb{A}$, $\mathbb{B}$] or by $\mathfrak{U}$B, of $\mathcal{FUNC}$ whose objects are the functors from $\mathbb{A}$ to $\mathbb{B}$ and whose arrows are the natural transformations connecting those functors, is a functor category.

In general, the above functor category $\mathbb{B}^{\mathbb{A}}$ is a quasi category, for the collection of functors from a category $\mathbb{A}$ to another category $\mathbb{B}$ is, in general, a conglomerate and so also is the case with the collection of natural transformations between even two functors.

Notice now the following.

Proposition 142 Let $\mathbb{A}$, $\mathbb{B}$, and $\mathbb{C}$ be categories. Then the functor category $(\mathbb{C}^{\mathbb{B}})^{\mathbb{A}}$ is isomorphic to the functor category $\mathbb{C}^{\mathbb{A}\times\mathbb{B}}$.

Proof Consider a functor $\mathfrak{F} : (\mathbb{C}^{\mathbb{B}})^{\mathbb{A}} \to \mathbb{C}^{\mathbb{A}\times\mathbb{B}}$ defined as follows:

$$(\mathfrak{F}(\mathfrak{A}))(A, B) = (\mathfrak{A}A)(B)$$

And if

$$(a, b) : (A, B) \to (C, D)$$

$$(\mathfrak{F}(\mathfrak{A}))(a, b) = \mathfrak{A}(C)(b) \circ \mathfrak{A}(\mathrm{a})_B = \mathfrak{A}(\mathrm{a})_B \circ \mathfrak{A}(A)(b)$$

and if $\eta : \mathfrak{A} \to \mathfrak{B}$, then

$\mathfrak{F}(\eta)(A, B) = (\eta_A)$.

We show that $\mathfrak{F}(\mathfrak{A})$, as defined above, is a functor from $\mathbb{A} \times \mathbb{B} \to \mathbb{C}$ and that $\mathfrak{F}(\eta)$ is a natural transformation from $\mathfrak{A} \to \mathfrak{B}$.

Clearly,

$$\mathfrak{F}(\mathrm{id}_{\mathfrak{A}})(A, B) = ((\mathrm{id}_{\mathfrak{A}})_A)_B = (\mathrm{id}_{\mathfrak{A}A})_B = \mathrm{id}_{\mathfrak{A}(A)(B)}.$$

Thus, $\mathfrak{F}$ preserves identities.

Now, let $\mathfrak{A} \xrightarrow{\eta} \mathfrak{B} \xrightarrow{\xi} \mathfrak{C}$. Then we have, for each $(A, B) \in \mathbb{A} \times \mathbb{B}$,

$$\mathfrak{F}(\xi \circ \eta)(A, B) = ((\xi \circ \eta)_A)_B = (\xi_A \circ \eta_A)_B = (\xi_A)_B \circ (\eta_A)_B$$
$$= \mathfrak{F}(\xi)(A, B) \circ \mathfrak{F}(\eta)(A, B) = (\mathfrak{F}(\xi) \circ \mathfrak{F}(\eta))(A, B).$$

Therefore, the defined $\mathfrak{F}$ is a functor.

Now, let $t : \mathbb{C}^{\mathbb{A}\times\mathbb{B}} \to (\mathbb{C}^{\mathbb{A}})^{\mathbb{B}}$ be defined as follows:

$\mathfrak{G}(\mathfrak{H})(A)$ = the right-associated functor $\mathfrak{H}(A, -) : \mathbb{B} \to \mathbb{C}$.

$(\mathfrak{G}(H)(g))_A = \mathfrak{H}(g, \mathrm{id}_g)$ if $g : A \to B$.

$(\mathfrak{G}(\tau_A))_B = \tau_{(A, B)}$ if $\tau : \mathfrak{H} \to \mathfrak{K}$.

We show that $\mathfrak{G}$ as defined above is a functor from $\mathbb{C}^{\mathbb{A}\times\mathbb{B}}$ to $(\mathbb{C}^{\mathbb{A}})^{\mathbb{B}}$. Firstly, we have

$$(\mathfrak{G}(\mathrm{id}_{\mathfrak{H}})_A)_B = (\mathrm{id}_{\mathfrak{H}})_{(A, B)} = (\mathrm{id}_{\mathfrak{H}})_{(A, B)} = \mathrm{id}_{(\mathfrak{G}(\mathfrak{H})A)(B)}.$$

Next, if $\mathfrak{H} \xrightarrow{\mu} \mathfrak{K} \xrightarrow{\nu} \mathfrak{L}$, then we have

$$(\mathfrak{G}(\nu \circ \mu)_A)_B = (\nu \circ \mu)_{(A, B)} = \nu_{(A, B)} \circ \mu_{(A,B)} = ((\mathfrak{G}\nu \circ \mathfrak{G}\mu)_A)_B.$$

Thus, $\mathfrak{G}$ preserves identities and compositions. Hence, η as defined is a functor from $\mathbb{C}^{\mathbb{A}\times\mathbb{B}}$ to $(\mathbb{C}^{\mathbb{A}})^{\mathbb{B}}$.

Now, let η be an arrow of the functor category $(\mathbb{C}^{\mathbb{A}})^{\mathbb{B}}$. Then

$$((\mathfrak{G} \circ \mathfrak{F}(\eta))_A)_B = (\mathfrak{G}(\mathfrak{F}(\eta))_A)_B = \mathfrak{F}(\eta)_{(A,B)} = ((\eta)_A)_B.$$

Therefore, we have

$$\mathfrak{G} \circ \mathfrak{F}(\eta) = \eta \Rightarrow \mathfrak{G} \circ \mathfrak{F} = \mathrm{id}_{[\mathbb{A}, [\mathbb{B}, \mathbb{C}]]}.$$

Furthermore, if σ is an arrow of category $\mathbb{C}^{\mathbb{A}\times\mathbb{B}}$, then we have

$$(\mathfrak{F} \circ \mathfrak{G}(\sigma))(A, B) = \mathfrak{F}(\mathfrak{G}(\sigma))(A, B) = \mathfrak{G}(\sigma)_A)_B = \sigma_{(A, B)}.$$

Therefore, we have for the functors $\mathfrak{F}$ and $\mathfrak{G}$,

$$\mathfrak{F} \circ \mathfrak{G}(\sigma) = \sigma \Rightarrow \mathfrak{F} \circ \mathfrak{G} = \mathrm{id}_{[\mathbb{A} \times \mathbb{B}, \mathbb{C}]}.$$

Therefore, we have an isofunctor $\mathfrak{F}$ between the categories $(\mathbb{C}^{\mathbb{A}})^{\mathbb{B}}$ and $\mathbb{C}^{\mathbb{A}\times\mathbb{B}}$. Hence, the functor categories $(\mathbb{C}^{\mathbb{A}})^{\mathbb{B}}$ and $\mathbb{C}^{\mathbb{A}\times\mathbb{B}}$ are isomorphic.

Now, we can also note the following.

Proposition 143 For any categories $\mathbb{A}$ and $\mathbb{B}$, the functor categories $\mathbb{B}^{\mathbb{A}}$ and $\mathbb{B}^{\mathrm{op}\mathbb{A}^{\mathrm{op}}}$ are dually equivalent.

It is now easy to see the following.

Proposition 144 Let $\mathbb{A} \underset{\mathfrak{G}}{\overset{\mathfrak{F}}{\rightrightarrows}} \mathbb{B}$ be functors and $\tau : \mathfrak{F} \to \mathfrak{F}$ be a natural transformation. Then τ is an isoarrow of the functor category $\mathbb{B}^{\mathbb{A}}$ iff it is a natural isomorphism of $\mathfrak{F}$ and $\mathfrak{G}$.

Proposition 145 Let ζ be any arrow of the functor category $\mathbb{B}^{\mathbb{A}}$. Then ζ is an isoarrow (monoarrow, bidirectional arrow, constant arrow, or zero arrow) if for each object $A \in \mathbb{A}$, ζ_A is an isoarrow (monoarrow, bidirectional arrow, constant arrow, or zero arrow).

Proof If each ζ_A is a monoarrow, then let ζ and δ be any arrows of $\mathbb{B}^{\mathbb{A}}$ such that $\zeta_A \circ \zeta_A = \zeta_A \circ \delta_A$ for each object $A \in \mathbb{B}$. Then for each such object A, we have $\zeta_A = \delta_A$, that is, $\zeta = \delta$, for ζ_A is given to be a monoarrow for each $A \in \mathbb{B}$. Hence, ζ is a monoarrow in $\mathbb{B}^{\mathbb{A}}$.

By duality, it follows that if each ζ_A is an epiarrow, then ζ is an epiarrow.

Then it follows that whenever each ζ_A is an isoarrow or a bidirectional arrow, ζ is an isoarrow or a bidirectional arrow.

Theorem 2 Let a category $\mathbb{B}$ possess the property that it has an initial object (it has a zero object, and it is pointed). Then for any category $\mathbb{A}$, the functor category $\mathbb{B}^{\mathbb{A}}$ also has the corresponding property that it has an initial object (it has a zero object, and it is pointed).

Proof Let I be the initial object of category $\mathbb{B}$ and consider a constant functor $\mathfrak{Const} : \mathbb{A} \to \mathbb{B}$ such that $\mathfrak{Const} : (f) = \mathrm{id}_I$ for all $f \in \mathbb{A}$. Then for any object $\mathfrak{G} \in \mathbb{B}^{\mathbb{A}}$, the natural transformation $\eta = \eta_A : \mathfrak{Const} \to \mathfrak{G}$ is defined as the unique arrow from I to $\mathfrak{G}A$ for any object $A \in \mathbb{A}$ because the square

$$\begin{array}{ccc}
\mathfrak{Const}\,(A) = I & \xrightarrow{\quad \eta_A \quad} & \mathfrak{G}A \\
{\scriptstyle \mathfrak{Const}\,(f) = \mathrm{id}_I}\Big\downarrow & & \Big\downarrow{\scriptstyle \mathrm{id}_B} \\
\mathfrak{Const}\,(B) = I & \xrightarrow[\quad \eta_B \quad]{} & \mathfrak{G}B
\end{array}$$

commutes because there is only one arrow to any other object from an initial object in a category. Clearly, η is then the only such natural transformation from $\mathfrak{Const}$ to $\mathfrak{G}$. Therefore, $\mathbb{B}^{\mathbb{A}}$ has an initial object.

By the duality principle, it follows that $\mathbb{B}^{\mathbb{A}}$ also has a terminal object if category $\mathbb{B}$ has a terminal object.

Then it also follows from the above that if category $\mathbb{B}$ has a zero object, then so is the case that $\mathbb{B}^{\mathbb{A}}$ has a zero object.

Now, let f_{KH} denote a zero arrow $K \xrightarrow{f} H$. Define for any pair of functors $\mathfrak{F}$, $\mathfrak{G}$: $\mathbb{A} \to \mathbb{B}$, a natural transformation $(\zeta_{\mathfrak{F}\mathfrak{G}}) = (\zeta_{\mathfrak{F}\mathfrak{G}})_A : \mathfrak{F} \to \mathfrak{G} = f_{\mathfrak{F}(A)} \circ {}_{\mathfrak{G}(A)}$. Consider a zero arrow from $\mathfrak{F}(A)$ to $\mathfrak{G}(B)$ in the pointed category $\mathbb{B}$. Then by the uniqueness of the zero arrow from $\mathfrak{F}(A)$ to $\mathfrak{G}(B)$ in the pointed category $\mathbb{B}$, the following square commutes:

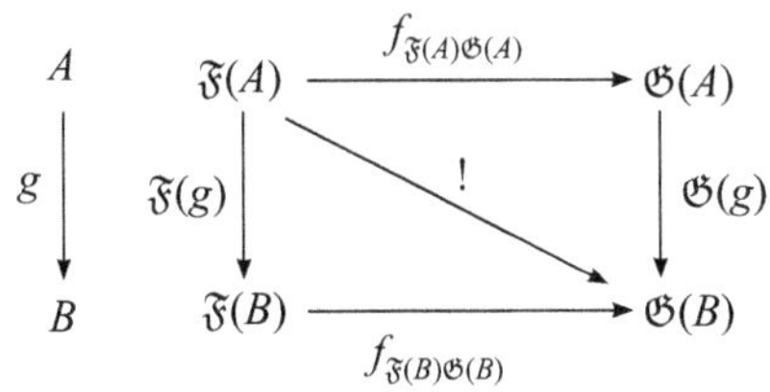

But in that case, every arrow $(\zeta_{\mathfrak{F}\mathfrak{G}})_A$ is necessarily a zero arrow. Hence, the functor category $\mathbb{B}^{\mathbb{A}}$ is a pointed category.

Therefore, whenever category $\mathbb{B}$ is a pointed category, we are assured that the functor category $\mathbb{B}^{\mathbb{A}}$ is also a pointed category. As we will see later, an appropriate additive structure is definable over the hom-collection of a pointed category having finite products. We note that measures are countably additive. This theorem will then play an important role in defining measures in the categorical context, as we will see in Chapter 9.

6.11 Evaluation Functor

Proposition 146 Given any categories $\mathbb{A}$ and $\mathbb{B}$, there exists a functor

$$\mathfrak{Ev} : \mathbb{B}^{\mathbb{A}} \times \mathbb{B} \to \mathbb{B}$$

defined by $\mathfrak{Ev} : (\mathfrak{F}, A) = \mathfrak{F}(A)$ and for each natural transformation $\eta : \mathfrak{F} \to \mathfrak{G}$ and an $\mathbb{A}$-arrow $f : A \to B$ by $\mathfrak{Ev}(\eta, f) = \mathfrak{G}(f) \circ \eta_A = \eta_B \circ \mathfrak{F}(f)$.

Functor $\mathfrak{Ev}$ is called an evaluation functor for $\mathbb{B}^{\mathbb{A}}$.

Proof Because the following square commutes

$$\begin{array}{ccccc} A & & \mathfrak{F}A & \xrightarrow{\eta_A} & \mathfrak{G}A \\ f\big\downarrow & & \mathfrak{F}f\big\downarrow & & \big\downarrow\mathfrak{G}f \\ B & & \mathfrak{F}B & \xrightarrow[\eta_B]{} & \mathfrak{G}B \end{array}$$

$\mathfrak{Ev}$ is well defined. We now show that $\mathfrak{Ev}$ preserves identities as well as the compositions.

Let the identity arrow in $\mathbb{B}^{\mathbb{A}} \times \mathbb{A}$ be $(\mathrm{id}_{\mathfrak{F}}, \mathrm{id}_A)$. Then

$$\mathfrak{Ev}(\mathrm{id}_{\mathfrak{F}}, \mathrm{id}_A) = \mathfrak{F}(\mathrm{id}_A) \circ \mathrm{id}_{\mathfrak{F}A} = \mathrm{id}_{\mathfrak{F}A} \circ \mathrm{id}_{\mathfrak{F}A} = \mathrm{id}_{\mathfrak{F}A} = \mathrm{id}_{\mathfrak{Ev}(\mathfrak{F},A)}.$$

Furthermore, let

$$(\mathfrak{F},A) \xrightarrow{(\eta,f)} (\mathfrak{G},B) \xrightarrow{(\eta,g)} (\mathfrak{H},C).$$

Then we have

$$\begin{aligned} \mathfrak{Ev}((\tau,g) \circ (n,f)) &= \mathfrak{Ev}(\tau\eta, gf) = \mathfrak{H}(gf) \circ (\tau\eta)_A = \mathfrak{H}g \circ (\mathfrak{H}f \circ \tau_A) \circ \eta_A \\ &= \mathfrak{H}g \circ (\tau_B \circ \mathfrak{G}f) \circ \eta_A = (\mathfrak{H}g \circ \tau_B) \circ (\mathfrak{G}f \circ \eta_A) \\ &= \mathfrak{Ev}((\tau,g) \circ \mathfrak{Ev}(n,f)). \end{aligned}$$

Therefore, $\mathfrak{Ev}$ is a functor.

Let $\mathfrak{Ev} : \mathbb{B}^{\mathbb{A}} \times \mathbb{A} \to \mathbb{B}$ be an evaluation functor.

Definition 124 (evaluation functor relative to object) Given the above evaluation functor, the left-associated functor

$$\mathfrak{Ev} : (-,A)\mathbb{B}^{\mathbb{A}} \to \mathbb{B}$$

is called the evaluation functor relative to the object A of category $\mathbb{A}$.

Clearly, we have

$$\mathfrak{Ev}(-,A)(\mathfrak{F}) = \mathfrak{F}(A) \text{ for each } \mathfrak{F} : \mathbb{A} \to \mathbb{B} \text{ and}$$

$$\mathfrak{Ev}(-,A)(\tau) = \tau_A \text{ for each } \tau * \mathrm{Mor}(\mathbb{B}^{\mathbb{A}}).$$

with the right-associated functor $\mathfrak{Ev}(\mathfrak{F}, -A)\ \mathbb{A} \to \mathbb{B}$ being functor $\mathfrak{F}$ for each functor $\mathfrak{F} : \mathbb{A} \to \mathbb{B}$.

6.12 "Constant Functor" Functor, or CF-Functor

Definition 125 (CF-functor) Given two categories $\mathbb{A}$ and $\mathbb{B}$, a functor $\mathfrak{Const} : \mathbb{B} \to \mathbb{B}^{\mathbb{A}}$ is called a CF-functor if:

- For each object B of category $\mathbb{B}$, $\mathfrak{Const}(B)$ is the constant functor from $\mathbb{A}$ to $\mathbb{B}$ whose value at each $\mathbb{A}$-object is B and whose value at each arrow of category $\mathbb{A}$ is the identity id_B.
- For each arrow $B \xrightarrow{f} C$ in category $\mathbb{B}$, $\mathfrak{Const}(f)$ is the natural transformation from $\mathfrak{Const}(B)$ to $\mathfrak{Const}(C)$ with $\mathfrak{Const}(f))_A = f$ for each object A of category $\mathbb{A}$.

We now observe:

Observation 26 For any nonempty categories $\mathbb{A}$ and $\mathbb{B}$, the CF-functor has a left-inverse and for each $\mathbb{A}$-object A, the evaluation functor relative to A has a right-invers

6.13 Limits and Colimits

In Chapter 5, we had encountered equalizers, products, and pullbacks. In each of these cases, there exists with a given diagram a distinguished object together with distinguished arrows, for example, a product object and arrows of projection in the case of a product, such that this distinguished object is connected with the rest of the objects of that diagram in a unique way. As noted before, we can treat such a distinguished object and arrows associated with it as a source (or, dually, as a sink).

In what follows, we generalize the above observation and call any such diagram as above as a limit diagram. Such constructions associate with a given diagram a certain source called the limit of the diagram. Any such limit can be completely characterized, up to isoarrows, by a purely categorical property.

In this context, we will, to begin with, introduce a slight change of perspective by defining a diagram in a different way, now using functors. This changed perspective of historical significance not only helps us with the generalizations of various categorical concepts but also shows the essential sameness of conceptions such as a product, a terminal object, or a pullback, each being a limit diagram.

Notice, in advance, that we could simply have started with the concept of a functor, then introduced diagrams and then limits, and then could have introduced various properties of arrows in a category as limits of various types. This is indeed an allowed perspective for category theory. It is an important change in the overall perspective as well.

Definition 126 The following nomenclature will be used:

- A diagram in a category $\mathbb{A}$ is a functor $\mathfrak{D} : \mathbb{I} \to \mathbb{A}$ with codomain $\mathbb{A}$. The domain, $\mathbb{I}$, is called the scheme of the diagram.
- A diagram with a small (or finite) scheme is called small (or finite).

Note 7 There is no technical difference whatsoever between a diagram and a functor or between a scheme and a category. Nonetheless, we use the alternate terminology of a diagram and its scheme when treating limits and colimits. For example, we often denote the image of an object $i \in \mathbb{I}$ under a diagram $\mathfrak{D}$ by $\mathfrak{D}_i$ rather than $\mathfrak{D}(i)$. Notice that this notation is more consistent with that for sources and sinks.

Example 38 A diagram in $\mathbb{A}$ with a discrete scheme is essentially nothing but a family of $\mathbb{A}$-objects.

Example 39 A diagram in $\mathbb{A}$ with the scheme $\bullet \rightrightarrows \bullet$ is essentially just a pair of $\mathbb{A}$-arrows with a common source and a common target.

Definition 127 (natural source and limit) Let $\mathfrak{D} : \mathbb{I} \to \mathbb{A}$ be a diagram.

- An $\mathbb{A}$-source $(A \xrightarrow{f_i} \mathfrak{D}_i)_{i \in \mathrm{Ob}(I)}$ is said to be natural for $\mathfrak{D}$ provided that for each $\mathbb{I}$-arrow $i \xrightarrow{d} j$, the following triangle commutes:

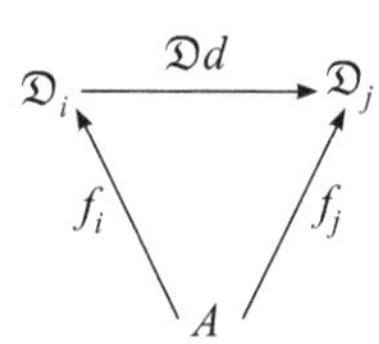

- A *limit of* $\mathfrak{D}$ is a natural source $(L \xrightarrow{\ell_i} \mathfrak{D}_i)_{i \in \mathrm{Ob}(I)}$ for $\mathfrak{D}$ with the (universal) property that each natural source $(A \xrightarrow{f_i} \mathfrak{D}_i)_{i \in \mathrm{Ob}(I)}$ for $\mathfrak{D}$ uniquely factors through it; that is, for every such source there exists a unique arrow $f : A \to L$ with $f_i = \ell_i \circ f$ for each $i \circ \mathrm{Ob}(I)$. That is to say, the following triangle commutes:

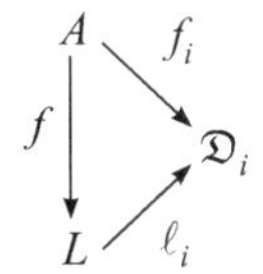

Limit and colimit situations

Example 40 Let the scheme $\mathbb{I} = \emptyset$ and $\mathfrak{D} : \mathbb{I} \to \mathbb{C}$ be a diagram. Then $(L, (\ell_i))$ is a limit of $\mathfrak{D}$ iff L is a terminal object and $(\ell_i) = \emptyset$. Similarly, $((\ell_i), L)$ is a colimit of a diagram $\mathfrak{D}$ iff L is an initial object and $(\ell_i) = \emptyset$.

In general, for a diagram $\mathfrak{D} : \mathbb{I} \to \mathbb{A}$ with a discrete scheme, every source with a codomain $(\mathfrak{D}_i)_{i \in \mathrm{Ob}(I)}$ is natural. A source is a limit of $\mathfrak{D}$ iff it is a product of the family $(\mathfrak{D}_i)_{i \in \mathrm{Ob}(I)}$. That is to say, products are limits of diagrams with discrete schemes. In particular, an object, considered as an empty source, is a limit of the empty diagram (i.e., the one with an empty scheme) iff it is a terminal object.

Example 41 For a pair of $\mathbb{A}$-arrows $A.28ex f 0.28ex g B$, considered as a diagram $\mathfrak{D}$ with a scheme $\bullet \rightrightarrows \bullet$, a source $(A \xrightarrow{e} C \xrightarrow{h} B)$ is natural provided $g \circ e = h = f \circ e$. Then in this case, h is determined by e. Hence, $C \xrightarrow{e} A$ is an equalizer of $A0.28ex\ f\ 0.28ex\ g\ B$ iff the source $(A \xrightarrow{e} C \xrightarrow{f \circ e} B)$ is a limit of $\mathfrak{D}$. Thus we may say (imprecisely) that equalizers are limits of diagrams with scheme $\bullet \rightrightarrows \bullet$. If, in the above scheme, the two arrows are replaced by an arbitrary set of arrows, then limits of diagrams with such schemes are, as seen in Chapter 5, called multiple equalizers.

Example 42 If I is a down-directed poset (as a category), then limits of diagrams with scheme I are called projective (or inverse) limits.

(Down-directed means that every pair of elements has a lower bound. The dual notion is up-directed.)

If, for example, $I = \mathbb{N}^{\mathrm{op}}$ is the poset of all non-negative integers with the opposite of the usual ordering, a diagram $\mathfrak{D} : I \to \mathbb{A}$ with this scheme is essentially a sequence

$$\cdots \xrightarrow{d_2} \mathfrak{D}_2 \xrightarrow{d_1} \mathfrak{D}_1 \xrightarrow{d_0} \mathfrak{D}_0$$

of $\mathbb{A}$-arrows (where $\mathfrak{D}(n+1 \to n) = \mathfrak{D}_{n+1} \xrightarrow{d_n} \mathfrak{D}_n$, $\mathfrak{D}(n+2 \to n) = d_n \circ d_{n+1}$, etc.). A natural source for $\mathfrak{D}$ is a source $(A \xrightarrow{f_n} D_n)_{n \in N}$, with $f_n = d_n \circ f_{n+1}$ for each n. In $\mathbb{S}$et, a projective limit of a diagram $\mathfrak{D}$ with a scheme $\mathbb{N}^{op}$ is a source $(L \xrightarrow{\ell_n} \mathfrak{D}_n)_{n \in N}$, where L is the set of all sequences $(x_n)n \in N$ with $x_n \in \mathfrak{D}_n)$ and $d_n(x_n + 1) = x_n$ for each $n \in \mathbb{N}$ and where each ℓ_m is a restriction of the mth projection $\pi_m : \prod_{n \in \mathbb{N}} \mathfrak{D}_n \to \mathfrak{D}_m$.

Example 43 Let $\mathbb{I} = \mathbb{C}$ and let diagram $\mathfrak{D} : \mathbb{I} \to \mathbb{C}$ be the identity functor. Then $(L, (\ell_A))$ is a limit of $\mathfrak{D}$ iff L is an initial object of $\mathbb{C}$ and for each object $A \in \mathbb{C}$, ℓ_A is the unique arrow from L to A. Similarly, $((\ell_A), L)$ is a colimit of $\mathfrak{D}$ iff L is a terminal object of $\mathbb{C}$ and for each $A \in \mathbb{C}$, ℓ_A is the unique arrow from A to L.

The sufficiency is obvious. For the necessity let $\mathfrak{L}$ be a limit of $\mathfrak{D}$ and let $L \xrightarrow{f} A$ be an arrow. By the naturality of $\mathfrak{L}$ for $\mathfrak{D}$ we obtain $f \circ \ell_L = \ell_A$. Application of this to $\ell_A \circ \ell_L = \ell_A = \ell_L \circ \mathrm{id}_L$ yields $f = \ell_A$ for each object A. Hence, by the uniqueness requirement in the definition of limit, $\ell_L = \mathrm{id}_L$. Consequently, $f = f \circ \mathrm{id}_L = f \circ \ell_L = \ell_A$. Thus $\hom(L, A) = \{\ell_A\}$ for each object A.

Example 44 If category $\mathbb{A}$ has an initial object i_0 with $\hom(i_0, i) = \{m_i\}$ for each $i \in \mathrm{Ob}(\mathbb{I})$, then every diagram $\mathfrak{D} : \mathbb{I} \to \mathbb{A}$ with scheme $\mathbb{I}$ has a limit:

$$(\mathfrak{D}_{i_0} \xrightarrow{\mathfrak{D}_{m_i}} \mathfrak{D}_i)_{i \in \mathrm{Ob}(\mathbb{I})}$$

Example 45 A square

$$\begin{array}{ccc} P & \xrightarrow{\bar{f}} & B \\ {\scriptstyle \bar{g}}\downarrow & & \downarrow{\scriptstyle g} \\ A & \xrightarrow[f]{} & C \end{array}$$

is a pullback square iff the 3-source $(P, (\bar{g}, f \circ \bar{g}, \bar{f}))$ is a limit of the 2-sink $A \xrightarrow{f} C \xleftarrow{g} B$, considered as a diagram in $\mathbb{A}$ with the scheme $\bullet \to \bullet \leftarrow \bullet$. In short, pullbacks are, loosely speaking, the limits of diagrams with scheme $\bullet \to \bullet \leftarrow \bullet$.

As can be noticed, the following are the special cases of limits:

- Product
- Terminal object

- Equalizer
- Pullback

We have already seen that a monoarrow is definable using a pullback diagram: it is an arrow whose pullback along itself is an identity arrow. Consequently, the concept of a limit and, dually, that of a colimit are the two properties of diagrams that, in general, can be considered to be the basis of categorical properties of arrows.

Proposition 147 Any limit $(L, (\ell_i))$ of a functor $\mathfrak{D} : \mathbb{I} \to \mathbb{C}$ is an (extremal) monosource.

Proof Let *X*0.28*ex r* 0.28*ex s L* be two $\mathbb{C}$-arrows with $\ell_i r = \ell_i s$ for each $i \in \text{Ob}(\mathbb{I})$. Then $(X, (\ell_1 r))$ is a natural source for $\mathfrak{D}$. There then exists, by the definition of a limit, a unique arrow $X \xrightarrow{x} L$ such that $\ell_i h = \ell_i r$ for each $i \in \text{Ob}(\mathbb{I})$. Clearly, each r and s is such x. Hence, $r = s$ and, therefore, $(L, \ell_i))$ is a monosource.

To prove that it is an extremal monosource, let its factorization using an epiarrow e be $L \xrightarrow{\ell_i} \mathfrak{D}_i = L \xrightarrow{e} K \xrightarrow{f_i} \mathfrak{D}_i$. As e is an epiarrow, $(K, (f_i))$ is natural for $\mathfrak{D}$. By the definition of a limit, there then exists a unique arrow $k: K \to L$ such that $f_i = \ell_i k$ for each $i \in \text{Ob}(\mathbb{I})$. Then we have $\ell_i ke = f_i e = \ell_i = \ell_i \circ \text{id}_L$. As $(L, (\ell_i))$ is a monosource, we have $ke = \text{id}_L$ and epiarrow e is a section. Hence, e is an isoarrow.

As a corollary, we then have:

Corollary 11 Every regular monoarrow is an extremal monoarrow, and every product is an extremal monosource.

We had seen that associated with an equalizer, a product, or a pullback, there is a certain uniqueness in the diagram. This uniqueness must also translate into the uniqueness associated with a limit diagram. It is then easy to see that this uniqueness associated with a limit diagram is the following.

Proposition 148 (uniqueness of limit) If $(L, (\ell_i))$ and $(K, (f_i))$ are the limits of a functor $\mathfrak{D} : \mathbb{I} \to \mathbb{C}$, then there exists a unique isoarrow $h : L \to K$ such that the triangle

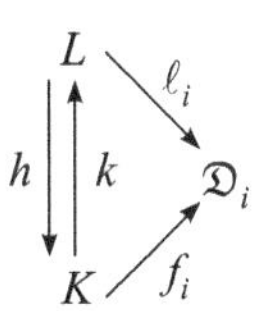

commutes for each $I \to \mathrm{Ob}(\mathbb{I})$.

Proof By the definition of a limit, there are unique arrows $k : K \to L$ and $h : L \to K$ such that for all $I \in \mathrm{Ob}(\mathbb{I})$, we have $\ell_i \circ k = f_i$ and $f_i \circ h = \ell_i$. But then $\ell_i \circ k \circ h = \ell_i = \ell_i \circ \mathrm{id}_L$ for $I \in \mathrm{Ob}(\mathbb{I})$ and, hence, $k \circ h = \mathrm{id}_L$. Similarly, $h \circ k = \mathrm{id}_K$. Thus, h and k are unique such isoarrows.

Because of this essential uniqueness of a limit, we can think of a generic limit, whenever it exists, of a diagram $\mathfrak{D}$ in any category. We will denote this generic limit by the symbol $\mathrm{Lim}\mathfrak{D}$. Similarly, we will denote the dual of a limit, the colimit, by the symbol $\mathrm{CoLim}\mathfrak{D}$. We then write for a specific instance $(L, (\ell_i))$ of a limit as $\mathrm{Lim}\mathfrak{D} \approx (L, (\ell_i))$ and that of a colimit as: $\mathrm{CoLim}\ \mathfrak{D} \approx ((\ell_i), L)$. By the abuse of terminology, we will also call the object L as a limit and write $\mathrm{Lim}\mathfrak{D} \approx L$ or as a colimit and write $\mathrm{CoLim}\mathfrak{D} \approx L$.

It is now easy to see that:

$\mathrm{Lim}\mathfrak{D}$ is

a terminal object when	$\mathbb{I}$ is an empty category.
a product when	$\mathbb{I}$ is a discrete category.
$\approx \mathrm{Eq}(\mathfrak{D}(p), \mathfrak{D}(q))$ when	$\mathbb{I}$ is represented by $\bullet \overset{p}{\underset{q}{\rightrightarrows}} \bullet$.
a pullback when	$\mathbb{I}$ is represented by $\bullet \to \bullet \leftarrow \bullet$.
a push-out when	$\mathbb{I}$ is represented by $\bullet \leftarrow \bullet \to \bullet$.

On the other hand, for the dual properties, $\mathrm{CoLim}\mathfrak{D}$ is

an initial object when	$\mathbb{I}$ is an empty category.
a coproduct when	$\mathbb{I}$ is a discrete category.
$\approx \mathrm{CoEq}(\mathfrak{D}(p), \mathfrak{D}(q))$ when	$\mathbb{I}$ is represented by $\bullet \overset{p}{\underset{q}{\rightrightarrows}} \bullet$.
a pullback when	$\mathbb{I}$ is represented by $\bullet \leftarrow \bullet \to \bullet$.
a push-out when	$\mathbb{I}$ is represented by $\bullet \to \bullet \leftarrow \bullet$.

Now, a diagram in a category $\mathbb{C}$ is a functor $\mathfrak{D} : \mathbb{I} \to \mathbb{D}$ and a limit $(L, (\ell_i))$ of diagram $\mathfrak{D}$ is a natural source for $\mathfrak{D}$ such that every natural source for $\mathfrak{D}$ uniquely factors through it. For some categories $\mathbb{A}$, it may happen that every functor $\mathfrak{D} : \mathbb{I} \to \mathbb{C}$ has a limit in $\mathbb{C}$. We adopt the following nomenclature for such situations involving a scheme $\mathbb{I}$.

Definition 128 ($\mathbb{I}$-complete category) A category $\mathbb{C}$ is said to be $\mathbb{I}$-complete when every diagram $\mathfrak{D} : \mathbb{I} \to \mathbb{C}$ has a limit in $\mathbb{C}$.

Definition 129 (complete category) A category $\mathbb{C}$ is said to be complete provided $\mathbb{C}$ is $\mathbb{I}$-complete for all the small schemes $\mathbb{I}$.

Definition 130 (finitely complete category) A category $\mathbb{C}$ is said to be finitely complete when $\mathbb{C}$ is $\mathbb{I}$-complete for all finite schemes $\mathbb{I}$.

We now note the following.

- All categories are $\mathbf{1}$-complete, $\mathbf{2}$-complete, as well as $\mathbf{3}$-complete.
- When $\mathbb{I}$ is $\emptyset$, any $\mathbb{I}$-complete category has a terminal object and also any $\mathbb{I}^{\text{op}}$-complete category has an initial object.
- When $\mathbb{I}$ is $\bullet \rightrightarrows \bullet$, any $\mathbb{I}$-complete category has equalizers and also any $\mathbb{I}^{\text{op}}$-complete category has coequalizers.
- When $\mathbb{I}$ is $\bullet\ \bullet$, that is, a discrete category having two objects, any $\mathbb{I}$-complete category has binary products and also any $\mathbb{I}^{\text{op}}$-complete category has binary coproducts.
- When $\mathbb{I}$ is $\bullet \to \bullet \leftarrow \bullet$, an $\mathbb{I}$-complete category has pullbacks and an $\mathbb{I}^{\text{op}}$-complete category has push-outs.

The following proposition establishes the existence of a unique functor from functor category $\mathbb{C}^{\mathbb{I}}$ to $\mathbb{C}$ when $\mathbb{C}$ is $\mathbb{I}$-complete.

Proposition 149 Let $\mathbb{I}$ be a scheme and $\mathbb{C}$ be an $\mathbb{I}$-complete category and for each diagram $\mathfrak{D} : \mathbb{I} \to \mathbb{C}$ let $(L_{\mathfrak{D}}, (\ell_i(\mathfrak{D}))$ be the limit of diagram $\mathfrak{D}$ in $\mathbb{I}$. Then there exists a unique functor $\mathfrak{Lim}_{\mathbb{I}} : \mathbb{C}^{\mathbb{I}} \to \mathbb{C}$, to be called an $\mathbb{I}$-limit functor, such that:

- For every object $\mathfrak{D} \in \mathbb{C}^{\mathbb{I}}$, $\mathfrak{Lim}_I(\mathfrak{D}) = L_{\mathfrak{D}}$.
- For every arrow (natural transformation) $\eta = (\eta_i) : \mathfrak{D} \to \mathfrak{G}$ in $\mathbb{C}^{\mathbb{I}}$ and for each $i \in \text{Ob}(\mathbb{I})$, the following square commutes:

$$\begin{array}{ccc}
\mathfrak{Lim}_I(\mathfrak{D}) = L_{\mathfrak{D}} & \xrightarrow{\ \ell_i(\mathfrak{D})\ } & \mathfrak{D}(i) \\
{\scriptstyle \mathfrak{Lim}_I(\eta)}\Big\downarrow & & \Big\downarrow{\scriptstyle \eta_i} \\
\mathfrak{Lim}_I(\mathfrak{G}) = L_{\mathfrak{G}} & \xrightarrow[\ \ell_i(\mathfrak{G})\ ]{} & \mathfrak{G}(i)
\end{array}$$

Proof As can be easily checked, the source $(L_{\mathfrak{D}}, \eta_i \circ \ell_i(\mathfrak{D}))$ is natural for $\mathfrak{G}$. Hence, there exists a unique natural transformation $\mathfrak{Lim}_I(\eta)$, an arrow in $\mathbb{C}^{\mathbb{I}}$, such that the given square commutes for all $i \in \mathrm{Ob}(\mathbb{I})$. It is then also easy to see that $\mathfrak{Lim}_{\mathbb{I}}$ defines the required functor.

Similarly, if category $\mathbb{C}$ is $\mathbb{I}$-co-complete, then there exists a functor, called an $\mathbb{I}$-colimit functor for $\mathbb{C}$, and is denoted by $\mathfrak{CoLim}_{\mathbb{I}} : \mathbb{C}^{\mathbb{I}} \to \mathbb{C}$.

Now, the existence of limits with a subcategory of a category having limits is also important. Therefore, we have:

Definition 131 A subcategory $\mathbb{D}$ of a category $\mathbb{C}$, with an embedding functor $\mathfrak{E} : \mathbb{D} \to \mathbb{C}$, is said to be closed under the formation of $\mathbb{I}$-limits in $\mathbb{C}$ when for every diagram $\mathfrak{D} : \mathbb{I} \to \mathbb{D}$ every $\mathbb{C}$-limit of $\mathfrak{E} \circ \mathfrak{D}$ is in $\mathbb{D}$.

Definition 132 A subcategory $\mathbb{D}$ of a category $\mathbb{C}$ that is closed under the formation of $\mathbb{I}$-limits for every small category $\mathbb{I}$ is called a complete subcategory of $\mathbb{C}$.

Note that, in general, if $\mathbb{D}$ is a complete subcategory of a category $\mathbb{C}$ with an embedding functor $\mathfrak{E} : \mathbb{D} \to \mathbb{C}$, if $\mathfrak{D} : \mathbb{I} \to \mathbb{D}$ is a diagram, and if $(L, (\ell_i))$ is a limit of $\mathfrak{E} \circ \mathfrak{D}$, then $(L, (\ell_i))$ need not be a limit of diagram $\mathfrak{D}$. (Recall that an embedding functor is not necessarily full, that is, onto on hom-collection restrictions, but is the one that is injective on arrows.) Therefore, a complete subcategory of a complete category need itself not be complete.

Nonetheless, we have the following.

Proposition 150 A complete subcategory of a complete category that is also a full subcategory of that complete category is itself complete.

Notice, however, that the converse of the above statement, that a complete category $\mathbb{D}$ that is also a full subcategory of a complete category $\mathbb{C}$ is necessarily a complete subcategory of $\mathbb{C}$, does not hold.

We will now state some propositions without proof.

Proposition 151 For any category $\mathbb{C}$, the following are equivalent (TFAE):

- $\mathbb{C}$ is finitely complete.
- $\mathbb{C}$ has finite products and pullbacks.
- $\mathbb{C}$ has finite products and inverse images.
- $\mathbb{C}$ has pullbacks and a terminal object.

- $\mathbb{C}$ has finite products and equalizers.
- $\mathbb{C}$ has finite products and finite intersections.
- $\mathbb{C}$ has finite products, equalizers, and finite intersections of regular subobjects.

Proposition 152 For any category $\mathbb{C}$, TFAE:

- $\mathbb{C}$ is complete.
- $\mathbb{C}$ has multiple pullbacks and a terminal object.
- $\mathbb{C}$ has products and pullbacks.
- $\mathbb{C}$ has products and finite intersections.
- $\mathbb{C}$ has products and equalizers.
- $\mathbb{C}$ has products, equalizers, and intersections of regular subobjects.
- $\mathbb{C}$ is finitely complete and has inverse limits.
- $\mathbb{C}$ has products and inverse images.

Proposition 153 For a small category $\mathbb{C}$, TFAE:

- $\mathbb{C}$ is complete.
- $\mathbb{C}$ is co-complete.
- $\mathbb{C}$ is equivalent to a complete lattice. (See Appendix 10.)

We now note that completeness and co-completeness are dual notions. Therefore, it may appear that the existence of one should have no implications for the existence of the other. However, the completeness of a category is equivalent to its co-completeness when the category is small! The appearance of the smallness of the category in this equivalence appears somewhat surprising here. The smallness condition in definition 3 has a crucial role in ensuring the internal consistency of the category as a mathematical structure. In a sense, certain constructions exist only when an appropriate smallness of categorical collections is ensured.

Note also that the schemes $\mathbb{I}$ and $\mathbb{I}^{\mathrm{op}}$ are, in general, not equivalent but always dually equivalent. Therefore, in general, the category of diagrams $\mathfrak{D} : \mathbb{I} \to \mathbb{C}$, that is, the functor category $\mathbb{C}^{\mathbb{I}}$ is distinct from the category of diagrams $\mathfrak{D} : \mathbb{I}^{\mathrm{op}} \to \mathbb{C}$, that is, from the functor category $\mathbb{C}^{\mathbb{I}^{\mathrm{op}}}$ $\mathbb{I}^{\mathrm{op}}$.

Now, in what follows, we note the conditions under which completeness is equivalent to co-completeness. Note, however, that a complete category is, in general, not necessarily co-complete.

Proposition 154 Every complete, well-powered, extremal co-well-powered category $\mathbb{C}$ has coequalizers.

Then we have as a corollary:

Corollary 12 If a category $\mathbb{C}$ is complete, well-powered, and co-well-powered, then TFAE:

- $\mathbb{C}$ is co-complete.
- $\mathbb{C}$ has coproducts.
- $\mathbb{C}$ has finite coproducts and direct limits.

There do exist certain furthermore categorical situations in which completeness is equivalent to co-completeness. We shall, however, not deal with them here. A reader interested in these situations is referred to the literature provided in the bibliography.

6.14 Functors Preserving and Reflecting Limits

A diagram in any category $\mathbb{A}$ is a functor $\mathfrak{D} : \mathbb{I} \to \mathbb{A}$. Any functor $\mathfrak{F} : \mathbb{A} \to \mathbb{B}$ will then have to be of some special type in order to preserve limits of diagrams $\mathfrak{D}$ in $\mathbb{A}$. The purpose of this section is to elaborate on the nature of functors that preserve and reflect limits. Let us therefore define first what we mean by the preservation and reflection of limits by functors.

Let $\mathbb{I}$ be a scheme, and let $\mathfrak{F} : \mathbb{A} \to \mathbb{B}$ be a functor.

Definition 133 (functors preserving limit) Functor $\mathfrak{F}$ is said to preserve $\mathbb{I}$-limits provided that whenever $\mathfrak{D} : \mathbb{I} \to \mathbb{A}$ is a diagram and $(L, (\ell_i))$ is a limit of diagram $\mathfrak{D}$ in category $\mathbb{A}$, $(\mathfrak{F}L, (\mathfrak{F}\ell_i))$ is a limit of diagram $\mathfrak{F} \circ \mathfrak{D}\mathbb{I} \to \mathbb{B}$ in category $\mathbb{B}$.

Definition 134 (functors reflecting limit) Similarly, a functor $\mathfrak{F} : \mathbb{A} \to \mathbb{B}$ is said to reflect $\mathbb{I}$-limits provided that whenever $\mathfrak{D} : \mathbb{I} \to \mathbb{B}$ is a diagram in category $\mathbb{A}$ and $(L, (\ell_i))$ is a source in category $\mathbb{A}$ such that $(\mathfrak{F}L, (\mathfrak{F}\ell_i))$ is a limit of diagram $\mathfrak{F} \circ \mathfrak{D}\mathbb{I} \to \mathbb{B}$ in category $\mathbb{B}$, then $(L, (\ell_i))$ is a limit of diagram $\mathfrak{D}$ in category $\mathbb{A}$.

We also adopt the following nomenclature:

Definition 135 (functor: preserves limits, preserves finite limits) Whenever a functor $\mathfrak{F}$ preserves $\mathbb{I}$-limits for every small (finite) scheme $\mathbb{I}$, we say that $\mathfrak{F}$ preserves (finite) limits.

Definition 136 (functor: reflects limits, reflects finite limits) Whenever a functor $\mathfrak{F}$ reflects $\mathbb{I}$-limits for every small (finite) scheme $\mathbb{I}$, we say that $\mathfrak{F}$ reflects (finite) limits.

A reader may have already noted that the above definitions are quite general and apply to the functorial preservation or reflection of a terminal object, product, pullback, equalizer, etc. These are also the special instances of the general definitions of functorial preservation and reflection of categorical property that were considered in Section 6.4.

Now, we state the following important results:

Proposition 155 Let a category $\mathbb{C}$ be finitely complete. Then for any functor $\mathfrak{F} : \mathbb{C} \to \mathbb{D}$, TFAE:

- $\mathfrak{F}$ preserves finite limits.
- $\mathfrak{F}$ preserves pullbacks and terminal objects.
- $\mathfrak{F}$ preserves finite products and pullbacks.
- $\mathfrak{F}$ preserves finite products and inverse images.
- $\mathfrak{F}$ preserves finite products and equalizers.
- $\mathfrak{F}$ preserves finite products and finite intersections.

Proposition 156 Let a category $\mathbb{C}$ be complete. Then for any given functor $\mathfrak{F} : \mathbb{C} \to \mathbb{D}$, TFAE:

- $\mathfrak{F}$ preserves limits.
- $\mathfrak{F}$ preserves multiple pullbacks and terminal objects.
- $\mathfrak{F}$ preserves products and pullbacks.
- $\mathfrak{F}$ preserves products and inverse images.
- $\mathfrak{F}$ preserves products and equalizers.
- $\mathfrak{F}$ preserves products and finite intersections.
- $\mathfrak{F}$ preserves finite limits and inverse limits.

Now, the following should also be evident:

Proposition 157 If a functor $\mathfrak{F}$ preserves pullbacks, then it preserves monoarrows.

Proposition 158 Let a category $\mathbb{C}$ have equalizers. Let a functor $\mathfrak{F}$: $\mathbb{C} \to \mathbb{D}$ preserve equalizers. Then functor $\mathfrak{F}$ is faithful iff $\mathfrak{F}$ reflects epiarrows.

Proposition 159 Let $\mathfrak{F} : \mathbb{C} \to \mathbb{D}$ be a faithful functor that reflects isoarrows. If, furthermore, category $\mathbb{C}$ has $\mathbb{I}$-limits (or $\mathbb{I}$-colimits) and $\mathfrak{F}$ preserves them, then $\mathfrak{F}$ also reflects them.

Proposition 160 Let $\mathbb{I}$ be a complete category, and let $\mathfrak{F} : \mathbb{C} \to \mathbb{D}$ preserve limits and reflect isoarrows. Then functor $\mathfrak{F}$ is faithful and reflecting limits, monoarrows, and epiarrows.

Proposition 161 Every full and faithful functor reflects $\mathbb{I}$-limits and $\mathbb{I}$-colimits for every scheme $\mathbb{I}$.

Proposition 162 If $\mathfrak{F}$ and $\mathfrak{G}$ are naturally isomorphic functors and $\mathbb{I}$ is any scheme, then $\mathfrak{F}$ preserves or reflects $\mathbb{I}$-limits or $\mathbb{I}$-colimits if and only $\mathfrak{G}$ also does so.

Proposition 163 Every equivalence functor preserves and reflects both $\mathbb{I}$-limits and $\mathbb{I}$-colimits for every scheme $\mathbb{I}$. In other words, completeness and co-completeness are categorical properties.

Now, of particular interest are the limits in functor categories. The following shows quite generally that any two types of limits commute. Therefore, we also see that every functor category $\mathbb{B}^{\mathbb{A}}$ inherits the completeness and co-completeness characteristics of its base category $\mathbb{B}$.

Let $\mathbb{I}$ and $\mathbb{J}$ be the two respective schemes of diagrams $\mathfrak{F}_{\mathbb{I}} : \mathbb{I} \to \mathbb{C}$ and $\mathfrak{D}_{\mathbb{I}} : \mathbb{J} \to \mathbb{C}$ in a category $\mathbb{C}$. Now, let $\mathfrak{D} : \mathbb{I} \times \mathbb{J} \to \mathbb{C}$ be a diagram in category $\mathbb{C}$ with the scheme $\mathbb{I} \times \mathbb{J}$. With the above data, we have the following result:

Theorem 3 Suppose for every object $i \in \mathbb{I}$, the right associated diagram (functor) $\mathfrak{D}(i, -) : \mathbb{J} \to \mathbb{C}$ has a limit $(L_i, (\ell^i_j)_{\mathbb{J}})$. Then:

- There exists a unique diagram (functor) $\mathfrak{F} : \mathbb{I} \to \mathbb{C}$ such that for each $i \in \mathbb{I}$, $\mathfrak{F}(i) = L_i$ and for each arrow $m : I \to i'$ in $\mathbb{I}$ and for each $j \in \mathbb{I}$, the following square commutes:

$$\begin{array}{cccc} i & \mathfrak{F}(i) = L_i & \xrightarrow{\ \ell^i_j\ } & D(i,j) \\ m\big\downarrow & \mathfrak{F}(m)\big\downarrow & & \big\downarrow \mathfrak{D}(m,j) \\ i' & \mathfrak{F}(i') = L_{i'} & \xrightarrow[\ \ell^{i'}_j\ ]{} & \mathfrak{D}(i',j) \end{array}$$

- Diagram $\mathfrak{D}$ has limit iff diagram $\mathfrak{F}$ has a limit, and any source $(L, (f_i)_{\mathbb{I}})$ is a limit of diagram $\mathfrak{F}$ iff $(L, (\ell^i_j \circ f_i)_{\mathbb{I}\times\mathbb{J}})$ is a limit of diagram $\mathfrak{D}$.

Proof Let the data be as provided.

- For each arrow $m : I \to i'$, let $\mathfrak{F}(m)$ be the arrow from L_i to $L_{i'}$.

Such an arrow exists by the fact that $(L_{i'}, (\ell^{i'}_j)_{\mathbb{J}})$ is a limit of $\mathfrak{D} : (i', -)\mathbb{J} \to \mathbb{C}$ and, furthermore, the source $(L_i, (\mathfrak{D}(m, j) \circ \ell^i_j)_{\mathbb{J}})$ is natural for $\mathfrak{D}(i', -)$. Now, $\mathfrak{D}(m, -)$ is a natural transformation. Consequently, $\mathfrak{F}$ preserves identities and compositions. Therefore, $\mathfrak{F}$ is a functor. Moreover, by the uniqueness in the definition of a limit, $\mathfrak{F} : \mathbb{I} \to \mathbb{C}$ can be defined in no other manner and, hence, is a unique functor.

- Now, let $(L, (f_i)_{\mathbb{I}})$ be the limit of diagram $\mathfrak{F}$. Then it can be seen that the source $(L, (\ell^i_j \circ f_i)_{\mathbb{I}\times\mathbb{J}})$ is natural for $\mathfrak{D}$. Let $(K, (k_{ij})_{\mathbb{I}\times\mathbb{J}})$ be another source that too is natural for $\mathfrak{D}$. Then for each $I \in \mathbb{J}$ and for each arrow $n : j \to j'$ in $\mathbb{J}$, we have $k_{ij} = \mathfrak{D}(i, n) \to k_{ij}$. The following diagram therefore commutes for each $i \in \mathbb{I}$:

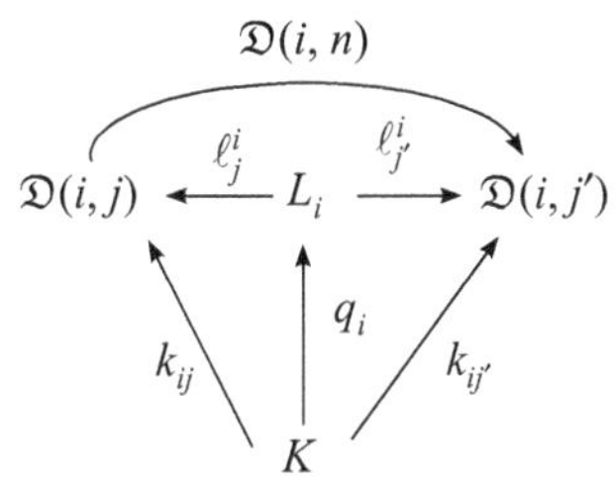

Here $q_i : K \to L_i$ is a unique arrow.
Now, for each $j \in \mathbb{J}$ and for each arrow $m : I \to i'$ in $\mathbb{I}$, we have

$$\ell^{i'}_j \circ \mathfrak{F}m \circ q_i = \mathfrak{D}(m, j) \circ \ell^i_j \circ q_i = \mathfrak{D}(m, j) \circ k_{ij} = k_{i'j} = \ell^{i'}_j \circ q_{i'} .$$

Then as the source $(L_i, (\ell^i_j)_{\mathbb{J}})$ is a monosource, we have $\mathfrak{F}m \circ q_i = q_{i'}$ for the unique functor $\mathfrak{F}$. But $(L, (f_i)_{\mathbb{I}})$ is given to be a limit of diagram $\mathfrak{F}$. There therefore exists a unique arrow $q : K \to L$ such that the following diagram commutes:

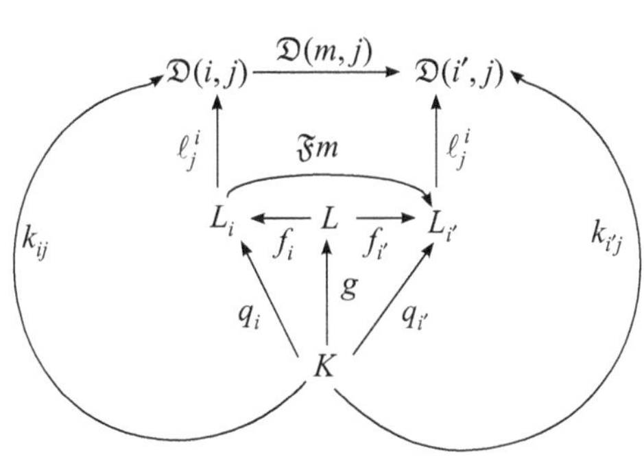

But as every $(L_i, (\ell^i_j)_{\mathbb{J}})$ is a monosource, arrow $q : K \to L$ is unique with respect to the property that for every $i \in \mathbb{I}$ and $j \in \mathbb{I}$, we always have $k_{ij} = \left(\ell^i_j \circ f_i\right) \circ g$. Therefore, the natural source $(L, (\ell^i_j \circ f_i)_{\mathbb{I}\times\mathbb{J}})$ is a limit of diagram $\mathfrak{D}$.

The converses also follow similarly. Hence, the result.

We note here that the above result is often referred to as the theorem of pointwise evaluation of limits.

Because of the above, we obtain:

Corollary 13 If a category $\mathbb{C}$ is $\mathbb{I}$-complete as well as $\mathbb{J}$-complete and if $\mathfrak{D} : \mathbb{I} \times \mathbb{J} \to \mathbb{C}$ is a diagram, then

$$\mathrm{Lim}_{\mathbb{I}}\ (\mathrm{Lim}_{\mathbb{J}}\ \mathfrak{D}(i, j) \approx \mathrm{Lim}_{\mathbb{I}\times\mathbb{J}}\ \mathfrak{D} \approx \mathrm{Lim}_{\mathbb{J}}\ (\mathrm{Lim}_{\mathbb{I}}\ \mathfrak{D}(i, j)).$$

That is to say, any two types of limits commute.

Furthermore, we also obtain:

Corollary 14 If a category $\mathbb{C}$ is $\mathbb{I}$-complete as well as $\mathbb{J}$-complete, then it is $(\mathbb{I} \times \mathbb{J})$-complete too.

As a consequence of the above results, we have:

- When both $\mathbb{I}$ and $\mathbb{J}$ are discrete schemes, we obtain the iteration of products, that is, products of products.
- When $\mathbb{I}$ is a discrete scheme and $\mathbb{J}$ is representable as $\bullet \rightrightarrows \bullet$, we have the result as the products of equalizers as equalizers of products.
- When both $\mathbb{I}$ and $\mathbb{J}$ are representable as $\bullet\!\to \bullet \leftarrow\!\bullet$, we have the result that the pullbacks can be pasted together.
- When $\mathbb{I}$ is a discrete scheme and $\mathbb{J}$ is representable as $\bullet\!\to \bullet \leftarrow\!\bullet$, we have the result that products of pullbacks are pullbacks of products.

Now, the following is an instance of the inheritance of properties of one category by the derived category.

Proposition 164 Let a diagram $\mathfrak{D} : \mathbb{I} \to \mathbb{B}^{\mathbb{A}}$ be given with the scheme $\mathbb{I}$. For each object $A \in \mathbb{A}$, let $\mathfrak{E}_A : \mathbb{B}^{\mathbb{A}} \to \mathbb{B}\mathbb{C}$ be the evaluation functor relative to A. If for each $A \in \mathbb{A}$, diagram $\mathfrak{E}_A \circ \mathfrak{D}\ \mathbb{I} \to \mathbb{B}$ has a limit $(L_A, (\ell^A_i)_{\mathbb{I}})$, then diagram $\mathfrak{D}$ also has a limit $(\mathfrak{F}, (f_i)_{\mathbb{I}})$. Here $\mathfrak{F} : \mathbb{A} \to \mathbb{B}$ is a functor such that $\mathfrak{F}(A) = L_A$ for each object $A \in \mathbb{A}$, and for each $i \in \mathbb{I}$, f_i is the natural transformation $(\ell^A_i)_{A\in\mathbb{A}} : \mathfrak{F} \to \mathfrak{D}(i)$.

Proof Firstly, we note that the categories $((\mathbb{B}^{\mathbb{A}})^{\mathbb{I}}$ and $\mathbb{B}^{\mathbb{I}\times\mathbb{A}}$ are isomorphic categories. Hence, diagram $\mathfrak{D} : \mathbb{I} \to \mathbb{B}^{\mathbb{A}}$ can be constructed as diagram $\mathfrak{D} : \mathbb{I} \times \mathbb{A} \to \mathbb{B}$. The above is then an appropriate restatement of the theorem of pointwise evaluation of limits.

Corollary 15 Whenever $\mathbb{B}$ is an $\mathbb{I}$-complete ($\mathbb{I}$-co-complete) category for the scheme $\mathbb{I}$, the functor category $\mathbb{B}^{\mathbb{A}}$ too is $\mathbb{I}$-complete ($\mathbb{I}$-co-complete) for any category $\mathbb{A}$.

6.15 Representable Functors

Definition 137 A functor $\mathfrak{F} : \mathbb{A} \to \mathbb{S}\text{et}$ is said to be representable (by an $\mathbb{A}$-object A) provided $\mathfrak{F}$ is naturally isomorphic to the hom-functor $\hom(A, -) : \mathbb{A} \to \mathbb{S}\text{et}$.

Note 8 Objects that represent the same functor (or two naturally isomorphic functors) are isomorphic.

Example 46 Forgetful functors are often representable.

Proposition 165 The following hold:

- Representable functors preserve monosources (i.e., if $\mathfrak{G} : \mathbb{A} \to \mathbb{S}\text{et}$ is a representable functor and S is a monosource in $\mathbb{A}$, then $\mathfrak{G}S$ is a monosource in $\mathbb{S}\text{et}$).
- Faithful functors reflect monosources (i.e., if $\mathfrak{G} : \mathbb{A} \to \mathbb{B}$ is a faithful functor, $S = (A, f_i)$ is a source in $\mathbb{A}$, and $\mathfrak{G}S = (\mathfrak{G}A, \mathfrak{G}f_i)$ is a monosource in $\mathbb{B}$, then S is a monosource in $\mathbb{A}$).

Proposition 166 The following hold:

- Hom-functors preserve limits.
- Representable functors preserve limits.
- If a functor preserves small limits, then it preserves monosources.
- If a functor preserves finite limits then it preserves monoarrows and regular monoarrows.
- If a functor preserves equalizers, then it also preserves regular monoarrows. (The converse is, however, false.)

6.16 Elements and Yoneda's Results

Now, it is often useful to think of an arrow in the manner of a variable belonging to its target. That is to say, an arrow $x : A \to B$ can be imagined as a variable x belonging to B and defined over A. We then write $x \in^A B$ to indicate that x is an arrow from A to B and will say that it is an element of B defined over A. We also call A the domain of variation of the element x. Clearly, every object A then has at least one element, id_A, the identity that is now called a generic element.

Then if $x \in^A B$ and $f : B \to C$, then $f \circ x \in^A C$. Then we can regard an arrow as a function taking elements to elements. When emphasizing this aspect, we can use the notation $f(x)$ for $f \circ x$.

However, it needs to be emphasized that the basic idea here is not that of introducing such variables everywhere in categorical considerations, for the main message of category theory is not to use any conceptions unless it is advantageous to do so.

In the above spirit, we note that many of the properties of arrows in a category can be described using elements. For example, an object is a terminal object of its category iff it has exactly one element for any domain of variation. As another example, any arrow $f : A \to B$ in category $\mathbb{C}$ is a monoarrow iff for every pair x, y of the elements of A defined on X, we have $f(x) = f(y)$, implying $x = y$, that is, f is injective (1-1) on elements defined on every object $X \in \mathrm{Ob}(\mathbb{C})$.

In the same spirit, an arrow $f : A \to B$ of category $\mathbb{C}$ is an isoarrow iff f is a bijection between the elements of A defined over X and the elements of B defined over X for all objects X in category $\mathbb{C}$.

As another example of the element view, a functor $\mathfrak{F} : \mathbb{A} \to \mathbb{B}$ takes any element of $A \in \mathrm{Ob}(\mathbb{A})$ to an element of $\mathfrak{F}A$ in such a way that generic elements are taken to generic elements and the action of $\mathfrak{F}$ commutes with the change of the domain of variation of the element.

Within this element view, consider subobjects. Then when (A_o, g) is a subobject of an object A and when an element $x \in^X A$ factors through A_o, that is, the triangle commutes, we say that the element x of A is an element of A_o and write $x \in_A^X A_o$. Often the subscript A is omitted when it is clear from the context that A_o is a subobject of A.

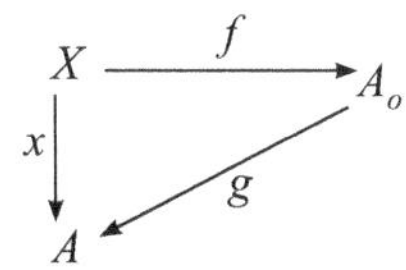

Proposition 167 If $g: A_o \to A$ and $g': A_{o'} \to A$ are two monoarrows in a category $\mathbb{C}$, then A_o and $A_{o'}$ have the same elements of A iff there exists an isoarrow $a: A_o \to A_{o'}$ such that the triangle on the right side in the following diagram commutes:

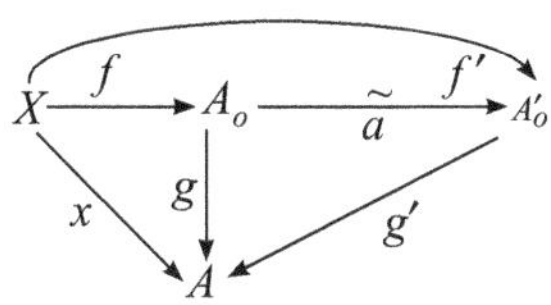

Proof If A_o and $A_{o'}$ have the same elements of A, then since the monoarrow $g \in_A^{A_o} A_o$, it factors through $A_{o'}$ and there exists an arrow $a: A_o \to A_{o'}$ such that $g = g' \circ a$. Similarly, for the monoarrow g', there exists an arrow $a': A_{o'} \to A_o$ such that $g = g \circ a'$. But g and g' are monoarrows. Hence, $\alpha \circ \alpha' = \mathrm{id}_{A'}$ and $\alpha' \circ \alpha = \mathrm{id}_{A'}$. The converse is also easy to establish.

Consequently, a subobject is completely determined by its elements:

Proposition 168 Two monoarrows having the same elements are equivalent as per definition 32, Chapter 5.

Now, we know from proposition 61, Chapter 5, that any arrow $f: T \to A$ having a terminal object T (whenever existing in a category) as a source is a monoarrow of that category and from proposition 52, Chapter 5, that any two terminal objects of a category are connected by a unique isoarrow. Consequently, the subobject (T, f) contains exactly one arrow and no more. Such an arrow is called a global element. An object is, however, not always determined by its global element. The following illustrates this and motivates the sheaf theory.

Consider a topological space X and the set $\mathbb{R}$ of real numbers. Let $\mathcal{O}(X)$ denote the category of open sets of X and arrows as the inclusion functions of one open set into another. Let $\mathfrak{I}: \mathcal{O}(\mathbb{X})^{\mathrm{op}} \to \mathbb{S}\mathrm{et}$ be a contravariant functor that takes each open set U to a set of real-valued continuous functions defined on U and each inclusion function of U into an open set V to a map from $\mathfrak{J}(V)$ to $\mathfrak{J}(U)$ that restricts a continuous function defined on V to U.

An important aspect of the restriction functions is that these are, in general, not surjective. In particular, there exist continuous functions defined on an open set that cannot be extended to a bigger open set. (Consider function $f(x) = 1/x$ as an example.)

Functor $\mathfrak{J}$ is an object of the functor category $[\mathcal{O}(\mathbb{X})^{\text{op}}, \mathbb{S}\text{et}]$ and the terminal object of this functor category is that functor that associates a singleton set to each of the open sets of the space X and the only possible monoarrow with each inclusion function of $\mathcal{O}(X)$. A global element of functor $\mathfrak{J}$ is then easily seen to be a continuous real-valued (global) function defined on *all* of the topological space X. (That is why we call an arrow from a terminal object a global element.)

Then an element of a functor $\mathfrak{F}$ is a natural transformation into that functor $\mathfrak{F}$. If $f : A \to B$ is any arrow in category $\mathbb{C}$, then arrow f induces a natural transformation from the hom-functor hom(B, –) to the hom-functor hom(A, –) by way of the composition of arrows in category $\mathbb{C}$: the component of this natural transformation at an object $C \in \text{Ob}(\mathbb{C})$ takes an arrow $g : B \to C$ to an arrow $g \circ f : A \to C$.

It is a straightforward (but worthwhile) computation to check that the above construction yields a contravariant functor $\mathfrak{Yoneda}$: $\mathbb{C} \to \mathbb{S}\text{et}^{\mathbb{C}}$. We call this functor Yoneda's embedding or map.

Let us denote the hom-functor in the functor category $\mathbb{B}^{\mathbb{A}}$ by the special symbol $\mathfrak{Nat}(\mathfrak{F},\mathfrak{G})$, where $\mathfrak{F},\mathfrak{G} : \mathbb{A} \to \mathbb{B}$. Notice that $\mathfrak{Nat}(-, -)$ is contravariant in the first variable.

Then a map that takes a functor $\mathfrak{F} : \mathbb{C} \to \mathbb{S}\text{et}$ to $\mathfrak{Nat}(\text{hom}(B, -), \mathfrak{F})$ and an object $B \in \text{Ob}(\mathbb{C})$ is a functor $\mathfrak{P}$ from $\mathbb{S}\text{et}^{\mathbb{C}} \times \mathbb{C}$ to $\mathbb{S}\text{et}$.

Another similar functor is the evaluation functor $\mathfrak{Ev} : \mathbb{S}\text{et}^{\mathbb{C}} \times \mathbb{C} \to \mathbb{S}\text{et}$, of Section 6.11, which takes $(\mathfrak{F}, B)$ to $\mathfrak{F}B$, and (τ, h) with arrow $h : B \to A$ and natural transformation $\tau : \mathfrak{F} \to \mathfrak{G}$ to $\mathfrak{G}h \circ \tau_B$.

The following is then a remarkable result due to Yoneda:

Lemma 2 (Yoneda's lemma) The map

$$\phi : \mathfrak{Nat}\,(\text{hom}(B, -), \mathfrak{F}), \to \mathfrak{F}B$$

defined by $\phi(\tau) = \tau_B(\text{id}_B)$ is a natural isomorphism of the evaluation functor $\mathfrak{Ev}$ and functor $\mathfrak{P}$, as have been defined above.

Proof We note that the inverse of the map ϕ takes an element x of an object $\mathfrak{F}B$ of category $\mathbb{C}$ to the natural transformation τ defined by $\tau_A(k) = \mathfrak{F}k(x)$, where $k \in \hom(B, A)$. It is then easy to verify that ϕ is the claimed natural isomorphism of the stated functors.

This lemma has many important consequences. As a first example, we have the following.

Theorem 4 (Yoneda's embedding theorem)

- The map that takes an arrow $f : A \to B$ in category $\mathbb{C}$ to the induced natural transformation $\hom(B, -) \to \hom(A, -)$ is a full and faithful contravariant functor from category $\mathbb{C}$ of arrow f to the functor category $\mathbb{S}\mathrm{et}^{\mathbb{C}}$.
- The map that takes an arrow $f : A \to B$ to the induced natural transformation $\hom(-, A) \to \hom(-, B)$ is a full and faithful functor from category $\mathbb{C}$ of arrow f to the functor category $\mathbb{S}\mathrm{et}^{\mathbb{C}^{\mathrm{op}}}$.

Proof It is easy to verify that the involved maps are functors as claimed. The fullness and faithfulness of these stated functors follows on using Yoneda's lemma, with functor $\mathfrak{F}$ being replaced by the hom-functor hom(A, –). Part 2 of Yoneda's embedding theorem is the dual of its first statement.

It then follows that Yoneda's embedding theorem essentially states that there exists a natural bijection between arrows $h : A \to B$ and the natural transformations $\hom(B, -) \to \hom(A, -)$. While claiming the existence of such and other arrows arising in this context, it is often the practice to say that the natural bijection or the arrow exists "by Yoneda."

Another consequence of Yoneda's lemma occurs when one of the elements of functor $\mathfrak{F}$ is a natural isomorphism defined on the hom-functor $\hom(A, -)$. Then we have a special element:

Definition 138 (universal element) When $\alpha : \hom(A, -) \to \mathfrak{F}$ is a natural isomorphism, the element $u \in \mathfrak{F}A$ corresponding to α is called a universal element for functor $\mathfrak{F}$ and functor $\mathfrak{F}$ is called the representable functor, it being represented by object A of category $\mathbb{C}$.

(Notice that this definition of a representable functor is the same as that given before in Section 6.15.)

The following proposition then provides a characterization of the universal elements of a functor:

Proposition 169 If $\mathfrak{F} : \mathbb{C} \to \mathbb{S}\mathrm{et}$ is a functor, then $u \in \mathfrak{F}A$ is a universal element iff for each object $B \in \mathrm{Ob}(\mathbb{C})$ and for each element $t \in \mathfrak{F}B$, there exists one and only one arrow $k : A \to B$ such that: $\mathfrak{F}k(u) = t$.

Proof If u is any such universal element corresponding to the natural isomorphism $\alpha : \hom(A, -) \to \mathfrak{F}$ and $t \in \mathfrak{F}A$, then the claimed arrow $k : A \to B$ is precisely the element $\alpha^{-1}B(t)$ in the hom-collection $[A, B]$.

Conversely, if $u \in \mathfrak{F}A$ satisfies the stated conditions, then there exists by Yoneda some natural isomorphism $\alpha : \hom(A, -) \to \mathfrak{F}$. It is then easy to verify that the map that takes $t \in \mathfrak{F}B$ to the only arrow $k : A \to B$ is an inverse in $\mathbb{S}\mathrm{et}^{\mathbb{C}}$ to αB and, hence, a universal element.

Notice that u can be an element of $\mathfrak{F}B$ for some B without it being a universal element in $\mathfrak{F}B$ if the conditions (on u and A) of the above lemma do not get satisfied.

Now, we can characterize pullbacks in terms of elements. For this purpose, consider the following diagram $\mathfrak{D}$ in category $\mathbb{C}$:

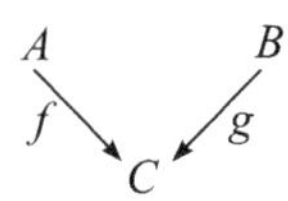

Such a diagram is often called a cone. Our interest now lies in finding elements x of A and elements y of B defined over the same object P of category $\mathbb{C}$ such that $f(x) = g(y)$. That is to say, we are interested in finding an object of category $\mathbb{C}$ whose elements are those pairs (x, y) with $f(x) = g(y)$. Obviously, the following commuting diagram is then implied:

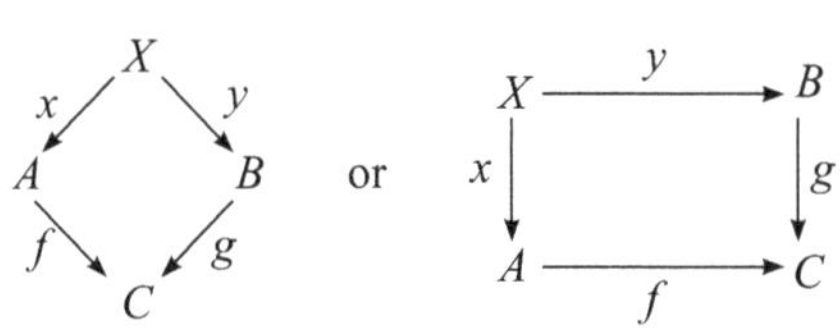

In this case, we call (X, x, y) a commuting or cap cone over $\mathfrak{D}$ based on X and regard it as an element of $\mathfrak{D}$ defined on X. Also, the collection of all the commuting or cap cones over $\mathfrak{D}$ based on X will then be denoted by the symbol $\mathfrak{Cone}(X, \mathfrak{D})$.

Our intention is to turn this construction into a functor $\mathfrak{Cone}$ $(-, \mathfrak{D})$ to category $\mathbb{S}\mathrm{et}$ in the following manner.

If $q : Y \to X$ is an arrow in $\mathbb{C}$ and (X, x, y) is a commutative cone over $\mathfrak{D}$ based on X, we define the functor as

$$\mathfrak{Cone}(q, \mathfrak{D})(X, x, y) = (Y, x \circ q, y \circ q),$$

which itself is a commutative cone over $\mathfrak{D}$ based on Y:

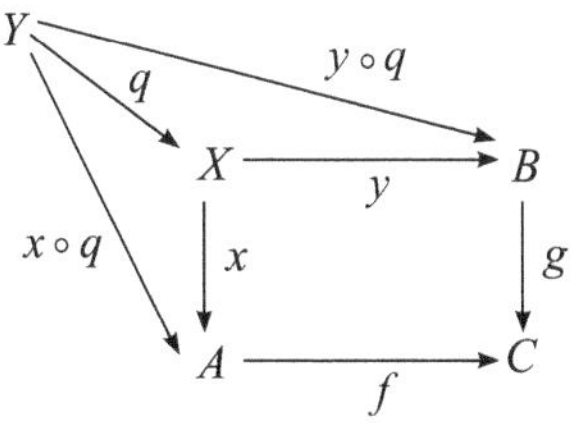

It is easy to check that this indeed defines a functor to $\mathbb{S}$et.

Now, the universal element, that we denote as (P, p_1, p_2), of $\mathfrak{Cone}(-, \mathfrak{D})$ is a pullback of diagram $\mathfrak{D}$ for which the following diagram commutes with the unique arrow $q : X \to P$:

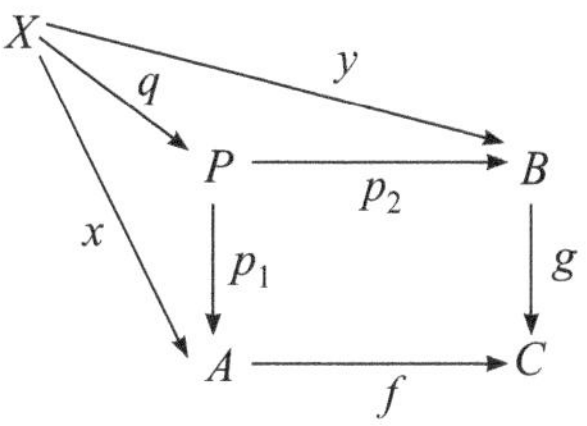

Functor $\mathfrak{Cone}(-, \mathfrak{D})$ is then representable. In other words, a pullback of a diagram $\mathfrak{D}$ is then a universal element of functor $\mathfrak{Cone}(-, \mathfrak{D})$, making that functor representable. Notice that the cone functor exists for every category but a particular cone may not have a pullback.

We can also turn the construction of a subobject into a functor $\mathfrak{Sub}$ by noting that a pullback of a monoarrow is a monoarrow in the same category, that is, arrow p_1 is also a monoarrow if arrow g is a monoarrow in the last diagram. Of course, we need to know more: we need to know that a pullback of a subobject is also a well-defined subobject.

To this end, let $\mathfrak{Sub}(A)$ denote the collection of all the subobjects of an object A of a category $\mathbb{C}$. (Notice that, in what follows, we are calling an equivalence class $[x]$ of the monoarrow x from object X to A as a subobject $(X, [x])$ of A and not just the pair (X, x), as we had done earlier.) Now, if $f: B \to A$ is a $\mathbb{C}$-arrow and $(S, [s])$ is a subobject of A, then let $\mathfrak{Sub}(f)(s)$ denote the pullback of any s in $[s]$ along f:

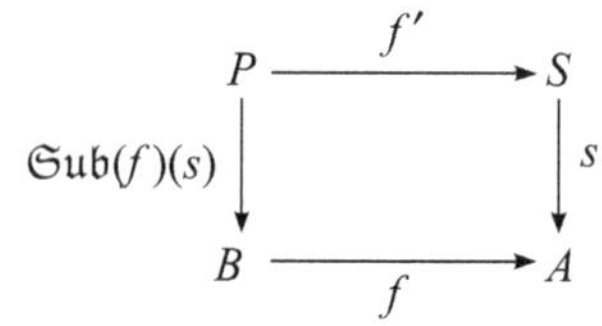

We then need to prove:

Proposition 170 If $s : S \to A$ and $t : S \to A$ are monoarrows in a category $\mathbb{C}$ determining the same subobject $(S, [s])$ *of* $A \in \mathrm{Ob}(\mathbb{C})$, then the pullbacks of s and t along $f : B \to A$ determine the same subobject of B.

Proof Firstly, a subobject is completely determined by its elements. (Recall that $x \in^P S$, that is, an element x of S means an arrow $x : P \to A$ and $f(y) \in^P_A S$ means an arrow $f \circ y : P \to A$ factoring through S.) Now, let the pullback of s along f be $y \mid f(y) \in^P_A S$ while the pullback of t along f be $z \mid f(z) \in^P_A S$. Collecting them, these two collections must be the same since a subobject $(S,[s])$ is entirely determined by its elements like f'.

It is then easy to verify that $\mathfrak{Sub}$ is indeed a functor.

Now, let a category $\mathbb{C}$ have finite limits. For a fixed object $A \in \mathrm{Ob}(\mathbb{C})$, let $- \times A$ be a functor from $\mathbb{C}$ to itself defined as follows: if $f: B \to B'$ is an arrow in $\mathbb{C}$, then $f \times A$ is arrow $(f \circ p_1, p_2) : B \times A \to B' \times A$, where p_1 and p_2 are the product projection arrows. By composition, we obtain a contravariant subobject functor $\mathfrak{Sub}(- \times A) : \mathbb{C} \to \mathbb{S}\mathrm{et}$. We then define:

Definition 139 (power object) The power object, denoted as PA of an object A in category $\mathbb{C}$, is an object, whenever it exists, that represents functor $\mathfrak{Sub}(- \times A)$ so that the functors $\hom(-, PA)$ and $\mathfrak{Sub}(- \times A)$ are naturally isomorphic.

This definition says precisely that given an arrow $f: B \to B'$ the following square commutes, with $\phi : \hom(-, PA) \to \mathfrak{Sub}(- \times A)$ being a natural transformation:

$$\begin{array}{ccc} \hom(B, \mathrm{P}A) & \xrightarrow{\phi(A,B)} & \mathfrak{Sub}(B \times A) \\ {\scriptstyle \hom(f,\, \mathrm{P}A)}\downarrow & & \downarrow{\scriptstyle \mathfrak{Sub}(f \times A)} \\ \hom(B', \mathrm{P}A) & \xrightarrow[\phi(A,B')]{} & \mathfrak{Sub}(B' \times A) \end{array}$$

In other words, the definition says that the elements of *PA* defined on *B* are essentially equivalent to subobjects of $B \times A$.

Now, if category $\mathbb{C}$ has a terminal object *T* and the power object *PT* exists, then functor $\mathfrak{Sub}(- \times A)$ is represented by *PT*, for $T \times A \simeq A$, that is, objects $T \times A$ and *A* are isomorphic.

Definition 140 A category $\mathbb{T}$ is called a topos if it has finite limits and each of its objects has a power object.

We would like to emphasize here that the aforementioned, deceptively simple, concept of a topos has quite rich, surprising, and powerful consequences. Many interesting categories are toposes. For example, category $\mathbb{S}$et is a topos. The category of a complete lattice is a topos. Set-valued functor categories are also important examples of toposes, or topoi.

In fact, category $\mathbb{S}$et provided the basic motivation for topos theoretical developments, which visualized a topos as a basis for mathematics. Nonetheless, we also note that a topos is not a generic category.

One then sees that the element view is helpful in many situations for grasping the underlying mathematical relationships in categories when it is advantageous to do so.

Functors, connecting categories, enjoy a pivotal position in category theory. Arrows and objects in one category can be studied in relation to arrows and objects of another category with the help of functors connecting them. It is to such studies that we now turn in the next chapter.

Chapter 7

Universal Associations by Functors

7.1 Introduction

Arrows and objects in categories may display some special associations in relation to functors connecting them. Such an association can be that of a source in the domain category of a functor with a source in the codomain category of that functor. For category theory, such functor-related properties of arrows and objects are obviously important for the analysis of the relationships of various categories.

Consider a functor $\mathfrak{F} : \mathbb{A} \to \mathbb{B}$. Let $A \in \mathrm{Ob}(\mathbb{A})$ be such that there exists an arrow $u : B \to \mathfrak{F}A$ for an object B in the target category $\mathbb{B}$ of $\mathfrak{F}$. We will call the pair (u, A) as an $\mathfrak{F}$-associate of B because it is functor $\mathfrak{F}$ that is creating such an association. (It is usually called an $\mathfrak{F}$-structured arrow with domain B. We however prefer the terminology adopted here.)

7.2 Objects and Arrows: Properties vis-à-vis Functors

7.2.1 Universal Associate and Universal Arrow

Definition 141 (universal associate and universal arrow) Let $\mathfrak{G} : \mathbb{A} \to \mathbb{B}$ be a functor and let $B \in \mathbb{B}$ be a $\mathbb{B}$-object. A $\mathfrak{G}$-associate

Subtlety in Relativity
Sanjay M. Wagh

ISBN 978-981-4774-44-4 (Hardcover), 978-1-351-23832-8 (eBook)
www.panstanford.com

(u, A) of B is then called a $\mathfrak{G}$-universal associate for B provided that for each $\mathbb{A}$-object A' and each $\mathbb{B}$-arrow u': $B \to \mathfrak{G}A'$, there exists a unique $\mathbb{A}$-arrow $f': A \to A'$ such that u' factors through $\mathfrak{G}A$, that is, $u' = \mathfrak{G}f' \circ u$ or the following triangle commutes:

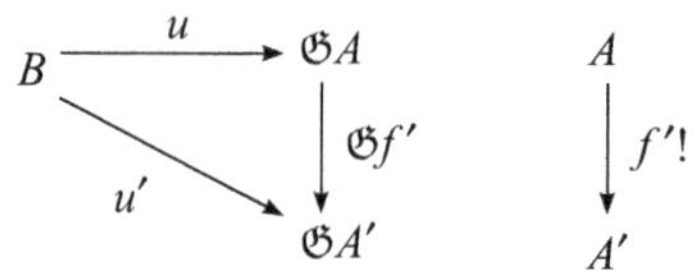

We then call the aforementioned arrow $u : B \to \mathfrak{G}A$ as a $\mathfrak{G}$-universal arrow for B.

Notice that not every functor $\mathfrak{G}$ would create such an association. Notice also that this universality is a property of functor $\mathfrak{G}$, for it is functor $\mathfrak{G}$ that is creating the association.

Example 47 Consider a subcategory $\mathbb{A}$ of category $\mathbb{B}$, and let $\mathfrak{I} : \mathbb{A} \to \mathbb{B}$ be the corresponding inclusion functor. Now, if B is any $\mathbb{B}$-object, then an $\mathbb{A}$-reflection for B is precisely an $\mathfrak{I}$-universal arrow for B.

Example 48 Let $\mathfrak{U} : \mathbb{A} \to \mathbb{S}\text{et}$ be a forgetful or underlying functor, which forgets the structure of the objects of $\mathbb{A}$ and associates with every object of $\mathbb{A}$ an underlying set from $\mathbb{S}\text{et}$. For each set B, there can then exist a $\mathfrak{U}$-universal arrow (u_B, A_B), with $u_B : B \to \mathfrak{U}(A_B)$ being the inclusion of set B into the underlying set $\mathfrak{U}(A_B)$ of object A_B. Object A_B is usually called as the $\mathfrak{U}$-free object, and the inclusion function u_B is called the insertion of the generator.

Example 49 In the case of a free object with one generator, u_B labels a single "distinguished" member—the generator—in the underlying set of A_B, with $B = \{*\}$ being a singleton set.

Example 50 Analogous results hold also for forgetful functors to categories other than category $\mathbb{S}\text{et}$.

Notions of a universal coassociate and a universal coarrow, which are dual to those of a universal associate and a universal arrow, respectively, can now be stated in the following manner:

7.2.2 Universal Coassociate and Universal Coarrow

Definition 142 (universal coassociate and universal coarrow) Consider a functor $\mathfrak{F} : \mathbb{B} \to \mathbb{A}$ and let $A \in \mathbb{A}$ be an $\mathbb{A}$-object. A pair $(B, \hat{u})$, with B being a $\mathbb{B}$-object and $\hat{u} : \mathfrak{F}B \to A$ being an arrow in $\mathbb{A}$, is called as an $\mathfrak{F}$-universal coassociate for A provided for each $\mathbb{A}$-object B' and each $\mathbb{A}$-arrow $f : \mathfrak{F}B' \to A$, there exists a unique $\mathbb{B}$-arrow $\hat{f} : B' \to B$ such that f factors through $\mathfrak{F}B$, that is, $f = \hat{u} \circ \mathfrak{F}\hat{f}$ or the following triangle commutes:

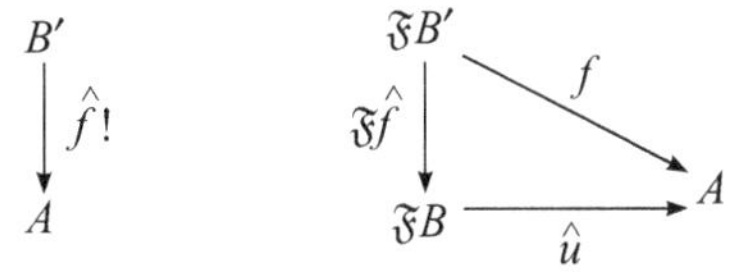

Notice that $(\hat{u}, B)$ is a universal associate for A with respect to functor $\mathfrak{F}^{op} : \mathbb{B}^{op} \to \mathbb{A}^{op}$. We call $\hat{u}$ a universal coarrow for B.

A co-reflection arrow is then an example of a universal coarrow.

It can now be also seen that the limits of diagrams $\mathfrak{D} : \mathbb{I} \to \mathbb{C}$ with the scheme $\mathbb{I}$ can be interpreted as $\mathfrak{G}$-universal coarrows relative to a suitable functor $\mathfrak{G} : \mathbb{C} \to \mathbb{C}^{\mathbb{I}}$, while the colimits are $\mathfrak{G}$-universal arrows of this kind.

7.2.3 Functor-Generation of Objects

Now, an equalizer diagram in the target category $\mathbb{B}$ can be associated by a functor $\mathfrak{G} : \mathbb{A} \to \mathbb{B}$ with a pair of arrows in the source category $\mathbb{A}$. This association can furthermore be such as to hold only when the pair of arrows of category $\mathbb{A}$ is equal. We then say that the equalizer arrow in the target category functor-generates that object.

Definition 143 Let $\mathbb{A}$ and $\mathbb{B}$ be categories, and let $\mathfrak{G} : \mathbb{A} \to \mathbb{B}$ be a functor. For an $\mathbb{A}$-object A and a $\mathbb{B}$-object B, let $g : B \to \mathfrak{G}A$ be a $\mathbb{B}$-arrow.

- Arrow g is said to $\mathfrak{G}$-generate A provided that whenever $A \underset{k}{\overset{h}{\rightrightarrows}} A'$ are two arrows in $\mathbb{A}$ such that $(\mathfrak{G}h) \circ g = (\mathfrak{G}k) \circ g$ then $h = k$. Diagrammatically, we have:

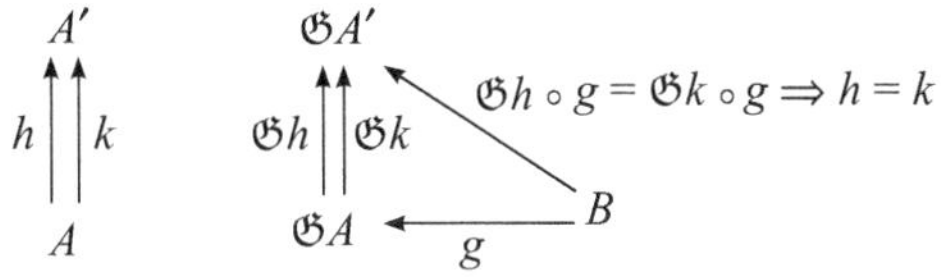

- Arrow g is said to extremally $\mathfrak{G}$-generate A provided
    - It $\mathfrak{G}$-generates A.

- (Extremal condition): Whenever there exists an $\mathbb{A}$-monoarrow m with target A, that is, $m : * \to A$, and a $\mathbb{B}$-arrow f such that $g = \mathfrak{G}m \circ f$, then m must be an isoarrow. Diagrammatically, we have:

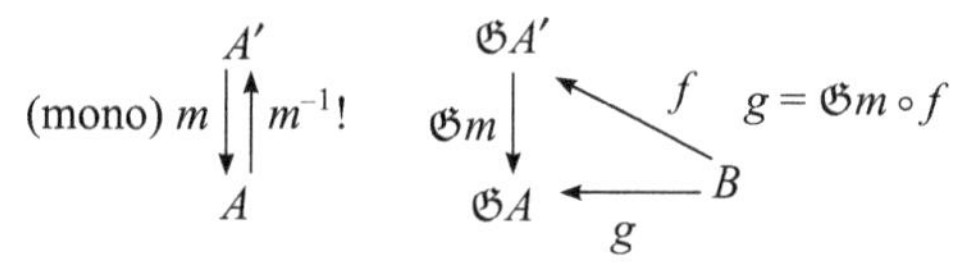

- A $\mathbb{B}$-object B is said to $\mathfrak{G}$-generate (extremally $\mathfrak{G}$-generate) an $\mathbb{A}$-object A whenever there exists some $\mathbb{B}$-arrow $g : B \to \mathfrak{G}A$ $\mathfrak{G}$-generating (extremally $\mathfrak{G}$-generating) object A.

Proposition 171 If a category $\mathbb{A}$ has equalizers and functor $\mathfrak{F} : \mathbb{A} \to \mathbb{B}$ preserves equalizers, then a $\mathbb{B}$-arrow $g : B \to \mathfrak{F}A$ extremally $\mathfrak{F}$-generates A if it satisfies the extremal condition (143) of definition 143.

Proof We need to show only that the given arrow g $\mathfrak{G}$-generates the object A. So, assume that the given arrow g satisfies the extremal condition (143) of definition 143. Let $A \underset{s}{\overset{r}{\rightrightarrows}} A'$ be $\mathbb{A}$-arrows such that $(\mathfrak{G}r) \circ g = (\mathfrak{G}s) \circ g$. As category $\mathbb{A}$ has equalizers, the pair (r, s) also has an equalizer that we denote by (K, k), where arrow k is a monoarrow in $\mathbb{A}$. Now, since it is given that functor $\mathfrak{G}$ preserves equalizers, we have $(\mathfrak{G}K, \mathfrak{G}k) \approx \mathrm{Eq}(\mathfrak{G}r, \mathfrak{G}s)$. From the definition of an equalizer, there then exists a unique arrow $h : B \to \mathfrak{G}\mathrm{K}$ such that $g = (\mathfrak{G}\mathrm{k}) \circ h$. But it is given that g satisfies the extremal condition. Hence, k is an isoarrow. As k is an equalizer of r and s, it then follows that $r = s$ and that g $\mathfrak{G}$-generates A.

Proposition 172 If (u, A) is a $\mathfrak{G}$-universal arrow for B, then arrow u extremal $\mathfrak{G}$-generates A.

Proof Here, (u, A) being a $\mathfrak{G}$-universal arrow means, diagrammatically, the left side, while arrow u (extremal) $\mathfrak{G}$-generates A means the right side of the following diagram:

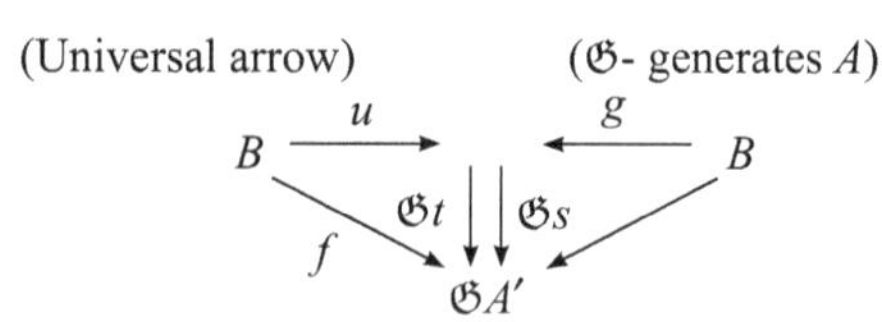

But if s and t are such that $\mathfrak{G}s \circ g = \mathfrak{G}t \circ g$, then there exists a unique $x : A \to A'$ such that $\mathfrak{G}x \circ u = \mathfrak{G}s \circ u$. Hence, $x = s = t$. This means that arrow u $\mathfrak{G}$-generates A.

It can be easily seen that the extremal condition is also satisfied.

Proposition 173 Universal arrows are essentially unique. That is to say, if $\mathfrak{F} : \mathbb{A} \to \mathbb{B}$ is a functor and (u, A) and (u', A') are both $\mathfrak{F}$-universal arrows for a $\mathbb{B}$-object B, then there exists a unique isoarrow $f : A \to A'$ in category $\mathbb{A}$ such that arrow u' factors through $\mathfrak{F}A$ in category $\mathbb{B}$, that is, the following triangle commutes:

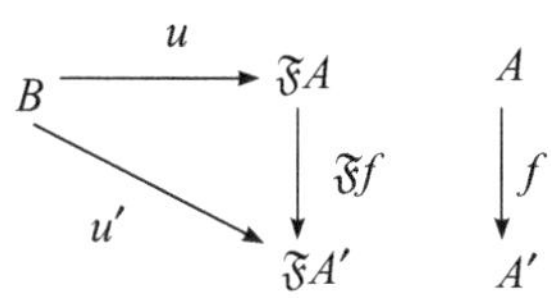

Proof We first show that u $\mathfrak{F}$-generates A. To this end, consider a pair of $\mathbb{A}$-arrows $A \underset{s}{\overset{r}{\rightrightarrows}} A'$ such that $(\mathfrak{F}r) \circ u = (\mathfrak{F}s) \circ u$.

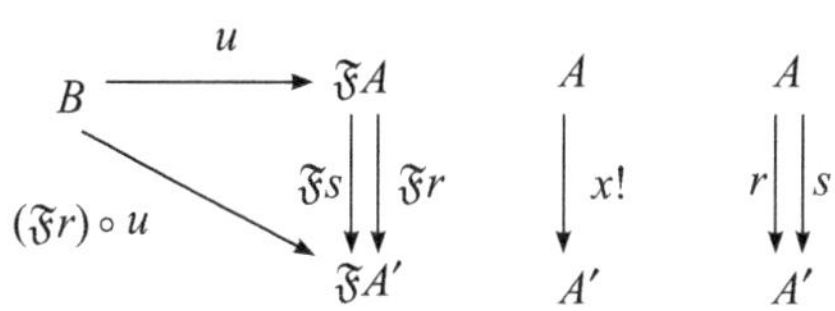

But u is an $\mathfrak{F}$-universal arrow for the $\mathbb{B}$-object B. Hence, there exists a unique $\mathbb{A}$-arrow $x : A \to A'$ such that $(\mathfrak{F}x) \circ u = (\mathfrak{F}r) \circ u = (\mathfrak{F}s) \circ u$. From this uniqueness, we have $x = r = s$. Thus, u $\mathfrak{F}$-generates A.

To prove that arrow u also satisfies the remaining condition, let there be an $\mathbb{A}$-monoarrow $m : A' \to A$ and a $\mathbb{B}$-arrow $f : B \to \mathfrak{F}A'$ such that $u = (\mathfrak{F}m) \circ f$.

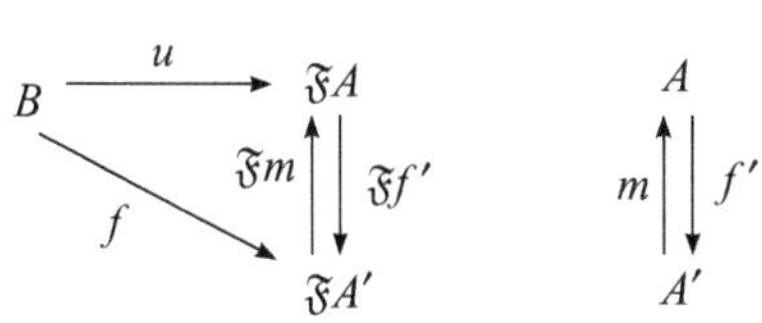

By the definition of a universal arrow, there also exists a unique $\mathbb{A}$-arrow $f' : A \to A'$ such that f factors through $\mathfrak{F}A$, that is, $f = (\mathfrak{F}f') \circ u$.

As a consequence of the above, the following diagram then commutes:

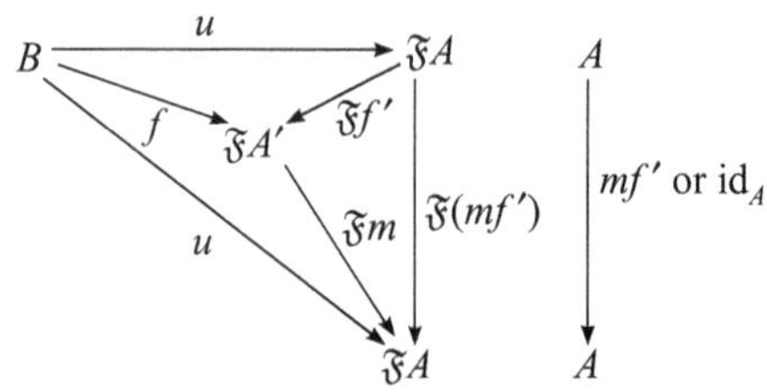

By the uniqueness in the definition of the universal arrow, we therefore have $mf' = \mathrm{id}_A$. Then the monoarrow m is also a retraction arrow, and, hence, it is an isoarrow.

Lemma 3 If $\mathfrak{G} : \mathbb{A} \to \mathbb{B}$ is a functor, $B_1, B_2 \in \mathrm{Ob}(\mathbb{B})$, $A_1, A_2 \in \mathrm{Ob}(\mathbb{A})$, (u_i, A_i) are $\mathfrak{G}$-universal arrows for $B_i (i = 1, 2)$, then for each of the $\mathbb{A}$-arrows $f : B_1 \to B_2$ there exists a unique $\mathbb{A}$-arrow $f' : A_1 \to A_2$ such that the following square commutes:

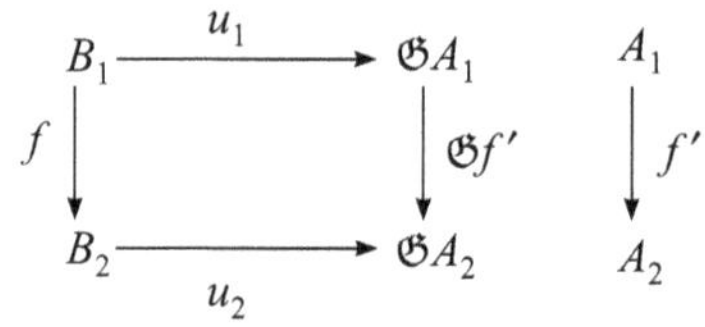

Proposition 174 Let $\mathfrak{H} : \mathbb{C} \to \mathbb{B}$ and $\mathfrak{G} : \mathbb{A} \to \mathbb{B}$ be functors. Let (u_C, A_C) be a $\mathfrak{G}$-universal arrow for object $\mathfrak{H}C$ for each $C \in \mathrm{Ob}(\mathbb{C})$.

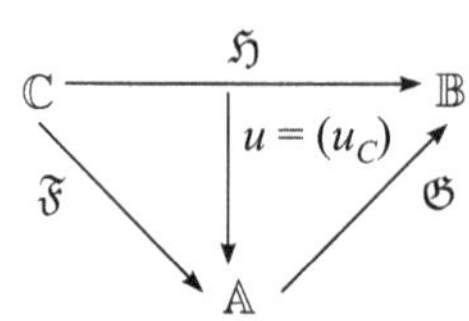

- Then there exists a unique functor $\mathfrak{F} : \mathbb{C} \to \mathbb{A}$ such that
  - For each $C \in \mathrm{Ob}(\mathbb{C})$, $\mathfrak{F}C = A_C$.
  - $u = (u_C)$ is a natural transformation from $\mathfrak{H}$ to $\mathfrak{G} \circ \mathfrak{F}$.

Proof Let $C \xrightarrow{f} C'$ and, then, $\mathfrak{H}C \xrightarrow{\mathfrak{H}f} \mathfrak{H}C'$ so that by lemma 3, there exists a unique arrow $f' : A_C \to A_{C'}$ such that the following square commutes:

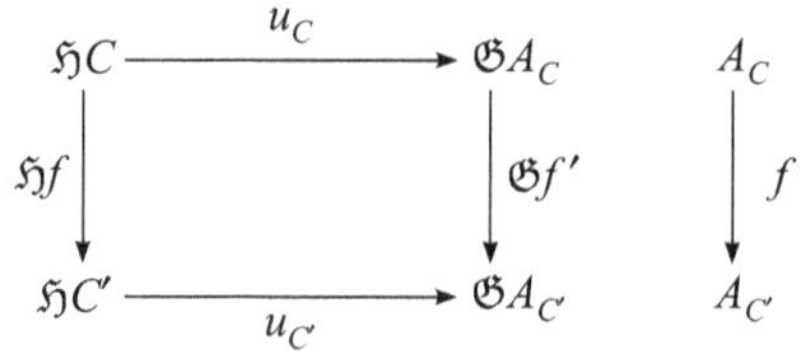

Moreover, if we let $C \xrightarrow{f} C' \xrightarrow{g} C''$, then there also exists a unique arrow $g' : A_{C'} \to A_{C''}$ and we can paste the squares as follows:

$$\begin{array}{ccccc} \mathfrak{H}C & \xrightarrow{u_C} & \mathfrak{G}A_C & A_C & C \\ {\scriptstyle \mathfrak{H}f}\downarrow & & \downarrow{\scriptstyle \mathfrak{G}f'} & \downarrow{\scriptstyle f'} & \downarrow{\scriptstyle f} \\ \mathfrak{H}C' & \xrightarrow{u_{C'}} & \mathfrak{G}A_{C'} & A_{C'} & C' \\ {\scriptstyle \mathfrak{H}g}\downarrow & & \downarrow{\scriptstyle \mathfrak{G}g'} & \downarrow{\scriptstyle g'} & \downarrow{\scriptstyle g} \\ \mathfrak{H}C'' & \xrightarrow{u_{C''}} & \mathfrak{G}A_{C''} & A_{C''} & C'' \end{array}$$

Now, let $\mathfrak{F}f = f'$ to generate a function $\mathfrak{F} : \mathrm{Mor}(\mathbb{C}) \to \mathrm{Mor}(\mathbb{A})$ such that $\mathfrak{F}(\mathrm{id}_C) = \mathrm{id}_{A_C}$. It is a unique function of this kind for which $\mathfrak{F}C = A_C$ and for which each square of the above type commutes.

From the above "pasted squares" it can be easily seen that the following square also commutes:

$$\begin{array}{cccc} \mathfrak{H}C & \xrightarrow{u_C} & \mathfrak{G}A_C & A_C \\ {\scriptstyle \mathfrak{H}(gf)}\downarrow & & \downarrow{\scriptstyle \mathfrak{G}x} & \downarrow{\scriptstyle x} \\ \mathfrak{H}C'' & \xrightarrow[u_{C''}]{} & \mathfrak{G}A_{C''} & A_{C''} \end{array}$$

But arrow u_C $\mathfrak{G}$-generates A_C. Hence, arrow x is unique. Therefore, we have $g' \circ f' = (gf)'$. Therefore, we also have $(\mathfrak{F}g) \circ (\mathfrak{F}g) = \mathfrak{F}(g \circ f)$. The aforementioned unique function $\mathfrak{F} : \mathrm{Mor}(\mathbb{C}) \to \mathrm{Mor}(\mathbb{A})$ is then a functor, and $u = (u_C)$ is a natural transformation from functor $\mathfrak{H}$ to functor $\mathfrak{G} \circ \mathfrak{F}$.

- Let $((k_C), K)$ be a colimit of $\mathfrak{H}$.
  - If $((k'_C), K')$ is a colimit of $\mathfrak{F}$, then there exists a unique $\mathbb{B}$-arrow $u' : K \to \mathfrak{G}K'$ such that, for each $C \in \mathrm{Ob}(\mathbb{C})$, the following square commutes:

$$\begin{array}{cccc} \mathfrak{H}C & \xrightarrow{u_C} & \mathfrak{G} \circ \mathfrak{F}(C) & \mathfrak{F}C \\ {\scriptstyle k_C}\downarrow & & \downarrow{\scriptstyle \mathfrak{G}k'_C} & \downarrow{\scriptstyle K'_C} \\ K & \xrightarrow[u']{} & \mathfrak{G}K' & K' \end{array}$$

In addition, (u', K') is a $\mathfrak{G}$-universal arrow for K.

- Conversely, if (u', K') is a $\mathfrak{G}$-universal arrow for K, then for each $C \in \text{Ob}(\mathbb{C})$, there exists a unique arrow $k'_C : \mathfrak{F}C \to K'$ such that the above square commutes and, in addition to this property, $((k'_C), K')$ is a colimit of $\mathfrak{F}$.

Proof From the first part of this proposition, we know that there exists a unique functor $\mathfrak{F} : \mathbb{C} \to \mathbb{A}$ for which each square of the considered type commutes. Then for each $g : \mathrm{C} \to \mathrm{C}'$, the following diagram also commutes:

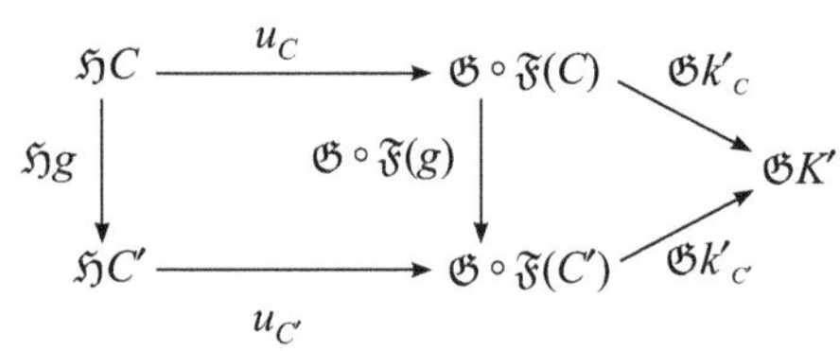

Therefore, $((\mathfrak{G}k'_C) \circ u_C, \mathfrak{G}K')$ is a natural sink for functor $\mathfrak{H}$. There then exists a unique arrow $u' : K \to \mathfrak{G}K'$ such that for each $C \in \text{Ob}(\mathbb{C})$, the following square commutes:

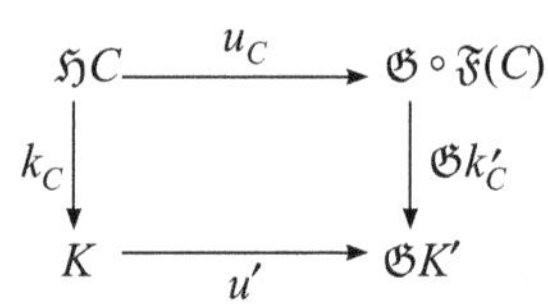

Now, to show that (u', K') is $\mathfrak{G}$-universal for K, let $f : K \to \mathfrak{G}A$. Note then that for each $C \in \text{Ob}(\mathbb{C})$, each $(u_C, \mathfrak{F}C)$ is a universal arrow. There therefore exists a unique $f_C : \mathfrak{F}C \to A$ such that $f \circ k_C = (\mathfrak{G}f_C) \circ u_C$, that is, the following triangle commutes:

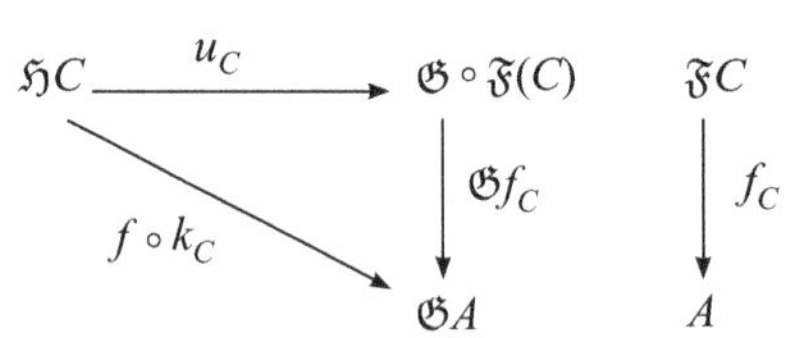

If $g : C \to C'$ is an arrow in category $\mathbb{C}$, then each of arrows f_C and $f_C \circ \mathfrak{F}g$ is an arrow $x : \mathfrak{F}\,C \to A$ for which the outer square of the following diagram commutes:

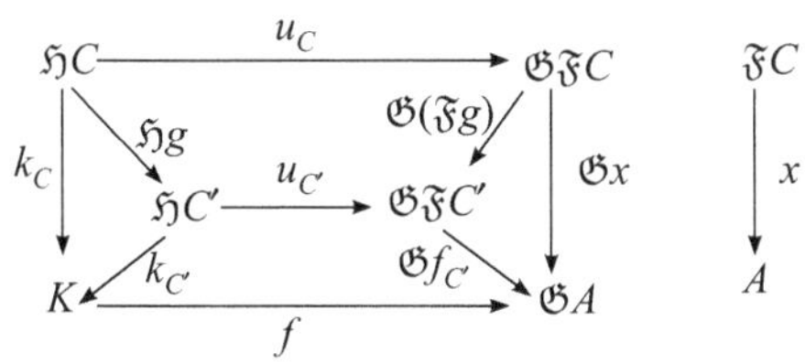

By the uniqueness in the above diagram, we then have $f_C = f_C \circ \mathfrak{F}(g)$. Then $((f_C), A)$ is a natural sink for functor $\mathfrak{F}$.

Now, as $((k'_C), K)$ is a colimit for diagram $\mathfrak{F}$, there exists a unique arrow $f' : K' \to A$ such that for each $C \in \mathrm{Ob}(\mathbb{C})$, we have f_C factoring through K' as $f_C = f' \circ k'_C$. Therefore, every part of the diagram

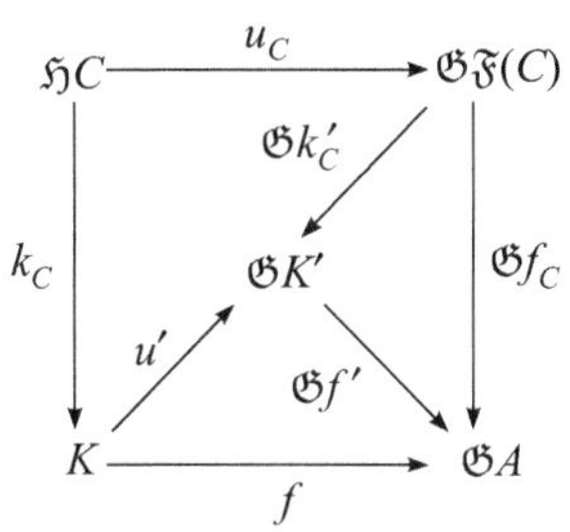

commutes, yielding thereby $f \circ k_C = (\mathfrak{G}f') \circ u' \circ k_C$ for each $C \in \mathrm{Ob}(\mathbb{C})$. Then as $((k'_C), K)$ is a colimit and consequently an episink, we have $f = (\mathfrak{G}f') \circ u'$. The uniqueness of f' with respect to this property in the diagram then follows from the above construction, the universal arrow $(u_C, \mathfrak{F}C)$, and the fact that f' $\mathfrak{G}$-generates $\mathfrak{F}C$. Hence, (u', K') is a $\mathfrak{G}$-universal arrow for K.

To prove the converse, that is, the second part of the proposition, let us assume that (u', K') is a $\mathfrak{G}$-universal arrow for K. Then as each $(u_C, \mathfrak{F}C)$ is $\mathfrak{G}$-universal for $\mathfrak{F}C$, for each $C \in \mathrm{Ob}(\mathbb{C})$, we have the existence of some arrow $k'_C : \mathfrak{F}C \to K'$ such that the following square commutes:

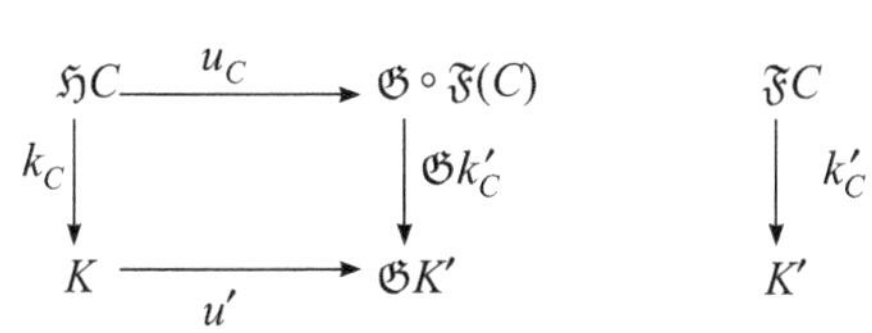

Now, $((k_C), K)$ is a colimit of diagram $\mathfrak{H}$ and $u : \mathfrak{H} \to \mathfrak{G}\mathfrak{F}$ is a natural transformation. Therefore, for every $g : C \to C'$, we have the commutativity of the following diagram:

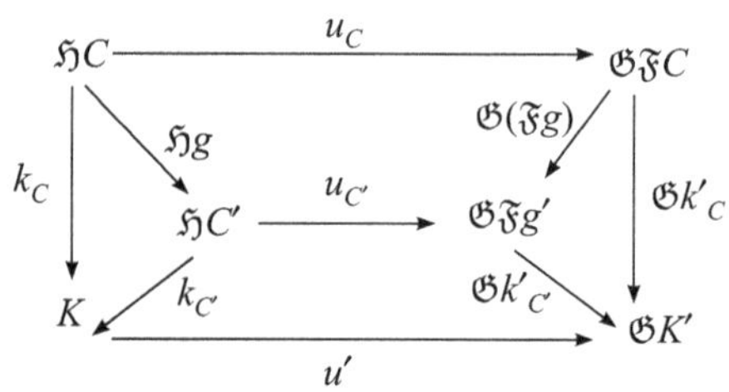

(An exception can occur for the rightmost triangle because the triangle functor $\mathfrak{G}$ is transforming need not commute in category $\mathbb{A}$, we note. This, however, does not concern the arguments that do, however, require the commutativity of the outer rectangle.)

As u_C $\mathfrak{G}$-generates $\mathfrak{F}C$, we have $k'_C = k'_C \circ \mathfrak{F}g$, and $((k'_C), K')$ is a natural sink therefore for functor $\mathfrak{F}$.

Let (p_C, P) be another natural sink for functor $\mathfrak{F}$. Then it can be seen that $((\mathfrak{G}p_C) \circ u_C, \mathfrak{G}P)$ is a natural sink for functor $\mathfrak{H}$. There then exists a unique arrow $q : K \to \mathfrak{G}P$ such that $(\mathfrak{G}p_C) \circ u_C = q \circ k_C$. That is to say, arrow $(\mathfrak{G}p_C) \circ u_C$ factors through K, or the following triangle commutes for every $C \in \mathrm{Ob}(\mathbb{C})$:

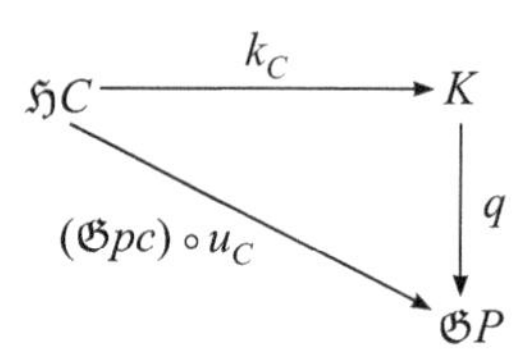

Furthermore, as (u', K') is a $\mathfrak{G}$-universal arrow for K, there exists an arrow $\tilde{q} : K' \to P$ such that the following triangle also commutes:

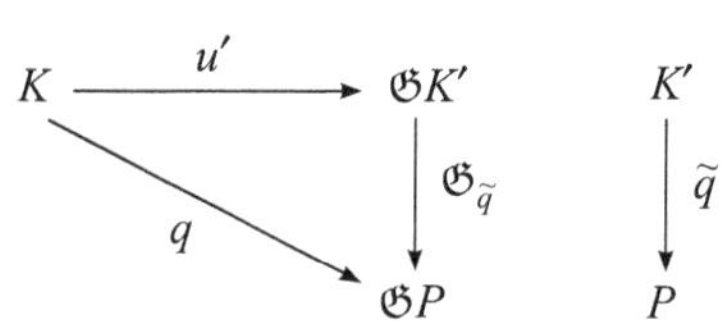

Clearly, each of $(\mathfrak{G}p_C) \circ u_C$ and $(\mathfrak{G}(\tilde{q} \circ k'_C)) \circ u_C$ factors through $\mathfrak{G}\mathfrak{F}C$, with $u_C : \mathfrak{H}C \to \mathfrak{G}\mathfrak{F}C$ being $\mathfrak{G}$-universal:

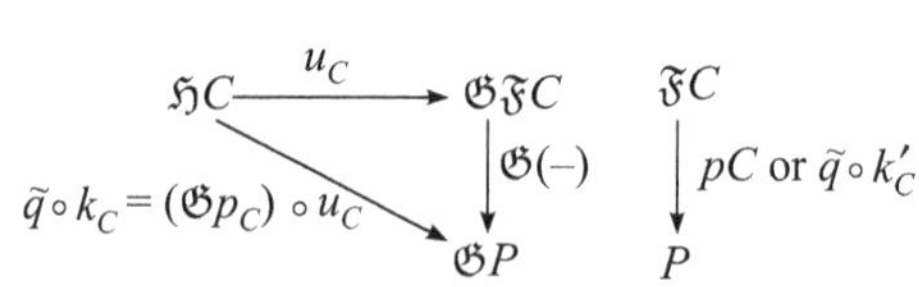

Then u_C $\mathfrak{G}$-generates $\mathfrak{F}C$ for every $C \in \text{Ob}(\mathbb{C})$. Therefore, we have $p_C = \tilde{q} \circ k'_C$ for every $C \in \text{Ob}(\mathbb{C})$. Arrow $\tilde{q}$ is unique with this property because (k_C, K) is an episink and (u', K') is $\mathfrak{G}$-universal. Hence, the result that (k'_C, K') is a colimit of functor $\mathfrak{F}$.

Corollary 16 Let $\mathbb{A}$ be a $\mathbb{C}$-co-complete category, and let $\mathfrak{G} : \mathbb{A} \to \mathbb{B}$ be a functor. Then the full subcategory $\mathbb{B}'$ consisting of all the $\mathbb{B}$-objects that have $\mathfrak{G}$-universal arrows is *closed* under the formation of $\mathbb{C}$-limits in category $\mathbb{B}$, that is, if diagram $\mathfrak{D} : \mathbb{C} \to \mathbb{B}$ has a colimit $((k'_C), K)$ in category $\mathbb{B}$ and if each $\mathfrak{D}C$ is in $\mathbb{B}'$, then K must also be in $\mathbb{B}'$.

Theorem 5 Let $\mathfrak{G} : \mathbb{A} \to \mathbb{B}$ be a functor such that for each $B \in \text{Ob}(\mathbb{C})$ there exists a $\mathfrak{G}$-universal arrow (η_B, A_B).

- Then there exists a unique functor $\mathfrak{F} : \mathbb{A} \to \mathbb{B}$ such that:
  - For each $B \in \text{Ob}(\mathbb{B})$, we have $\mathfrak{F}B = A_B$.
  - $\eta = (\eta_B) : \text{id}_{\mathbb{B}} \to \mathfrak{G} \circ \mathfrak{F}$ is a natural transformation.

Proof This part follows from proposition 174, with functor $\mathfrak{H}$ being replaced by the identity functor, $\text{id}_{\mathbb{B}} : \mathbb{B} \to \mathbb{B}$, of category $\mathbb{B}$.

- Moreover, functor $\mathfrak{F}$ preserves $\mathbb{C}$-colimits for each category $\mathbb{C}$.

Proof Let $\mathfrak{D}$: $\mathbb{C} \to \mathbb{B}$ be a diagram with colimit $((k_C), K)$. By assumption, the $\mathbb{B}$-object K has an associated $\mathfrak{G}$-universal arrow (η_K, A_K). By proposition 174, there exists for every $\mathbb{C}$-object C an arrow $k'_C : \mathfrak{F}\mathfrak{D}C \to A_K$ such that $((k'_C), K')$ is a colimit of $\mathfrak{F}\mathfrak{D}$ and each square of the following type commutes:

$$\begin{array}{ccccc} \mathfrak{D}C & \xrightarrow{\eta_{\mathfrak{D}C}} & \mathfrak{G}\mathfrak{F}\mathfrak{D}C & \qquad & \mathfrak{F}\mathfrak{D}C \\ \downarrow{\scriptstyle k_C} & & \downarrow{\scriptstyle \mathfrak{G}k'_C} & & \downarrow{\scriptstyle k'_C} \\ K & \xrightarrow[\eta_K]{} & \mathfrak{G}A_K = \mathfrak{G}\mathfrak{F}\text{K} & & A_K = \mathfrak{F}\text{K} \end{array}$$

But $\eta : \text{id}_{\mathbb{B}} \to \mathfrak{G}\mathfrak{F}$ is a natural transformation. Therefore, all arrows of the form $\mathfrak{F}k_C : \mathfrak{F}\mathfrak{D}C \to \mathfrak{F}K$ also make the above square commute. As $\eta_{\mathfrak{D}C}$ $\mathfrak{G}$-generates $\mathfrak{F}\mathfrak{D}C$, we have $\mathfrak{F}k_C = k'_C$ for every $C \in \text{Ob}(\mathbb{C})$. Then $((k'_C), A_K) = ((\mathfrak{F}k_C), \mathfrak{F}K)$ is a colimit of functor $\mathfrak{F}\mathfrak{D}$. Hence, functor $\mathfrak{F}$ preserves $\mathbb{C}$-colimits.

- There exists a unique natural transformation $\varepsilon : \mathfrak{F} \circ \mathfrak{G} \to \text{id}_{\mathbb{A}}$ such that

$$\mathfrak{G} \xrightarrow{\eta \star \mathfrak{G}} \mathfrak{G} \circ \mathfrak{F} \circ \mathfrak{G} \xrightarrow{\mathfrak{G} \star \varepsilon} \mathfrak{G} = \mathfrak{G} \xrightarrow{\mathrm{id}_{\mathfrak{G}}} \mathfrak{G}$$

and

$$\mathfrak{F} \xrightarrow{\mathfrak{F} \star \eta} \mathfrak{F} \circ \mathfrak{G} \circ \mathfrak{F} \xrightarrow{\varepsilon \star \mathfrak{F}} \mathfrak{F} = \mathfrak{F} \xrightarrow{\mathrm{id}_{\mathfrak{F}}} \mathfrak{F}.$$

That is to say, for each $A \in \mathrm{Ob}(\mathbb{A})$, we have

$$\mathfrak{G}(\varepsilon_A) \circ \eta_{\mathfrak{G}A} = \mathrm{id}_{\mathfrak{G}A}$$

and for each $B \in \mathrm{Ob}(\mathbb{B})$, we have

$$\varepsilon_{\mathfrak{F}B} \circ \mathfrak{F}(\eta_B) = \mathrm{id}_{\mathfrak{F}B}.$$

Proof Let $(\eta_{\mathfrak{G}A}, \mathfrak{F}\mathfrak{G}A)$ be the $\mathfrak{G}$-universal arrow for the $\mathbb{B}$-object $\mathfrak{G}A$ for each $\mathbb{A}$-object A. By the definition of a universal arrow, there then exists a unique arrow $\varepsilon_A : \mathfrak{F}\mathfrak{G}A \to A$ such that the following triangle commutes:

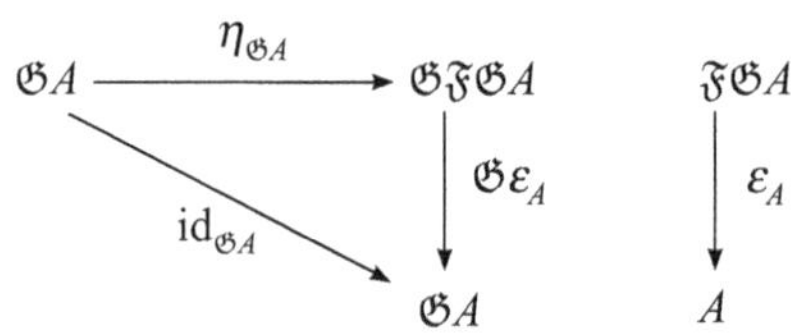

Let $\varepsilon = (\varepsilon_A)$. We show that ε is a natural transformation. To this end, let $f : A \to A'$. Then we have

$$\begin{aligned}\mathfrak{G}(f \circ \varepsilon_A) \circ \eta_{\mathfrak{G}A} &= \mathfrak{G}f \circ \mathfrak{G}\varepsilon_A \circ \eta_{\mathfrak{G}A}\\ &= \mathfrak{G}f \circ \mathrm{id}_{\mathfrak{G}A}\\ &= \mathrm{id}_{\mathfrak{G}A} \circ \mathfrak{G}f\\ &= \mathfrak{G}(\varepsilon_{A'}) \circ \eta_{\mathfrak{G}A'} \circ \mathfrak{G}f.\end{aligned}$$

However, $\eta : \mathrm{id}_{\mathbb{B}} \to \mathfrak{G}\mathfrak{F}$ is a natural transformation. Therefore, we also have

$$\begin{aligned}\mathfrak{G}(f \circ \varepsilon_A) \circ \eta_{\mathfrak{G}A} &= \mathfrak{G}\varepsilon_{A'} \circ \mathfrak{G}\mathfrak{F}\mathfrak{G}f \circ \eta_{\mathfrak{G}A}\\ &= \mathfrak{G}(\varepsilon_{A'} \circ \mathfrak{F}\mathfrak{G}f) \circ \eta_{\mathfrak{G}A}.\end{aligned}$$

As $\eta_{\mathfrak{G}A}$ $\mathfrak{G}$-generates $\mathfrak{F}\mathfrak{G}A$, we have $f \circ \varepsilon_A = \varepsilon_{A'} \circ \mathfrak{F}\mathfrak{G}f$. Thus, we have that $\varepsilon = (\varepsilon_A) : \mathrm{id}_A \to \mathfrak{F}\mathfrak{G}$ is a natural transformation. From the commutativity of the above triangle for every $A \in \mathrm{Ob}(\mathbb{A})$, it then follows that

$$\mathrm{id}_{\mathfrak{G}} = (\mathfrak{G} * \varepsilon) \circ (\eta * \mathfrak{G}).$$

Now, for every $B \in \text{Ob}(\mathbb{B})$, we have

$$\mathfrak{G}(\varepsilon_{\mathfrak{F}B} \circ \mathfrak{F}(\eta_B)) \circ \eta_B = \mathfrak{G}(\varepsilon_{\mathfrak{F}B}) \circ (\mathfrak{G}\mathfrak{F}(\eta_B) \circ \eta_B).$$

However, as η_B is a natural transformation, we have

$$\mathfrak{G}(\varepsilon_{\mathfrak{F}B}) = \eta_{\mathfrak{G}\mathfrak{F}B} \circ \eta_B = \mathfrak{G}(\text{id}_{\mathfrak{F}B}) \circ \eta_B.$$

Since, for each $B \in \text{Ob}(\mathbb{B})$, η_B also $\mathfrak{G}$-generates $\mathfrak{F}B$, we have

$$\varepsilon_{\mathfrak{F}B} \circ \mathfrak{F}(\eta_B) = \text{id}_{\mathfrak{F}B}.$$

Therefore, we also have

$$(\varepsilon * \mathfrak{F}) \circ (\mathfrak{F} * \eta) = \text{id}_{\mathfrak{F}}.$$

Hence, the result.

Proposition 175 Let $\mathfrak{G} : \mathbb{A} \to \mathbb{B}$ be a functor such that for each object $B \in \text{Ob}(\mathbb{B})$, each of (η_B, A_B) and (η'_B, A'_B) is a $\mathfrak{G}$-universal arrow for B. If $\mathfrak{F}$ and $\mathfrak{F}$ are the corresponding functors (whose existence is guaranteed by theorem 5) such that $\mathfrak{F}'B = A'_B$, then $\mathfrak{F}$ is naturally isomorphic to $\mathfrak{F}$.

Proof This proposition follows from the essential uniqueness of the universal arrows.

Proposition 176 Let $\mathfrak{G} : \mathbb{A} \to \mathbb{B}$ be a functor and let $B \in \text{Ob}(\mathbb{B})$.

- If B is an initial object of $\mathbb{B}$ and $u : B \to \mathfrak{G}A$ is the unique $\mathbb{B}$-arrow, then TFAE:
    - (u, A) is a $\mathfrak{G}$-universal arrow for B.
    - A is an initial object of category $\mathbb{A}$.
- If $\mathfrak{B} : 1 \to \mathbb{B}$ is a functor whose value at the single object * is B, then TFAE:
    - (u, A) is a $\mathfrak{G}$-universal arrow for B.
    - $(*, u, A)$ is an initial object of the comma category $(\mathfrak{B}, \mathfrak{G})$.

Functors are arrows connecting two categories. Therefore, the properties of arrows that we considered in Chapter 5 become relevant also for the functors when we consider the category of categories. Furthermore, functors provide us the association of the properties of the arrows of their source category with those of the arrows in their target category. There can be universal associations such as these. Such universal associations are important to the analysis of the structures forming categories. We consider such situations, adjoint situations, next.

Chapter 8

Adjoint Functors and Adjoint Situations

8.1 Introduction

In analyzing relationships between categories, functors play an obvious role. An isomorphism of categories, as an example, is a functor that identifies exactly identical categories. However, this is an unduly strong notion of the similarity of categories. An equivalence of categories, on the other hand, is provided by an equivalence functor that identifies for us the similarity of the properties of the arrows in categories it connects.

Nonetheless, in Chapter 7, we saw the existence of a functor $\mathfrak{F} : \mathbb{B} \to \mathbb{A}$ to correspond to a functor $\mathfrak{G} : \mathbb{B} \to \mathbb{A}$, for which every $\mathbb{B}$-object has a $\mathfrak{G}$-universal arrow (η_B, A_B), in such a way that for each object $A \in \text{Ob}(\mathbb{A})$, we have

$$(\mathfrak{G} * \varepsilon) \circ (\eta * \mathfrak{G}) = \text{id}_{\mathfrak{G}}$$

and for each object $B \in \text{Ob}(\mathbb{B})$, we have

$$(\varepsilon * \mathfrak{F}) \circ (\mathfrak{F} * \eta) = \text{id}_{\mathfrak{F}}.$$

The above can be seen to hold when $\mathfrak{F}$ and $\mathfrak{G}$ are either isofunctors or equivalence functors. However, neither $\mathfrak{G}$ nor $\mathfrak{F}$ is required to be an isofunctor or an equivalence functor for the above to hold. Therefore, an isofunctor and an equivalence are both unduly strong notions to capture the concept of the universality of the above type.

Subtlety in Relativity
Sanjay M. Wagh

ISBN 978-981-4774-44-4 (Hardcover), 978-1-351-23832-8 (eBook)
www.panstanford.com

The situation that is under consideration here is, however, an extremely important part of category theory in that it also abounds in mathematics. Whenever the aforementioned situation occurs for functors $\mathfrak{F}$ and $\mathfrak{G}$, we say that we have an adjunction or the adjoint situation with the categories $\mathbb{A}$ and $\mathbb{B}$. Quite unexpectedly such adjoint situations have been found to exist. With the existence of adjunction between two categories, we are able to analyze one category in relation to the other, for adjunction identifies the similarity of the universality structures of the categories. Therefore, adjunction imparts category theory its analytical powers that we will explore in the present chapter.

8.2 Adjoint and Coadjoint Functors

Definition 144 (adjoint functor) A functor $\mathfrak{G} : \mathbb{A} \to \mathbb{B}$ is called an adjoint functor provided that for every $\mathbb{B}$-object B there exists a $\mathfrak{G}$-universal arrow with source or domain B.

(Coadjoint functor) Dually to the above, a functor $\mathfrak{F} : \mathbb{B} \to \mathbb{A}$ is said to be a coadjoint functor provided that for every $\mathbb{B}$-object B there exists an $\mathfrak{F}$-couniversal arrow with target or codomain B.

Examples The following are immediately noticeable examples of adjoint and coadjoint functors:

- A subcategory $\mathbb{A}$ of a category $\mathbb{B}$ is reflective (co-reflective) in $\mathbb{B}$ if and only if (iff) the associated inclusion functor $\mathbb{A} \rightarrowtail \mathbb{B}$ is an adjoint (coadjoint) functor.
- Every equivalence functor is an adjoint as well as a coadjoint functor.
- Every isofunctor is an adjoint as well as a coadjoint functor.

8.3 Adjoint Situations

Definition 145 (adjunction or adjoint situation) Given categories $\mathbb{A}$ and $\mathbb{B}$, functors $\mathfrak{G} : \mathbb{A} \to \mathbb{B}$ and $\mathfrak{F} : \mathbb{B} \to \mathbb{A}$, as well as two natural transformations $\eta : \mathrm{id}_{\mathbb{B}} \to \mathfrak{G} \circ \mathfrak{F}$ and $\varepsilon : \mathfrak{F} \circ \mathfrak{G} \to \mathrm{id}_{\mathbb{A}}$ such that

$$\mathfrak{G} \xrightarrow{\eta \star \mathfrak{G}} \mathfrak{G} \circ \mathfrak{F} \circ \mathfrak{G} \xrightarrow{\mathfrak{G} \star \varepsilon} \mathfrak{G} = \mathfrak{G} \xrightarrow{\mathrm{id}_{\mathfrak{G}}} \mathfrak{G}$$

and

$$\mathfrak{F} \xrightarrow{\mathfrak{F}\star\eta} \mathfrak{F}\circ\mathfrak{G}\circ\mathfrak{F} \xrightarrow{\varepsilon\star\mathfrak{F}} \mathfrak{F} = \mathfrak{F} \xrightarrow{\mathrm{id}_{\mathfrak{F}}} \mathfrak{F},$$

we say that we have an adjunction or the adjoint situation. We denote the adjunction by any of the following:

$$(\eta, \varepsilon) : \mathfrak{F} \dashv : \mathfrak{G}(\mathbb{A}, \mathbb{B})$$

$$(\eta, \varepsilon) : \mathfrak{F} \dashv \mathfrak{G}$$

$$\mathfrak{F} \dashv \mathfrak{G}$$

We call functor $\mathfrak{F}$ as the left adjoint of functor $\mathfrak{G}$, which is called the right adjoint of functor $\mathfrak{F}$, and we write for it $\mathfrak{G} \vdash \mathfrak{F}$.

Definition 146 We call the natural transformation η in definition 145 of adjunction the unit of the adjunction and the natural transformation ε as the counit of the adjunction.

Definition 147 Moreover, a functor is said to have a left adjoint provided that it is a right adjoint of some functor and is said to have a right adjoint provided that it is a left adjoint of some functor.

Proposition 177 If $(\mathfrak{F}, \mathfrak{G}, \eta, \varepsilon)$ is an equivalence situation, then we have

$$(\eta, \varepsilon) : \mathfrak{F} \dashv \mathfrak{G}$$

and

$$(\varepsilon^{-1}, \eta^{-1}) : \mathfrak{G} \dashv \mathfrak{F}.$$

Furthermore, if $\mathfrak{G} : \mathbb{A} \to \mathbb{B}$ is an equivalence, then there exists functor $\mathfrak{F} : \mathbb{B} \to \mathbb{A}$, which is simultaneously a left adjoint and a right adjoint of $\mathfrak{G}$.

Now, according to theorem 5, Chapter 7, whenever a functor $\mathfrak{G} : \mathbb{A} \to \mathbb{B}$ has the property that every $\mathbb{B}$-object has a $\mathfrak{G}$-universal arrow, we have an adjoint situation. The following theorem now proves that every adjoint situation arises in this manner. That is to say, every adjoint situation actually corresponds to a functor $\mathfrak{G} : \mathbb{A} \to \mathbb{B}$ for which each $\mathbb{B}$-object has a $\mathfrak{G}$-universal arrow.

Theorem 6 Let $\mathfrak{G} : \mathbb{A} \to \mathbb{B}$ be a functor. Then the following hold:

- If each $B \in \mathrm{Ob}(\mathbb{B})$ has a $\mathfrak{G}$-universal arrow (η_B, A_B), then there exists a unique adjoint situation $(\eta, \varepsilon) : \mathfrak{F} \dashv \mathfrak{G}$ such that $\eta = (\eta_B)$ and we have for each $B \in \mathrm{Ob}(\mathbb{B})$, $\mathfrak{F}B = A_B$.

- Conversely, if we are given an adjoint situation $(\eta, \varepsilon) : \mathfrak{F} \dashv \mathfrak{G}$, then we have, for every $B \in \text{Ob}(\mathbb{B})$, $(\eta_B, \mathfrak{F}_B)$ as being $\mathfrak{G}$-universal for B.

Proof The first part of this theorem is seen to be an immediate consequence of theorem 5, Chapter 7.

To prove the second part, let us consider an object $B \in \text{Ob}(\mathbb{B})$ and an arrow $f : B \to \mathfrak{G}A$. We then need to show that there exists a $\mathfrak{G}$-universal arrow for B, that is, the existence of a unique arrow $f' : \mathfrak{F}B \to A$ such that the following triangle commutes:

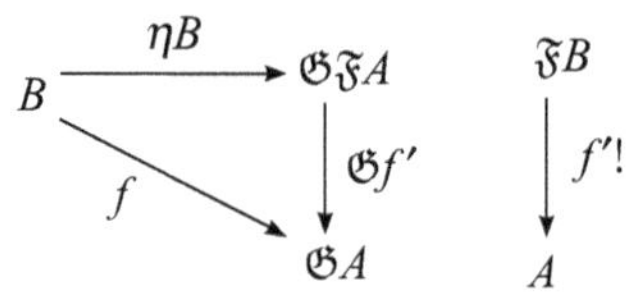

To this end, let $f' = \varepsilon_A \circ \mathfrak{F}f$. Then

$(\mathfrak{G}f') \circ \eta_B = (\mathfrak{G}\varepsilon_A) \circ (\mathfrak{G}\mathfrak{F})f \circ \eta_B.$

However, $\eta : \text{id}_{\mathbb{B}} \to \mathfrak{G}\mathfrak{F}$ is a natural transformation and, therefore, we have

$(\mathfrak{G} * \varepsilon) \circ (\eta * \mathfrak{G}) = \text{id}_{\mathfrak{G}}.$

Therefore, we have

$\mathfrak{G}\,(\varepsilon_A) \circ \eta_B \circ f' = \text{id}_{\mathfrak{G}A} \circ f' = f'.$

Thus, arrow $f' = \varepsilon_A \circ \mathfrak{F}f$ makes the above triangle commute.

To show the uniqueness of the above arrow $f' = \varepsilon_A \circ \mathfrak{F}f$, let $h : \mathfrak{F}B \to A$ be another arrow making the above triangle commute, that is, with $f = (\mathfrak{G}h) \circ \eta_B$. We then show that $h = \varepsilon_A \circ \mathfrak{F}f$. We then note that ε is a natural transformation for which we also have

$(\varepsilon * \mathfrak{F}) \circ (\mathfrak{F} * \eta) = \text{id}_{\mathfrak{F}}.$

As a result, we have the following commutative diagram:

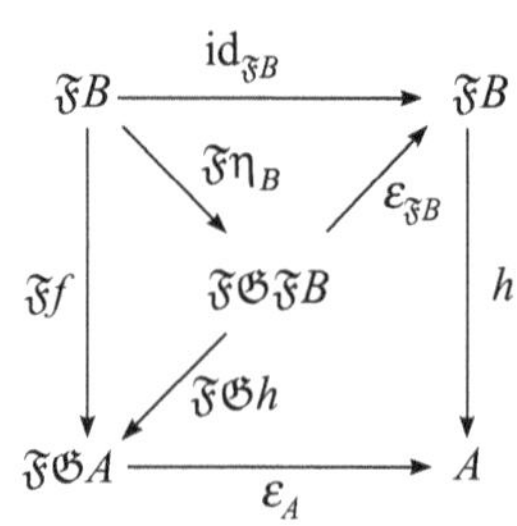

Hence, $h = \varepsilon_A \circ \mathfrak{F}f$. Therefore, $f' = \varepsilon_A \circ \mathfrak{F}f$ is a unique arrow with the required property. Then $(\eta_B, \mathfrak{F}B)$ is $\mathfrak{G}$-universal for $B \in \mathrm{Ob}(\mathbb{B})$.

As a corollary to the above, we have:

Corollary 17 If $\mathfrak{F}$ and $\mathfrak{F}'$ are both left adjoint to functor $\mathfrak{G}$, then $\mathfrak{F}$ is naturally isomorphic to $\mathfrak{F}'$.

We therefore say that adjoint situations are essentially unique.

An example of the adjoint situation is provided by a category $\mathbb{C}$ having $\mathbb{I}$-colimits and the functor category $\mathbb{C}^{\mathbb{I}}$, with the left adjoint being functor $\mathfrak{CoLim} : \mathbb{C}^{\mathbb{I}} \to \mathbb{C}$ and the right adjoint being the "constant functor" functor, or the CF-functor. Similarly, if the CF-functor is treated as a left adjoint, then the right adjoint is functor $\mathfrak{Lim}_{\mathbb{I}} : \mathbb{C} \to \mathbb{C}$ for the corresponding adjoint situation.

We can also establish the following duality:

Proposition 178 If $\mathfrak{G} : \mathbb{A} \to \mathbb{B}$, $\mathfrak{F} : \mathbb{B} \to \mathbb{A}$, $\eta : \mathrm{id}_{\mathbb{B}} \to \mathfrak{G}\mathfrak{F}$, and $\varepsilon : \mathfrak{G}\mathfrak{F} \to \mathrm{id}_{\mathbb{A}}$. Then the following are equivalent (TFAE):

- $(\eta, \varepsilon) : \mathfrak{F} \longrightarrow\!\!| : \mathfrak{G}(\mathbb{A}, \mathbb{B})$.
- $(\varepsilon, \eta) : \mathfrak{G}^{\mathrm{op}} \longrightarrow\!\!| \ \mathfrak{F}^{\mathrm{op}} : (\mathbb{B}^{\mathrm{op}}, \mathbb{A}^{\mathrm{op}})$.

Proof Given an adjoint situation of the form $(\eta, \varepsilon) : \mathfrak{F} \longrightarrow\!\!| \ \mathfrak{G} : (\mathbb{A}, \mathbb{B})$, we first translate it into a statement in terms of $\mathbb{A}^{\mathrm{op}}$ and $\mathbb{B}^{\mathrm{op}}$. We therefore have the functors as $\mathfrak{G}^{\mathrm{op}} : \mathbb{A}^{\mathrm{op}} \to \mathbb{B}^{\mathrm{op}}$ and $\mathfrak{F}^{\mathrm{op}} : \mathbb{B}^{\mathrm{op}} \to \mathbb{A}^{\mathrm{op}}$. As $\eta : \mathrm{id}_{\mathbb{B}} \to \mathfrak{G}\mathfrak{F}$ is an arrow in $[\mathbb{B}, \mathbb{B}]$, it is also an arrow in $[\mathbb{B}, \mathbb{B}]^{\mathrm{op}} \simeq [\mathbb{B}^{\mathrm{op}}, \mathbb{B}^{\mathrm{op}}]$. Therefore, we have $\eta : \mathfrak{G}^{\mathrm{op}} \circ \mathfrak{F}^{\mathrm{op}} \to \mathrm{id}_{\mathbb{B}^{\mathrm{op}}}$ in category $[\mathbb{B}^{\mathrm{op}}, \mathbb{B}^{\mathrm{op}}]$. Similarly, we have $\varepsilon : \mathrm{id}_{\mathbb{A}^{\mathrm{op}}} \to \mathfrak{F}^{\mathrm{op}} \circ \mathfrak{G}^{\mathrm{op}}$.

Now, translating the statement $\mathfrak{G} \xrightarrow{\eta \star \mathfrak{G}} \mathfrak{G}\mathfrak{F}\mathfrak{G} \xrightarrow{\mathfrak{G} \star \varepsilon} \mathfrak{G} = \mathfrak{G} \xrightarrow{\mathrm{id}_{\mathfrak{G}}} \mathfrak{G}$ from $[\mathbb{A}, \mathbb{B}]$ to $[\mathbb{A}^{\mathrm{op}}, \mathbb{B}^{\mathrm{op}}]$ involves reversing the directions of all the involved arrows. Therefore, we have

$$\mathfrak{G}^{\mathrm{op}} \xleftarrow{\eta \star \mathfrak{G}^{\mathrm{op}}} \mathfrak{G}^{\mathrm{op}}\mathfrak{F}^{\mathrm{op}}\mathfrak{G}^{\mathrm{op}} \xleftarrow{\mathfrak{G}^{\mathrm{op}} \star \varepsilon} \mathfrak{G}^{\mathrm{op}} = \mathfrak{G}^{\mathrm{op}} \xleftarrow{\mathrm{id}_{\mathfrak{G}^{\mathrm{op}}}} \mathfrak{G}^{\mathrm{op}}.$$

Similarly, we have

$$\mathfrak{F}^{\mathrm{op}} \xleftarrow{\mathfrak{F}^{\mathrm{op}} \star \eta} \mathfrak{F}^{\mathrm{op}}\mathfrak{G}^{\mathrm{op}}\mathfrak{F}^{\mathrm{op}} \xleftarrow{\varepsilon \star \mathfrak{F}^{\mathrm{op}}} \mathfrak{F}^{\mathrm{op}} = \mathfrak{F}^{\mathrm{op}} \xleftarrow{\mathrm{id}_{\mathfrak{F}^{\mathrm{op}}}} \mathfrak{F}^{\mathrm{op}}.$$

We therefore have the adjoint situation

$$(\varepsilon, \eta) : \mathfrak{G}^{\mathrm{op}} \longrightarrow\!\!| \ \mathfrak{F}^{\mathrm{op}} : (\mathbb{B}^{\mathrm{op}}, \mathbb{A}^{\mathrm{op}}).$$

It is also obvious that applying the same method to the adjoint situation $(\varepsilon, \eta) : \mathfrak{G}^{\mathrm{op}} \longrightarrow\!\!| \ \mathfrak{F}^{\mathrm{op}} : (\mathbb{B}^{\mathrm{op}}, \mathbb{A}^{\mathrm{op}})$ leads to the adjoint situation

$(\eta, \varepsilon) : \mathfrak{F} \dashv \mathfrak{G} : (\mathbb{A}, \mathbb{B})$.

Hence, the result.

Proposition 179 In an adjoint situation $(\eta, \varepsilon) : \mathfrak{F} \dashv \mathfrak{G} : (\mathbb{A}, \mathbb{B})$, functor $\mathfrak{F}$ preserves colimits and functor $\mathfrak{G}$ preserves limits.

Proof By theorem 6, we know that $(\eta_B, \eta B)$ is a $\mathfrak{G}$-universal arrow for each $B \in \text{Ob}(\mathbb{B})$. Then by theorem 5, Chapter 7, we know that $\mathfrak{F}$ preserves colimits.

Now, by proposition 178, we have the adjoint situation

$(\varepsilon, \eta) : \mathfrak{G}^{\text{op}} \dashv \mathfrak{F}^{\text{op}}$.

Then by theorem 5, Chapter 7, we know that $\mathfrak{G}^{\text{op}}$ preserves colimits and, therefore, $\mathfrak{G}$ preserves limits.

Proposition 180 Adjoint situations can be composed. That is to say, if we have the adjoint situations

$(\eta, \varepsilon) : \mathfrak{F} \dashv \mathfrak{G} : (\mathbb{A}, \mathbb{B})$

and

$(\beta, \delta) : \mathfrak{H} \dashv \mathfrak{K} : (\mathbb{B}, \mathbb{C})$

then we also have the adjoint situation

$((\mathfrak{K} * \eta * \mathfrak{H})\beta, \varepsilon \circ (\mathfrak{F} * \delta * \mathfrak{G})) : \mathfrak{F} \circ \mathfrak{H} \dashv \mathfrak{K} \circ \mathfrak{G}: (\mathbb{A}, \mathbb{C})$.

Proof As can be easily verified, this result is an immediate consequence of the application of Godement's rules.

Note 8 If the domain of a functor is a product of two categories, then it is often called a bifunctor. Similarly, for a trifunctor and functor of n-variables.

Proposition 181 Given functors $\mathfrak{G} : \mathbb{A} \to \mathbb{B}$ and $\mathfrak{F} : \mathbb{B} \to \mathbb{A}$, TFAE:

- $\mathfrak{F}$ is a left adjoint of $\mathfrak{G}$. That is to say, there exist natural transformations η and ε such that $(\eta, \varepsilon) : \mathfrak{F} \dashv \mathfrak{G}$.
- The associated set-valued bifunctors $\hom(\mathfrak{F}-, -) : \mathbb{B}^{\text{op}} \times \mathbb{A} \to \mathbb{S}\text{et}$, and $\hom(-, \mathfrak{G}-) : \mathbb{B}^{\text{op}} \times \mathbb{A} \to \mathbb{S}\text{et}$ are naturally isomorphic.

Proof For the first implication, let there be an adjunction $(\eta, \varepsilon) : \mathfrak{F} \dashv \mathfrak{G}$. Let us also define a mapping as $\tau : \hom(\mathfrak{F}-, -) \to \hom(-, \mathfrak{G}-)$ by the assignment $\tau_{BA} = \mathfrak{G}(f) \circ \eta_B$. Now, if $\tau_{BA}(f) = \tau BA(g)$, then we have $\mathfrak{G}(f) \circ \eta_B = \mathfrak{G}(g) \circ \eta_B$, implying $f = g$ by the uniqueness condition for the universal arrow for $(\eta_B, \mathfrak{F}B)$ is $\mathfrak{G}$-universal for $B \in \text{Ob}(\mathbb{B})$. Therefore, τ_{BA} is an injective function.

Now, let $f: B \to \mathfrak{G}A$. Then by the uniqueness in the definition of a universal arrow, there exists an arrow $f' : \mathfrak{F}B \to A$ such that $\tau_{BA}(f') = \mathfrak{G}(f') \circ \eta_B = f$. Therefore, τ_{BA} is bijective as well.

Next, to establish that τ is a natural transformation, we need to show that if $g : B' \to B$ and $f : A \to A'$, then the following square commutes:

$$\begin{array}{ccc}
\hom(\mathfrak{F}B, A) & \xrightarrow{\tau_{BA}} & \hom(B, \mathfrak{G}A) \\
{\scriptstyle \hom(\mathfrak{F}g, f)} \downarrow & & \downarrow {\scriptstyle \hom(g, \mathfrak{G}f)} \\
\hom(\mathfrak{A}B', A') & \xrightarrow[\tau_{B'A'}]{} & \hom(B', \mathfrak{G}A')
\end{array}$$

To this end, let $x \in \hom(\mathfrak{F}B, A)$. Then we have

$$\begin{aligned}
(\tau_{B'A'} \circ \hom(\mathfrak{F}g, f))(x) &= \tau_{B'A'}(f \circ x \circ \mathfrak{F}g) \\
&= \mathfrak{G}(f \circ x \circ \mathfrak{F}g) \circ \eta_{B'} \\
&= \mathfrak{G}(fx) \circ \mathfrak{G}\mathfrak{F}(g) \circ \eta_{B'}
\end{aligned}$$

However, $\eta : \mathrm{id}_{\mathbb{B}} \to \mathfrak{G}\mathfrak{F}$ is a natural transformation. Consequently, we also have

$$\begin{aligned}
\mathfrak{G}(fx) \circ \eta_B \circ g &= \mathfrak{G}f \circ (\mathfrak{G}x \circ \eta_B) \circ g \\
&= \hom(\mathrm{g}, \mathfrak{G}f)(\mathfrak{G}x \circ \eta_B) \\
&= (\hom(\mathrm{g}, \mathfrak{G}f) \circ \tau_{BA})(x)
\end{aligned}$$

Hence, the square commutes.

To prove the second part of the implication, let

$$\lambda : \hom(\mathfrak{F}-, -) \to \hom(-, \mathfrak{G}-)$$

be a natural transformation and let

$$\eta_B = \lambda_{B\mathfrak{F}B}(\mathrm{id}_{\mathfrak{F}B})$$

for each $B \in \mathrm{Ob}(\mathbb{B})$. As a consequence of theorem 6, we only need to show that $(\eta_B\ , \mathfrak{F}B)$ is a $\mathfrak{G}$-universal arrow for every $B \in \mathrm{Ob}(\mathbb{B})$.

8.4 Describing Adjunction

As a consequence of various theorems of this chapter, any adjoint situation can now be described in the following equivalent ways:

Given functors $\mathfrak{G} : \mathbb{A} \to \mathbb{B}$ and $\mathfrak{F} : \mathbb{B} \to \mathbb{A}$, the description that functor $\mathfrak{F}$ is a left adjoint of functor $\mathfrak{G}$ can be achieved by showing that for every object $B \in \mathrm{Ob}(\mathbb{B})$ and for every object $A \in \mathrm{Ob}(\mathbb{A})$:

- A family $(\eta_B, \mathfrak{F}B)$ is that of $\mathfrak{G}$-universal arrows.
- A family $(\mathfrak{G}A, \varepsilon_A)$ is that of $\mathfrak{G}$-couniversal arrows.
- There exist natural transformations

  $\eta = (\eta_{\mathrm{B}}) : \mathrm{id}_{\mathbb{B}} \to \mathfrak{G}\mathfrak{F}$

  and

  $\varepsilon = (\varepsilon_{\mathrm{A}}) : \mathfrak{F}\mathfrak{G} \to \mathrm{id}_{\mathbb{A}}$

  such that

  $$\mathfrak{G} \xrightarrow{\eta\star\mathfrak{G}} \mathfrak{G}\circ\mathfrak{F}\circ\mathfrak{G} \xrightarrow{\mathfrak{G}\star\varepsilon} \mathfrak{G} = \mathfrak{G} \xrightarrow{\mathrm{id}_{\mathfrak{G}}} \mathfrak{G}$$

  and

  $$\mathfrak{F} \xrightarrow{\mathfrak{F}\star\eta} \mathfrak{F}\circ\mathfrak{G}\circ\mathfrak{F} \xrightarrow{\varepsilon\star\mathfrak{F}} \mathfrak{F} = \mathfrak{F} \xrightarrow{\mathrm{id}_{\mathfrak{F}}} \mathfrak{F}\,.$$
- There exists a natural transformation $\lambda = (\lambda_{BA}) : \mathrm{hom}(\mathfrak{F}-, -) \to \mathrm{hom}(-, \mathfrak{G}-)$.

It is usual to describe adjunction by providing the natural transformation λ above, for it is easier and by far the quickest one to state after the definition of a category. Therefore, in the literature on category theory, it is this that is quite often the stated definition of adjunction.

As seen in Chapter 7, the notion of a universal arrow is that of the relation of a source in one category with a source in the other category. This relation, which is determined up to natural isomorphisms of functors, is in such a manner that an arrow in the source of the codomain category of a functor is uniquely related to a commuting triangle involving the arrows from the source of the domain category of that functor. Therefore, it is the first or second of the above ways of describing adjoint situations that is the principal conception of the underlying universality and is, in many cases, also the easier one to establish.

An adjunct situation between two categories shows us how a source (a pair of objects and a family of arrows, with that object as the domain) of the first category is related to a source in the second category. When an adjoint situation occurs, the two involved categories possess an essentially identical structure for mathematical relationships for sources in them. With respect to the nature of sources in them, the two categories are therefore similar. Analyzing the structure of sources of one of these two categories is then sufficient to also provide us the structure of sources in the other category.

Chapter 9

Categorical Measures

9.1 Introduction

Of all the real-valued functions defined on a set, certain functions are special and these special functions are called measures. To begin with, let us recall how measures arise in set theory.

A collection $\mathcal{A}$ of the subsets of a set X, which is closed under countable unions and taking complements, is called a σ-algebra of the subsets of X. An example is the set of all subsets of X, the power set of X.

A real-valued function μ defined on the collection ε of sets is said to be countably additive when for any sequence $\{E_i\}$ of pairwise disjoint sets in ε; whose union is also in ε; we have

$$\mu\left(\biguplus_i E_i\right)=\sum_i \mu(E_i).$$

The standard or usual theory of measures calls an extended real-valued, non-negative, countably additive set-function μ over σ-algebra $\mathcal{A}$ with $\mu(\phi) = 0$ as a real-valued measure, ϕ being the null set. Countable additivity is the basic or defining property of a measure.

Our purpose is to extend this notion of measures to any category. An object-free definition of a category is always possible. So, to

Subtlety in Relativity
Sanjay M. Wagh

ISBN 978-981-4774-44-4 (Hardcover), 978-1-351-23832-8 (eBook)
www.panstanford.com

define measures over any category, an appropriate additive structure should exist over the arrows or morphisms of any arbitrary category.

In category theory, an additivity can then be defined only over the hom-collections. Additivity function + may associate a unique arrow c with any pair of arrows (a, b) from the hom-collection $[A, B]$. This association then needs to satisfy certain properties; in particular, a suitable arrow, a zero arrow 0, needs to exist within every hom-collection as an additive identity: $0 + a \to a$. A category is then said to be additive. A group of addition (of real numbers) can be viewed as an additive category.

However, every category is not additive, for arrows to satisfy conditions of associations of the definition of additivity may not exist, in general. In particular, a zero arrow may not exist.

This difficulty can be overcome by making a suitable construction over a category. The strategy for this construction can be as follows: We form families of objects of a category because the operation of concatenation of families has the additivity. From the families of the objects of any category, an additive category can always be constructed.

For a category $\mathbb{C}$, let us denote the additive category of the families of $\mathbb{C}$-objects as $p\mathbb{F}\text{amily}(\mathbb{C})$, with a zero arrow existing in every of its hom-collections. Since a family of only one object can be formed, every object of category $\mathbb{C}$ is in $p\mathbb{F}\text{amily}(\mathbb{C})$. Therefore, category $\mathbb{C}$ is naturally a subcategory of category $p\mathbb{F}\text{amily}(\mathbb{C})$. As categories are connected by functors, we have an inclusion functor, $\mathfrak{I}_{\mathbb{C}} : \mathbb{C} \rightarrowtail p\mathbb{F}\text{amily}(\mathbb{C})$, corresponding to the inclusion of category $\mathbb{C}$ in category $p\mathbb{F}\text{amily}(\mathbb{C})$ as a subcategory.

Then for any functor $\mathfrak{F} : p\mathbb{F}\text{amily}(\mathbb{C}) \to \mathbb{A}$, where $\mathbb{A}$ is any additive category, we obtain a functor from category $\mathbb{C}$ to that additive category $\mathbb{A}$ by the functor composition $\mathfrak{F} \circ \mathfrak{I}_{\mathbb{C}}\ \mathbb{C} \to \mathbb{A}$. Such a composition of functors can now provide us various additive quantities over that generic category $\mathbb{C}$. This construction then leads us to such measure functors. Measure functors generalize the usual notion of measures to the categorical context.

Before we describe the aforementioned construction in all its details, let us first consider certain preliminaries to develop the background against which it can be more clearly perceived.

9.2 Preliminaries

Let us initiate ideas by recalling the following here: A real-valued function μ defined on a collection e of sets is said to be additive if, whenever $E, F \in \varepsilon$, $E \cup F \in \varepsilon$, and $E \cap F \in \phi$, then

$$\mu(E \uplus F) = \mu(E) + \mu(F),$$

where $\uplus$ denotes a disjoint union of the involved sets. It is finitely additive when ε is any finite collection of pairwise disjoint sets and countably additive when ε is any sequence $\{E_i\}$ of pairwise disjoint sets in ε, whose union is also in collection ε, and $\mu\left(\uplus_i E_i\right) = \sum_i \mu(E_i)$.

Next, a nonempty collection S of subsets of a set X is called a ring of subsets of X if it contains the union and the difference of any two of its members. It is called a *σ-ring* if it also contains the union of any sequence of its members.

A ring is called an algebra $\mathcal{A}$ of the subsets of X if and only if (iff) $X \in \mathcal{A}$. An algebra closed under countable unions is called a σ-algebra. So, for any nonempty set X, a collection $\mathcal{A}$ of subsets of X that is closed under a countable union and taking complements is a σ-algebra of subsets of X. Any σ-algebra is either finite or of cardinality at least of the continuum, that is, **c**.

Now, a measure can be defined as an extended real-valued, non-negative, countably additive set-function μ for σ-algebra $\mathcal{A}$ of the subsets of a set X with $\mu(\phi) = 0$.

We call $(X, \mathcal{A})$ a measurable space. The triple (X, A, μ) is called a measure space, and the sets belonging to $\mathcal{A}$ are called μ-measurable. When every subset of a zero μ-measure set belongs to $\mathcal{A}$, that is, when the sets of a zero μ-measure form a σ-ideal, the measure space is called complete.

We generalize this notion of a measure to that of a signed measure for the measurable space (X, A). We also generalize it by replacing the real line $\mathbb{R}$ with an abstract additive group.

Now, the pair $(X, \mathcal{A})$, where X is a set and $\mathcal{A}$ is a σ-algebra of the subsets of X generated by the topology on X, is called the Borel structure and $\mathcal{A}$ is called as the Borel σ-algebra. Members of $\mathcal{A}$ are called the Borel sets. Borel set theory further provides analytical foundations for this usual measure theory that, as is well known, provides the basis for the theory of integration. Almost all of these aforementioned concepts, as with many other mathematical notions, have their origins in the properties of the real line. This usual measure theory is based on the notion of the complement of a set due to the very definition of a σ-algebra.

Even when the categorical concept of a subobject of an object is analogous to that of the subset of a set, the notion of a subobject is not naturally available within the most general framework of category theory. That is to say, a given object may, in general, not have any subobjects. Moreover, no natural notion of the complement of a set is available within the general settings of category theory. Therefore, attempts to define the notion of a measure using an analogue of the σ-algebra for the subobjects of an object cannot, in general, succeed.

To provide categorical foundations to measure theory, we need to first identify appropriate categorical aspects of the usual, aforementioned, construction of the measures.

In this context, notice that the real line $\mathbb{R}$ can be viewed as a category $\mathbb{R}$, specifically, the additive group of reals as a category. For the countable additivity of real-valued measures, this category is essential.

Now, in the context of sets and functions, measures are the "extended real-valued, countably additive set-functions" in the sense already stated. To extend this sense to an arbitrary category, as is the present goal, we need to extract the general sense of this property of measures. The following such sense arises for measures on some set A.

Consider a set A and its power set $\mathcal{P}(A)$. Let us index the members of the power set $\mathcal{P}(A)$ by an indexing set I. For $i \in I$, we can form the collection $\mathcal{C}(A)$, to be called the extended power set of set A, of pairwise disjoint sets of the form $\{i\} \times A_i \equiv B_i$, where A_i is a subset of A with index i. We can also form the Cartesian product $\mathcal{C}(A) \times \mathcal{C}(A)$ and consider a function $f : \mathcal{C}(A) \to \mathcal{C}(A) \times \mathcal{C}(A)$. These are both free constructions.

Consider next a function $(\mathcal{C}(A) \times \mathcal{C}(A), \mu, \mathbb{R} \times \mathbb{R})$, where any rule μ assigns to each member (B_i, B_j) of $\mathcal{C}(A) \times \mathcal{C}(A)$ a unique element $(r, s) \in \mathbb{R} \times \mathbb{R}$ with $((B_i, B_j), (r, s)) \in \mu$, with μ being a subcollection of

$$\{((B_i, B_j), (r, s)) \mid B_i, B_j \in \mathcal{C}(A) \text{ and } r, s \in \mathbb{R}\}.$$

Furthermore, we will demand that $\mu : (\phi, \phi) \mapsto (0, 0)$. We also extend the subcollection μ if necessary.

Consider now the addition function $+ : \mathbb{R} \times \mathbb{R} \to \mathbb{R}$, with 0 as the additive identity of the reals $\mathbb{R}$. Any two elements r and s are related by the operation of the group additivity as $+(r, s) = r + s = r'$, with r' being some other (unique) element in $\mathbb{R}$.

Now, let

$$(+ \circ \mu)(B_i, \phi) = +(r, 0) = r \text{ and } (+ \circ \mu)(B_k, \phi) = +(s, 0) = s.$$

Then the group additivity, being the property of the function +, is also displayed by the composite function $+ \circ \mu$:

$$(+ \circ \mu)(B_i, B_k) = + \circ (\mu\,(B_i, B_k)) = +(r, s) = r + s = r'$$

For some $p \in I$, we may let $(r', 0) = \mu(B_p, \phi)$. Then $B_p \in \mathcal{C}(A)$ can be taken to be $B_i \uplus B_k$.

Clearly, every extended real-valued countably additive set-function on set A, that is, a measure on set A, satisfies the above construction. On the other hand, every function of the kind $+ \circ \mu$, considered above, is also an extended real-valued countably additive set-function on set A, and such a function is, therefore, a measure on set A.

The above is the essential meaning of the standard real-valued measures definable on set A.

In this construction, the crucial role is that of the addition function +, which has the group property, once the construction of the extended power set $\mathcal{C}(A)$ has been made. It is worth emphasizing that we are *not* considering set $\mathbb{R}$ of reals but the additive group or, more appropriately, only the commutative monoid (of addition of the reals). We then also replace category $\mathbb{R}$ by category $\mathbb{G}$ of the (semi) abelian group for generalization.

Next, we may be able to endow set X with the structure of a category $\mathbb{X}$ by providing appropriate arrows for its members. (The same set X can, of course, be viewed as a category in different possible ways.) Then the set function μ should be replaced by a suitable functor, a measure functor, from category $\mathbb{X}$ to category $\mathbb{G}$.

We notice, furthermore, that the subsets of a given set form a family of sets that can be considered as objects of an appropriate category. Thus, we are naturally led to consider families of objects in category $\mathbb{X}$ for the countable additivity of the special functor μ.

Fundamental to the countable additivity of measures is now the pairwise disjointedness of considered member sets of the σ-algebra and their union—the disjoint union. The concept of a coproduct in the case of the category of sets and functions is that of this disjoint union. We would therefore expect that the free coproduct completion of a given category plays an important role in the general definition of measures in abstract category theory. This will be seen to be the situation.

Now, let us first analyze the concept of additivity that can arise within the categorical context.

9.3 Additivity

To begin with, we define the following (levels of) additive structures using a function + on the hom-collections $[A, B]$ of arrows from any object A to any object B in a category $\mathbb{A}$.

Definition 148 A *notional additivity* structure on any category $\mathbb{A}$ is a function + that associates with each pair $A \underset{g}{\overset{f}{\rightrightarrows}} B$ of $\mathbb{A}$-arrows with a common source or domain A and a common target or codomain B another $\mathbb{A}$-arrow, denoted by $f + g : A \to B$ or by $A \xrightarrow{f+g} B$, such that

(N1) For each pair (A, B) of $\mathbb{A}$-objects, the function + induces on $[A, B]$ the structure of a commutative ring of addition without identity.

(N2) The composition of arrows in $\mathbb{A}$ is left and right distributive over +, that is, whenever $A \xrightarrow{k} B \underset{h}{\overset{g}{\rightrightarrows}} C \xrightarrow{f} D$ are $\mathbb{A}$-arrows, then $f \circ (g + h) = (f \circ g) + (f \circ h)$ and $(g + h) \circ k = (g \circ k) + (h \circ k)$, that is, the composition functions $[B, C] \times [A, B] \to [A, C]$ are bilinear.

Definition 149 A *monoid additivity* structure on any *pointed* category $\mathbb{A}$ is a notional additivity structure on the arrows of $\mathbb{A}$ such that

(M1) The property (N1) of the notional additivity holds with the commutative ring of addition having an identity or it being a commutative monoid.

(M2) The property (N2) of the notional additivity holds for the arrows in $\mathbb{A}$.

(M3) The zero arrows or morphisms of $\mathbb{A}$ act as monoid identities with respect to +, that is, for each $\mathbb{A}$-arrow f, $0 + f = f + 0 = f$. That is to say, the identity elements of the monoid behave as zero arrows or morphisms whenever the compositions are defined.

Definition 150 A *full* or a *group additivity* structure on any *pointed* category $\mathbb{A}$ is a monoid additivity structure such that the commutative monoid of the monoid additivity is an abelian group.

Definition 151 A *notional* (or *monoid* or *full*) *additive category* is defined to be a pair $(\mathbb{A}, +)$ with a (pointed) category $\mathbb{A}$, together with the structure of a commutative ring without identity (or commutative monoid or an abelian group), on each of its hom-collections $[A, B]$.

By a small abuse of the terminology, we will often say that only category $\mathbb{A}$ is itself a notional (or monoid or full) additive category. It follows that a notional additive category need not be a monoid additive or a full additive category; that a monoid additive category is a notional additive category but not necessarily a full additive category and that a full additive category is always both a notional additive and a monoid additive category. Notice also that the category needs to be pointed for it to possess either the monoid or full additivity structures.

Then we have categorically expressed the concept of the additivity of arrows or morphisms of a category.

The existence of additivity in a category has consequences for the properties of its objects and arrows. Any monoid or full additive category has the following properties for its objects and arrows.

Let $\left(A_k\right)_{k=1}^{n} \equiv \{A_1, A_2, \cdots, A_n\}$ be a finite family of objects of either a monoid or full additive category $\mathbb{A}$. Then we have the following result:

Proposition 182 A collection $(\phi_i : A_i \rightarrow A)$; $i = 1, \ldots, n)$ of arrows in any monoid or full additive category $\mathbb{A}$ serves for the construction of the coproduct A of the members of that family. In other words, the arrows of the collection $(\phi_i : A_i \rightarrow A)$; $i = 1, \ldots, n)$ are the corresponding coproduct injections iff there exists another collection $(\mathfrak{p}_i : A \rightarrow A_i)$ such that $\mathfrak{p}_i \phi_j = \delta_{ij}$ and $\sum_{k=1}^{n} \phi_k \mathfrak{p}_k = 1_A$. Here $\delta_{ij} = 0$ for $i \neq j$ and $\delta_{ii} = 1_{A_i}$ and $(\mathfrak{p}_i)_{i=1}^{n}$ are the product projections.

Proof Suppose A is a coproduct of the family, $(A_k)_{k=1}^n$ with coproduct injections $\phi_k : A_k \to A$. Fix an index i, and if there is an arrow $\mathfrak{p}_j : A_i \to A_j$ for all j such that $j \neq i$, then we take $\mathfrak{p}_i = 1_{A_i}$ to obtain an arrow $\mathfrak{p} : A \to A_i$, with $\mathfrak{p}\phi_i = 1_{A_i}$. Then $\mathfrak{p}_i$ is a retraction, and this is always true when $\mathbb{A}$ has a zero object. Then the arrows $\mathfrak{p}_i$ satisfying $\mathfrak{p}_i\phi_j = \delta_{ij}$ exist (and $\mathfrak{p}_i$ are projections). Next, for each i,

$$\left(\sum_{k=1}^{n} \phi_k \mathfrak{p}_k\right)\phi_i = \sum_{k=1}^{n}(\phi_k \mathfrak{p}_k \phi_i) = \sum_{k=1}^{n} \phi_k \delta_{ki} = \phi_i = 1_A \phi_i$$

and, hence, from the definition of a coproduct, we have that $\sum_{k=1}^{n} \phi_k \mathfrak{p}_k = 1_A$.

On the other hand, if there exist arrows $\mathfrak{p}_i$ satisfying conditions $\mathfrak{p}_i\phi_j = \delta_{ij}$ and $\sum_{k=1}^{n} \phi_k \mathfrak{p}_k = 1_A$, then, for $i = 1, \ldots, n$ and for any other object A' in $\mathbb{A}$, let $g_i : A_i \to A'$ and call $\sum_{k=1}^{n} g_k \mathfrak{p}_k = g : A \to A'$. Then we have

$$g\phi_i = \left(\sum_{k=1}^{n} g_k \mathfrak{p}_k\right)\phi_i = \sum_{k=1}^{n} g_k \delta_{ki} = g_i \,.$$

Now, for any other arrow g' such that $g'\phi_i = g_i$ for all i, we have

$$g' = g' 1_A = g'\left(\sum_{k=1}^{n} \phi_k \mathfrak{p}_k\right) = \sum_{k=1}^{n} g' \phi_k \mathfrak{p}_k = \sum_{k=1}^{n} g_k \mathfrak{p}_k = g \,.$$

Therefore, g is a unique such arrow and A is the coproduct of the family $(A_k)_{k=1}^n$ of objects $A_i \in \mathbb{A}\not\triangle$ with the collection of arrows $(\phi_i : A_i \to A)_{i=1}^n$ being the corresponding coproduct injections.

As a consequence of the above proposition, we then have the following corollaries:

Corollary 18 In any monoid or full additive category $\mathbb{A}$, a product (or a coproduct) of a family $(A_i)_I$ of its objects can be completed in a unique manner to a biproduct.

Corollary 19 For any monoid or full additive category $\mathbb{A}$, every (product) coproduct of a finite family $\{A_i\}_{i \in I}$ of its objects is also a biproduct of that family in that category.

Corollary 20 Either in a monoid or in a full additive category $\mathbb{A}$, the following are equivalent (TFAE) statements: (1) $\mathbb{A}$ has finite products, (2) $\mathbb{A}$ has finite coproducts, and (3) $\mathbb{A}$ has finite biproducts.

We also have the following proposition:

Proposition 183 If any pointed category $\mathbb{A}$ has finite biproducts, then there exists on $\mathbb{A}$ a unique monoid additive structure, that is, a function + on the arrows of $\mathbb{A}$, defined by the following:

If $A \underset{g}{\overset{f}{\rightrightarrows}} B$ are any two arrows with a common source A and a common target B, then

$$A \xrightarrow{f+g} B = \begin{cases} A \xrightarrow{\alpha} A\oplus A \xrightarrow{[f,g]} B = A \xrightarrow{<f,g>} B\oplus B \xrightarrow{\beta} B = \\ A \xrightarrow{\alpha} A\oplus A \xrightarrow{f\oplus g} B\oplus B \xrightarrow{\beta} B \end{cases}$$

.

The function + is then equivalently given by any of the compositions $[f, g] \circ \alpha$, $\beta \circ <f, g>$, and $\beta \circ (f \oplus g) \circ \alpha$ occurring in the above proposition. Here $[f, g]$ is the unique arrow from the definition of a coproduct, $<f, g>$ is the unique arrow from the definition of a product, and $f \oplus g$ denotes the unique arrow from the definition of a biproduct, all being identical for the situation under considerations.

Proof Let the biproducts be given by

$(\phi_A, \phi_{A'}, \mathrm{A} \oplus A, \pi_A, \pi_{A'})$ and $(\phi_B, \phi_{B'}, \mathrm{B} \oplus B, \pi_B, \pi_{B'})$.

The first part of the proof follows. If in a pointed category $\mathbb{A}$, having finite biproducts, an additive structure + exists, then the equalities of the aforementioned proposition hold because of the distributivity of the composition of arrows over the addition while using $\phi_A\pi_A + \phi_{A'}\pi_{A'} = 1_A$ and $\phi_B\pi_B + \phi_{B'}\pi_{B'} = 1_B$ to establish $[f, g] \circ \alpha = f + g$, $\beta \circ <f, g> = f + g$, and $\beta \circ (f \oplus g) \circ \alpha = f + g$.

To prove the converse part, let us define two operations as

$f + g = [f, g] \circ \alpha$ and $f * g = \beta \circ <f, g>$.

from the compositions occurring in these equalities. Our aim is to show that zero morphisms act as left and right identities with respect to them both, the distributivity of the composition of arrows in $\mathbb{A}$ holds for them both, associativity holds for them both, and these operations are the same. Having shown the above, we will then have shown the existence of an additive structure for category $\mathbb{A}$.

To this end, since $\mathbb{A}$ has a zero morphism in any collection $[A, B]$ of arrows from object A to object B in $\mathbb{A}$, for $f : A \to B$, we have $[f, 0] \circ \phi_A = f = f \circ \pi_A \circ \phi_A$ and $[f, 0] \circ \phi_B = 0 = f \circ \pi_A \circ \phi_B$. But the definition of coproduct implies that $[f, 0] = f \circ \pi_A$. Similarly, we have $[0, f] = f \circ \pi_B$. Hence, we have $f + 0 = [f, 0] \circ \alpha = f \circ \pi_A \circ \alpha = f = f \circ \pi_B \circ \alpha = [0, f] \circ \alpha = 0 + f$. Thus, $f + 0 = 0 + f$. By similar arguments, we have $f * 0 = 0 * f$.

Thus, the zero morphism 0 acts as left and right identities with respect to both aforementioned operations of * and +.

To prove the distributivity property, let us consider any arrows $p : C \to A$ and $q : B \to D$. Then we have

$$(f * g) \circ p = \alpha \circ \langle f, g\rangle \circ p = \alpha \circ \langle f \circ p, g \circ p\rangle = f \circ p + g \circ p.$$

Similarly, we have

$$q \circ (f + g) = q \circ [f, g] \circ \alpha = [q \circ f, q \circ g] \circ \alpha = q \circ f + q \circ g.$$

Then for the arrows $f, g, h, k : A \to B$, we will have with the * operation

$$\begin{aligned}(f + g) * (h + k) &= \beta \circ \langle f + g, h + k\rangle \\ &= \beta \circ \langle [f, g] \circ \alpha, [h, k] \circ \alpha\rangle \\ &= \beta \circ \langle [f, g], [h, k]\rangle \circ \alpha.\end{aligned}$$

But it also follows that

$$\begin{aligned}\beta \circ \langle [f, g], [h, k]\rangle \circ \alpha &= [\beta \circ \langle f, h\rangle, \beta \circ \langle h, k\rangle] \circ \alpha \\ &= [f * h, g * k] \circ \alpha \\ &= (f * h) + (g * k)\end{aligned}$$

Hence, we have

$$(f + g) * (h + k) = (f * h) + (g * k).$$

Then if g and h are the zero morphism, it immediately follows that

$$f * k = f + k \Rightarrow * = +.$$

Now, if only g is the zero morphism, then we establish the associativity of +. On the other hand, if we let f and k be the zero morphism, we establish the commutativity since we then have $g + h = h + g$. Operation + is then the required additive structure on $\mathbb{A}$.

The uniqueness of the additive structure is then also established. The additive operation induces on every collection $[A, B]$ of arrows in $\mathbb{A}$ the structure of a commutative monoid.

The following proposition follows as a consequence of various of above results:

Proposition 184 Given a pointed category $\mathbb{A}$ having finite products or finite coproducts, TFAE statements:

- There exists a unique monoid additive structure on $\mathbb{A}$.
- $\mathbb{A}$ has finite biproducts.

Now, if category $\mathbb{A}$ is any monoid or full additive category, then so also are the dual category $\mathbb{A}^{op}$ and the functor category $\mathbb{A}^{\mathbb{C}}$ monoid or full additive categories, where $\mathbb{C}$ is any pointed category or a small category. Also, all full subcategories and all product categories of monoid or full additive categories are similarly monoid or full additive categories. Moreover, if R is any ring (with identity) then R can be considered as a monoid additive category with exactly one object. Conversely, any monoid or full additive category with only one object can be considered as a ring with identity.

Every functor between two notional (monoid, full) additive categories $\mathbb{A}$ and $\mathbb{B}$ need not induce a ring (monoid, group) homomorphism of their respective additive structures. Consequently, we will have a subcollection of such functors that induce a ring (monoid, group) homomorphism of their respective additive structures.

Definition 152 If $\mathbb{A}$ and $\mathbb{B}$ are two notional (monoid, full) additive categories, then the functor $\mathfrak{F} : \mathbb{A} \to \mathbb{B}$ is called a notional (monoid, full) additive similarity functor provided that for each pair (A, B) of $\mathbb{A}$-objects, $\mathfrak{F}$ induces a ring (monoid, group) homomorphism from the collection $[A, B]$ of arrows in $\mathbb{A}$ to the collection $[\mathfrak{F}(A), \mathfrak{F}(B)]$ of arrows in $\mathbb{B}$.

The following theorem for monoid and full additive categories then follows.

Theorem 7 If $\mathfrak{F} : \mathbb{A} \to \mathbb{B}$ is a functor, where $\mathbb{A}$ and $\mathbb{B}$ are monoid (full) additive categories, and if $\mathbb{A}$ has finite products, then every monoid (full) additive similarity functor preserves products, coproducts, and biproducts. Under the stated conditions, any functor $\mathfrak{F}$ preserving products, coproducts, or biproducts is a monoid (full) additive similarity functor.

Consider monoid (full) additive categories $\mathbb{A}$ and $\mathbb{B}$ and the functor category $\mathbb{B}^{\mathbb{A}}$ and let the category of all the additive similarity functors from $\mathbb{A}$ to $\mathbb{B}$ be $\mathbb{A}$dd$\mathbb{S}$im$\mathbb{F}$unct[$\mathbb{A}$,$\mathbb{B}$]. In general, $\mathbb{A}$dd$\mathbb{S}$im$\mathbb{F}$unct[$\mathbb{A}$,$\mathbb{B}$] is, by the definition of a monoid (full) additive similarity functor, a full subcategory of $\mathbb{B}^{\mathbb{A}}$. Additive similarity functors indicate the similarity of the additive structures of the involved categories.

Now, to define measures over any category to the fullest extent, the freely obtainable suitable category, apart from it being a full additive category (it is then a pointed category), must also possess the notion of the disjoint union for its arrows so that it can be considered the analogue of the σ-algebra of a set from the standard theory of measures. Thus, our strategy demands that we freely construct from an arbitrary category a suitable category that not only is full additive but also possesses a suitable notion (generalization) of the pairwise disjointedness for its arrows.

A coproduct generalizes to any category the pairwise disjoint union of sets in the context of sets and functions. We could then construct a free coproduct completion category of an arbitrary category and endow it with an appropriate additive structure. Within this category, any object should be a coproduct of some other objects of that category. Within this category, we should be able to form the coproducts of coproducts of its objects. We are therefore naturally led to the considerations of families of objects of an arbitrary category $\mathbb{C}$. Such construction is therefore a free construction, a generalization of pairwise disjointness, for the objects of an arbitrary category.

9.4 Category: $\mathbb{F}$amily($\mathbb{C}$)

To use a coproduct for defining measures in the categorical context, we therefore construct for any arbitrary category $\mathbb{C}$, its free coproduct completion category by the following considerations of families of its objects. Such are then free constructions.

To this end, let us denote by A_I the family $(A_i)_{i\in I}$ of objects in an arbitrary category $\mathbb{C}$, indexed by some collection I. Category $\mathbb{F}$amily($\mathbb{C}$) of the families of objects in $\mathbb{C}$ can then be defined in the following manner:

Definition 153 Category $\mathbb{F}$amily($\mathbb{C}$):

- Objects in category $\mathbb{F}$amily($\mathbb{C}$) are the families A_I, B_J, C_K, . . . of objects in $\mathbb{C}$.

- An arrow from A_I to B_J in category $\mathbb{F}$amily($\mathbb{C}$) then involves a map of indexing sets I and J to associate index $i \in I$ to index $j \in J$ and a family of arrows from members $A_i \in A_I$ to members $B_{j(i)} \in B_J$ for each $i \in I$. An arrow from A_I to B_J is then a pair $(f, \mathfrak{f})$ with $f : I \to J$ as a map of sets and $\mathfrak{f}$ as a family $\mathfrak{f} : \left(A_i \xrightarrow{\mathfrak{f}_i} B_{j(i)} \right)_{i \in I}$ of arrows in $\mathbb{C}$.
- For arrows $(f, \mathfrak{f}) : A_I \to B_J$ and $(g, \mathfrak{g}) : B_J \to C_K$ in category $\mathbb{F}$amily($\mathbb{C}$), the composition $(g, \mathfrak{g}) \circ (f, \mathfrak{f})$ is defined as the arrow $(h, \mathfrak{h}) : A_I \to C_K$ such that $h = gf$ and $\mathfrak{h}_i = \mathfrak{g}_{f(i)}\, \mathfrak{f}_i$.
- An identity arrow for A_I in $\mathbb{F}$amily($\mathbb{C}$) is the identity map $\mathrm{id}_I : I \to I$ and the family, $\{\mathrm{id}_{A_i}\}_{i \in I}$, of identity arrows for $\mathbb{C}$-objects A_i, $i \in I$. It is easy to verify that $\mathbb{F}$amily($\mathbb{C}$) is, as stated above, indeed a category[1] by checking that the identity arrows of $\mathbb{F}$amily($\mathbb{C}$) act also as identities with respect to the above composition that is easily seen to be associative. Note also that a family can be empty. Since we can regard objects in $\mathbb{C}$ as one-member families, $\mathbb{C}$ is a subcategory of $\mathbb{F}$amily($\mathbb{C}$). Subcategory $\mathbb{C}$ is a full subcategory of $\mathbb{F}$amily($\mathbb{C}$) by the above construction.

For $A \in \mathbb{C}$, considered as a one-member family object in $\mathbb{F}$amily($\mathbb{C}$)and B_I in $\mathbb{F}$amily($\mathbb{C}$), an arrow $(f, \mathfrak{f}) : A \to B_I$ gets completely determined by a single element $i \in I$ that is the unique element in the image of f and the unique member of the family $\mathfrak{f}$ that we denote by $\mathfrak{f}_u$, which need not be the identity of A in $\mathbb{C}$, and we write in this case $(f, \mathfrak{f}) \equiv (i, \mathfrak{f}_u)$.

Then for any arrow $(g, \mathfrak{g}) : A_I \to B_J$, we have $(g, \mathfrak{g}) \circ (i, \mathfrak{f}_u) = (g(i), \mathfrak{g}_i \mathfrak{f}_u)$. Moreover, given an arrow $\alpha : X \to A$ in $\mathbb{F}$amily($\mathbb{C}$), where X is a one-member family object, we have $(i, \mathfrak{f}_u) \circ \alpha = (i, \mathfrak{f}_u\, \alpha)$. Therefore, given an object A_I in $\mathbb{F}$amily($\mathbb{C}$) and $i \in I$, we have a special arrow from A_i to A_I, namely, $(i, 1_{A_i})$, that we shall denote by $\mathbb{I}_i$.

Every object A_I in $\mathbb{F}$amily($\mathbb{C}$) is the coproduct of the same collection $(A_i)_{i \in I}$ considered as a family of objects in category $\mathbb{F}$amily($\mathbb{C}$), and the morphisms $\mathbb{I}_I : A_i \to (A_i)_{i \in I}$ are the coproduct injections. Any coproduct of coproducts of families of objects is

[1]In general, it is a quasi category. We will, however, freely use these concepts even when we simply mention here $\mathbb{F}$amily($\mathbb{C}$) only as a category.

the same as the coproduct of all the objects from all the families under consideration. Category $\mathbb{F}$amily($\mathbb{C}$) is then seen to be the free coproduct completion of category $\mathbb{C}$ as it admits arbitrary (small) coproducts. Every object in $\mathbb{F}$amily($\mathbb{C}$) is then a coproduct of some objects in it.

In the case of sets and functions, there exists a monoid additive structure (unions) over the σ-algebra of the sets. We therefore expect an additive structure to exist for category $\mathbb{F}$amily($\mathbb{C}$) as well. But for the monoid additivity, $\mathbb{F}$amily($\mathbb{C}$) needs to be a pointed category.

Notice, however, that category $\mathbb{F}$amily($\mathbb{C}$) has an initial object ϕ that is an empty family of objects of category $\mathbb{C}$ and an arrow from it to any object of $\mathbb{F}$amily($\mathbb{C}$) is the empty function arrow. However, this object is not the terminal object, which is a singleton, that is, a one-member family of empty; and the corresponding arrow from any object of $\mathbb{F}$amily($\mathbb{C}$) to it is an arrow (function) with an empty domain. Therefore, category $\mathbb{F}$amily($\mathbb{C}$) is not a pointed category.

This minor difficulty can now be resolved in the following manner: this resolution involves a rearrangement of the arrows of category $\mathbb{F}$amily($\mathbb{C}$) in a manner that leads to the resultant category being pointed, as required for the monoid additivity.

9.5 Category: *p*$\mathbb{F}$amily($\mathbb{C}$)

To construct a pointed category *p*$\mathbb{F}$amily($\mathbb{C}$) from category $\mathbb{F}$amily($\mathbb{C}$), we shall adopt the following procedure.

To begin with this construct, consider therefore a pair (A_I, A_i) where A_I is an object of $\mathbb{F}$amily($\mathbb{C}$) indexed by I, with $A_I \in A_i$ being a base object of that family A_I. Let (B_j, B_J) be another such pair where B_J is indexed by J, with base $B_j \in B_J$.

Now, consider from the arrows of $\mathbb{F}$amily($\mathbb{C}$) an arrow of the form $(f, \mathfrak{f}) : A_I \to B_J$ such that we also have $(f, \mathfrak{f})A_i = B_j$, with A_i and B_j being taken as one-member families. That is to say, we have $f(i) = j$ and there always exists an arrow $\mathfrak{f}_i : A_i \to B_j$ in collection $\mathfrak{f}$. We say that any such arrow preserves the bases of the families. We then form a collection of all such arrows in $\mathbb{F}$amily($\mathbb{C}$) and denote it by $[(A_I, A_i), (B_J, B_j)]$.

An identity arrow $\left(\mathrm{id}_I, \{\mathrm{id}_{A_i}\}_{i\in I}\right) : (A_I, A_i) \to (A_I, A_i)$ also exists in the aforementioned collection with obviously $\left(\mathrm{id}_I, \{\mathrm{id}_{A_i}\}_{i\in I}\right)A_i = A_i$, and this identity arrow also acts as an identity with respect to the composition of arrows, which is defined using the composition of arrows from the underlying category $\mathbb{F}$amily($\mathbb{C}$) with that composition also being associative. Therefore, $p\mathbb{F}$amily($\mathbb{C}$) is a category with (A_I, A_i) as objects and, as the hom-collection of arrows, the collection of arrows $[(A_I, A_i), (B_J, B_j)] = \{(f, \mathfrak{f}) : A_I \to B_J \text{ and } f(i) = j\}$.

Recall that category $\mathbb{F}$amily($\mathbb{C}$) has an object ϕ that is an empty family of the objects of category $\mathbb{C}$. Consider therefore an object $(\phi_\mathrm{I}, \phi_\mathrm{i})$ of category $p\mathbb{F}$amily($\mathbb{C}$), where ϕ_i denotes the family of empty and ϕ_i denotes that object of $\mathbb{F}$amily($\mathbb{C}$), which is an empty family. Since any arrow $(f, \mathfrak{f} : f(i) = j) : (\phi_\mathrm{I}, \phi_\mathrm{i}) \to (A_J, A_j)$ has $f(i) = j$ and as there always exists an empty morphism from ϕ_i to any A_j, there exists exactly one arrow in the collection $[(\phi_\mathrm{I}, \phi_\mathrm{i}), (A_J, A_j)]$. Consequently, $(\phi_\mathrm{I}, \phi_\mathrm{i})$ is the initial object of category $p\mathbb{F}$amily($\mathbb{C}$).

Similarly, it follows that any arrow $(g, \mathfrak{g} : g(i) = j) : (A_J, A_j) \to (\phi_\mathrm{I}, \phi_\mathrm{i})$ has $g(j) = i$ and as there always exists a morphism from any A_j to ϕ_i, that is an arrow from A_j with an empty domain, there exists exactly one arrow in the collection $[(A_J, A_j), (\phi_\mathrm{I}, \phi_\mathrm{i})]$. Consequently, $(\phi_\mathrm{I}, \phi_\mathrm{i})$ is the terminal object of category $p\mathbb{F}$amily($\mathbb{C}$).

Category $p\mathbb{F}$amily($\mathbb{C}$) therefore has an object $(\phi_\mathrm{I}, \phi_\mathrm{i})$ that is simultaneously an initial object as well as a terminal object. This object is therefore a zero object of category $p\mathbb{F}$amily($\mathbb{C}$). The zero morphisms of $p\mathbb{F}$amily($\mathbb{C}$) are then the arrows with empty domains, and these are its only zero morphisms. Category $p\mathbb{F}$amily($\mathbb{C}$) is therefore a pointed category.

It should be evident that this nature of $p\mathbb{F}$amily($\mathbb{C}$) arises from it being generated using finite families of objects of category $\mathbb{C}$. This construction is therefore analogous to the category of finite sets: just as we construct the category of pointed sets from the category of (finite) sets, category $p\mathbb{F}$amily($\mathbb{C}$) is constructible as a pointed category.

Since by virtue of proposition 184, category $p\mathbb{F}$amily($\mathbb{C}$) has finite coproducts, it also has finite biproducts. By virtue of proposition 183, there then exists a unique monoid additive structure + on the arrows of category $p\mathbb{F}$amily($\mathbb{C}$).

Category $p\mathbb{F}$amily($\mathbb{C}$) has all of its objects as the coproducts of some other objects in it. An empty family of objects of $p\mathbb{F}$amily($\mathbb{C}$) has a biproduct in $p\mathbb{F}$amily($\mathbb{C}$) since $p\mathbb{F}$amily($\mathbb{C}$) has a zero object.

Now, the category of a monoid, that we denote by $\mathbb{M}$, has a single object M, with the monoid identity being the only isomorphism of $\mathbb{M}$. It is then easy to see that $\mathbb{M}$ is a monoid additive category with its only isomorphism also being its zero morphism. Furthermore, category $\mathbb{G}$ of an (abelian) group also has a single object G but all of its arrows are isomorphisms. It is also easy to see that category $\mathbb{G}$ is a full additive category with its identity isomorphism also being its zero morphism. Of course, the single object is also a zero object of the respective category in either of these cases.

Consider now category $\mathbb{M}^{p\mathbb{F}\text{amily}(\mathbb{C})}$. An object of this category is a functor $\mathfrak{F} : p\mathbb{F}\text{amily}(\mathbb{C}) \to \mathbb{M}$, and any of its arrows is a natural transformation of two such functors $\mathfrak{F}$ and $\mathfrak{G}$. A natural transformation τ from $\mathfrak{F}$ to $\mathfrak{G}$ (denoted by $\tau : \mathfrak{F} \to \mathfrak{G}$ or by $\mathfrak{F} \xrightarrow{\tau} \mathfrak{G}$) is a function that assigns to each $p\mathbb{F}\text{amily}(\mathbb{C})$-object (A_I, A_i) an $\mathbb{M}$-morphism $\tau\,(A_I, A_i) : \mathfrak{F}(A_I, A_i) \to \mathfrak{G}(A_I, A_i)$ in such a way that for each $p\mathbb{F}\text{amily}(\mathbb{C})$-morphism $(A_I, A_i) \xrightarrow{(f,\mathfrak{f})} (B_J, B_j)$, the following square commutes:

$$\mathfrak{F}(A_I, A_i)\tau_{(A_I,A_i)}\mathfrak{G}(A_I, A_i)\mathfrak{F}(f,\mathfrak{f})\mathfrak{G}(f,\mathfrak{f})\mathfrak{F}(B_J, B_j)\tau_{(B_J,B_j)}\mathfrak{G}(B_J, B_j)$$

But in this case, $\mathfrak{F}(A_I, A_i)$, $\mathfrak{F}(B_J, B_j)$, $\mathfrak{G}(A_I, A_i)$, and $\mathfrak{G}(B_J, B_j)$ are the same object M of the monoid as a category. This diagram is then a diagram of category $\mathbb{M}$. As a monoid category has only its identity morphism as an isomorphism, no arbitrary arrow in the above diagram can be reversed keeping the resultant diagram commutative. So, it is easy to see that the collection of all the natural transformations from $\mathfrak{F}$ to $\mathfrak{G}$, that is, the collection $[\mathfrak{F}, \mathfrak{G}]$, has the structure of a commutative monoid. Therefore, the functor category $\mathbb{M}^{p\mathbb{F}\text{amily}(\mathbb{C})}$ is a monoid additive category.

Consider next category $\mathbb{G}^{p\mathbb{F}\text{amily}(\mathbb{C})}$. An object of this category is a functor $\mathfrak{F} : p\mathbb{F}\text{amily}(\mathbb{C}) \to \mathbb{G}$, and any of its arrows is a natural transformation of two such functors $\mathfrak{F}$ and $\mathfrak{G}$, a function, that assigns to each $p\mathbb{F}\text{amily}(\mathbb{C})$-object (A_I, A_i) a $\mathbb{G}$-morphism $\tau_{(A_I,A_i)} : \mathfrak{F}(A_I, A_i) \to \mathfrak{G}(A_I, A_i)$ in such a way that for each $p\mathbb{F}\text{amily}(\mathbb{C})$-morphism $(A_J, A_j) \xrightarrow{(f,\mathfrak{f})} (B_J, B_j)$, the aforementioned diagram, or square, commutes. In this case also, $\mathfrak{F}(A_I, A_i)$, $\mathfrak{F}(B_J, B_j)$, $\mathfrak{G}(A_I, A_i)$, and $\mathfrak{G}(B_J, B_j)$ are the same object G of the group as a category. This diagram is then a diagram of category $\mathbb{G}$ in which every morphism is an isomorphism. Consequently, in the above diagram, every arrow occurring in it can be reversed and still the resultant diagram would commute. Therefore, there always exists an inverse of any natural transformation $\tau_{(A_I, A_i)}$ above. It is easy to see that the collection of all natural transformations $[\mathfrak{F}, \mathfrak{G}]$ has the structure of a full abelian group. Therefore, category $\mathbb{G}^{p\mathbb{F}\text{amily}(\mathbb{C})}$ is always a full additive category.

We have therefore established the results that category $\mathbb{M}^{p\mathbb{F}\text{amily}(\mathbb{C})}$ is always a monoid additive category and that category $\mathbb{G}^{p\mathbb{F}\text{amily}(\mathbb{C})}$ is always a full additive category.

9.6 Additivity and Measures

We now encounter various additive structures in the context of measures to be defined within the premise of abstract category theory.

The first additive structure is the additive structure of the category, say $\mathbb{A}$, over which measures are defined (e.g., that of the reals). Another such additive structure is that of the pointed free coproduct completion category $p\mathbb{F}$amily($\mathbb{C}$) of any arbitrary category $\mathbb{C}$ as defined above. As a coproduct is analogous to disjoint union of sets, this additive structure of $p\mathbb{F}$amily($\mathbb{C}$) is analogous to the additive structure, by way of the disjoint unions of sets, of the σ-algebra of sets.

Functors relate any two categories. Every additive similarity functor then associates a collection of arrows (forming a commutative monoid or an abelian group) of one additive category with a collection of arrows (forming a commutative monoid or an abelian group) of another additive category while inducing a homomorphism (from one monoid or group to the other) between them. An additive similarity functor is therefore only an association identifying the similarity of additive structures of the two additive categories. For additive categories $p\mathbb{F}$amily($\mathbb{C}$) and $\mathbb{A}$, additive similarity functors establish therefore, simply, the similarity of their additive structures. However, this association of only the similarity of additive structures does not define measures in the categorical context. We therefore require another additive structure to define measures in the categorical context. Still, the additive structures of

$p\mathbb{F}\text{amily}(\mathbb{C})$ and $\mathbb{A}$ are both necessary for the categorical definition of measures.

Now, the additivity property of measures in the categorical context can be expected to be central to their definition. This is similar to the countable additivity of the usual measures. The next additive structure should be that of the additivity of functors relating the categories $p\mathbb{F}\text{amily}(\mathbb{C})$ and $\mathbb{A}$, each possessing an internal additive structure. This additivity of functors from $p\mathbb{F}\text{amily}(\mathbb{C})$ to $\mathbb{A}$ is the additive structure on the arrows, natural transformations, connecting these functors. As seen earlier, category $\mathbb{A}^{p\mathbb{F}\text{amily}(\mathbb{C})}$ is also an additive category because $p\mathbb{F}\text{amily}(\mathbb{C})$ and $\mathbb{A}$ are both additive categories. The additive structure of $\mathbb{A}^{p\mathbb{F}\text{amily}(\mathbb{C})}$ is precisely that of the natural transformations of functors from $p\mathbb{F}\text{amily}(\mathbb{C})$ to $\mathbb{A}$.

Therefore, $\mathbb{A}$-valued measures on an arbitrary category $\mathbb{C}$ are now defined to be the objects of the functor category $\mathbb{A}^{p\mathbb{F}\text{amily}(\mathbb{C})}$. The additive structure of category $\mathbb{A}^{p\mathbb{F}\text{amily}(\mathbb{C})}$ defines then the countable additivity of $\mathbb{A}$-valued measures on any arbitrary category $\mathbb{C}$.

In particular, the real-valued measures on category $\mathbb{C}$ are then $\mathbb{R}^{p\mathbb{F}\text{amily}(\mathbb{C})}$, where $\mathbb{R}$ is the category of the additive group of the real numbers. Non-negative real-valued measures on category $\mathbb{C}$ arise from a monoid of positive real numbers for category $\mathbb{R}_+{}^{p\mathbb{F}\text{amily}(\mathbb{C})}$. Now, if $\mathbb{C}$ is the category of a set, then we recover standard real-valued measures on a set as $\mathbb{R}^{p\mathbb{F}\text{amily}(\mathbb{C})}$. (Note here the completeness of category $\mathbb{C}$ with respect to objects of measure zero.)

Natural numbers form a commutative monoid under addition. Consequently, category $\mathbb{N}^{p\mathbb{F}\text{amily}(\mathbb{C})}$ is the category of counting measures on category $\mathbb{C}$. Also, $[0, 1]^{p\mathbb{F}\text{amily}(\mathbb{C})}$ is the category of probability measures on category $\mathbb{C}$, with [0, 1] as the category of the unit interval.

Therefore, measures on a category $\mathbb{C}$ are the objects of category $A^{p\mathbb{F}\text{amily}(\mathbb{C})}$, where $\mathbb{A}$ is any additive category. Notably, what we required for the definition of measures in the categorical context is the additivity of category $\mathbb{A}$ over which the measures are defined and the additivity of category $p\mathbb{F}\text{amily}(\mathbb{C})$ on which the measures are defined. The countable additivity of measures trivially followed because category $A^{p\mathbb{F}\text{amily}(\mathbb{C})}$ is then ensured to be an appropriately additive category.

9.7 Summary

In summary, we began with an arbitrary category $\mathbb{C}$ and freely constructed a pointed category $p\mathbb{F}\text{amily}(\mathbb{C})$ of the families of its objects. We showed that $p\mathbb{F}\text{amily}(\mathbb{C})$ is a monoid additive category. Given any monoid (full) additive category $\mathbb{A}$, we then showed that the functor category $A^{p\mathbb{F}\text{amily}(\mathbb{C})}$ is also a monoid (full) additive category. We then argued that measures in the categorical context are the objects of this freely constructible functor category.

On a closing note, the above is therefore the essence of the categorical foundations of measure theory. This categorical definition of measures can also be freed from the set theoretical restrictions, if any, by quasi-categoric considerations, if needed.

Chapter 10

The Meaning of Universal Relativity

Einstein's general principle of relativity contemplates physical laws to be independent of the states of motion of bodies of nature. The universality of this principle means that the laws of physics should be independent also of the mathematical structure we associate with natural bodies. Such physical laws can only arise within the overall framework of category theory.

10.1 Introduction

As seen earlier, in Newton's theory an extensionless point, a singleton, of Euclidean geometry is associated with a body to study its motion as a change in its distance from a reference body, a binary relation over the collection of such points of the Euclidean geometry.

A SU(2) spinor—a singleton of the collection of spinors—may also represent a natural body. Motion of that natural body now means the motion also of this spinor associated with it relative to the spinor associated with the reference physical body.

The general principle of relativity states that the laws of physics are such that they have the same (mathematical) form irrespective of the state of (motion of) the system of reference, any physical body. Then the universality of this principle of relativity in respect of the mathematical structure that we may associate with natural bodies

Subtlety in Relativity
Sanjay M. Wagh

ISBN 978-981-4774-44-4 (Hardcover), 978-1-351-23832-8 (eBook)
www.panstanford.com

requires those mathematical laws, to be the laws of physics, which do not depend on even the mathematical structure that we may associate with them.

Any mathematical structure (e.g., an extensionless point or a spinor) can be an object of a collection of mathematical structures similar to it. If we let collection mean a gathering of *all* the elements having some defining property such that their gathering does not have that defining property, then set theoretical paradoxes do not arise in its considerations. We will always use the word "collection" in this sense.

The object-free structure is fundamental to *all* mathematical structures, as we can always form a collection of mathematical structures similar to any chosen one and consider all changes of one to others of that collection (as a category). Notably, categorical can be the mathematical structure of the collection of all categories then.

Now, when we form a compound body from some physical bodies, values of certain of their properties (inertia, energy, charge, etc.) can be added to obtain values for the properties of the compound body. This needs to hold for representation of physical bodies by any mathematical structure. This suggests an additive function or structure over the generic category, whose objects, of any mathematical structure, represent bodies.

In general, a generic category need not possess an additive structure over its collection of arrows, addition being a very specific function on the collection of categorical arrows.

But a generic category, a collection of arrows, always has a subcollection of identity arrows of its partial binary algebra of arrows. Then we can always form families, including an empty family, of the objects (identity arrows) of any category. Intuitively, when we combine families to form a larger family, we have the conception of addition implicit in this operation that is analogous to forming a compound physical body.

This can also be implemented within the overall premise of category theory by constructing an appropriate derived category of the families of the objects of the generic category.

Functors to additive categories, for example, of a group (monoid) of addition of real numbers from a generic category, via the above derived category, provide then each a unique association of its object with an element of the concerned additive category. Such functors are therefore (additive category-valued) categorical measures, definable for every category.

We can define distance, change in distance, time, inertia, electric charge, etc., as categorical measures over any category. See later.

An endo-functor of a category is its one-to-one (1-1) map onto itself. Association of an object with an element of the additive category due to a measure functor can then be different on composition with such an endo-functor. Barring an identity endo-functor, any other endo-functor changes the associations of categorical measures, as it maps one object to a different another. Changes to all the measures definable over a category arise only in this manner.

Studies of such changes to categorical measures and their interrelationships are the categorical dynamics, which incorporate relationships of all the possible changes to notions like distance.

These categorical relations are exactly what we want to mathematically implement the universality of Einstein's general principle of relativity: functor relations are independent of the mathematical structure that we associate with any physical body and therefore free of any physical system of reference that can be any physical body and in any motion whatsoever.

Then we will have deduced the laws of physics without specifying their relation to the observable universe. Nevertheless, these laws of physics would be the best possible of our efforts at formulating them.

This approach has support in this statement by Einstein:

> Certainly it would be logically more correct to begin with the whole of the laws and to apply the "stipulation of meaning" to this whole first, i.e. to put the unambiguous relation to the world of experience last instead of already fulfilling it in an imperfect form for an artificially isolated part. . . .

Now, before we describe such laws of physics, let us in brief recapitulate the categorical method of arriving at them.

10.2 Categorical Basis

We denote objects by $A, B, \ldots$; arrows by $a, b, \ldots$; categories by $\mathbb{A}$, $\mathbb{C}, \ldots$; and functors by $\mathfrak{C}, \mathfrak{F}, \ldots$ A collection of all the arrows of a category $\mathbb{C}$ will be denoted by $\mathcal{C}(\mathbb{C})$.

Let $[A, B]$ denote a hom-collection of any category $\mathbb{C}$. Then an additivity over category $\mathbb{C}$ is an addition function $+ : [A, B] \times [A, B] \to [A, B]$, for any pair (A, B) of $\mathbb{C}$-objects, such that a zero arrow is additive identity $0 + a = a = a + 0$, and $a \circ (b + c) = a \circ b + a \circ c$, $(p + q) \circ r = p \circ r + q \circ r$ whenever compositions are defined.

Every category is not additive. Nevertheless, the additivity of the families of the objects of a generic category is captured by the following construction of a derived category.

10.2.1 Categorical Measures

A pointed free coproduct completion category, denoted by $p\mathbb{F}\text{amily}(\mathbb{C})$, of any category $\mathbb{C}$ is constructed with its objects as pairs (A_I, A_i), where $A_I = (A_n)_{n\in I}$ is a family of $\mathbb{C}$-objects, $A_i \in A_I$ for $i \in I$ is a fixed object, and I is an indexing collection.

An arrow from (A_I, A_i) to (B_J, B_j) is then a pair $(f, \mathfrak{f})$—with $f : I \to J$ as a map of collections and $\mathfrak{f}$ as a family $\mathfrak{f} = \{\mathfrak{f}_n\} : \left(A_n \xrightarrow{\mathfrak{f}_n} B_{j(n)}\right)_{n\in I}$ of $\mathbb{C}$-arrows—such that we also have $(f, \mathfrak{f})\, A_i = B_j$, with A_i and B_j being considered as one-member families. That is, we have $f(i) = j$, and there always exists an arrow $\mathfrak{f}_i : A_i \to B_j$ in collection $\mathfrak{f}$.

For any arrows $(f, \mathfrak{f}) : (A_I, A_i) \to (B_J, B_j)$ and $(g, \mathfrak{g}) : (B_J, B_j) \to (C_K, C_k)$ in category $p\mathbb{F}\text{amily}(\mathbb{C})$, their composition $(g, \mathfrak{g}) \circ (f, \mathfrak{f})$ in category $p\mathbb{F}\text{amily}(\mathbb{C})$ is defined as arrow $(h, \mathfrak{h}) : (A_I, A_i) \to (C_K, C_k)$ such that $h = gf$ and $\mathfrak{h}_n = \mathfrak{g}_{f(n)}\mathfrak{f}_n$, with $\mathfrak{h}_i = \mathfrak{g}_{f(i)}\, \mathfrak{f}_i : A_i \to C_k$.

An identity arrow $\left(\mathrm{id}_I, \{\mathrm{id}_{A_i}\}_{i\in I}\right) : (A_I, A_i) \to (A_I, A_i)$ of any object (A_I, A_i) of $p\mathbb{F}\text{amily}(\mathbb{C})$ then, obviously, has $\left(\mathrm{id}_I, \{\mathrm{id}_{A_i}\}_{i\in I}\right) A_i = A_i$. Also, every object (A_I, A_i) of category $p\mathbb{F}\text{amily}(\mathbb{C})$ is a coproduct of objects $(\{A_i\}, A_i)_{i\in I}$, with $\P_i = (i, \mathrm{id}_{A_i}) : (\{A_i\}, A_i) \to (A_I, A_i)$ being coproduct injections. We shall say that objects $(\{A_i\}, A_i)_{i\in I}$ are the constituents of an object (A_I, A_i).

Category $p\mathbb{F}\text{amily}(\mathbb{C})$ has a zero object $(\{\emptyset\}, \emptyset)$, where $\emptyset$ is an empty family of $\mathbb{C}$-objects and $\{\emptyset\}$ denotes a singleton (family) of that empty family. The (only) zero arrows of $p\mathbb{F}\text{amily}(\mathbb{C})$ are those with empty domains.

Now, any hom-collection $[(A_I, A_i), (B_J, B_j)]$ of category $p\mathbb{F}\text{amily}(\mathbb{C})$ contains an arrow with an empty source, a zero arrow. This arrow is then an additive identity for the addition function

$$+ : [(A_I, A_i), (B_J, B_j)] \times [(A_I, A_i), (B_J, B_j)] \to [(A_I, A_i), (B_J, B_j)].$$

With $(g, \mathfrak{g})$ and $(h, \mathfrak{h})$ as any arrows from hom-collection $[(A_I, A_i), (A_I, A_i)]$, to establish the existence of an addition function + over this hom-collection, we need to establish the equality

$$(f, \mathfrak{f}) \circ [(g, \mathfrak{g}) + (h, \mathfrak{h})] = (f, \mathfrak{f}) \circ (g, \mathfrak{g}) + (f, \mathfrak{f}) \circ (h, \mathfrak{h})$$
$$= (g, \mathfrak{g}) \circ [(f, \mathfrak{f}) + (h, \mathfrak{h}) \circ (f, \mathfrak{f})$$
$$= (g, \mathfrak{g}) + (h, \mathfrak{h})] \circ (f, \mathfrak{f})$$

for any arrow $(f, \mathfrak{f}) : (A_I, A_i) \rightarrow (A_I, A_i)$.

Then any composition $(f, \mathfrak{f}) \circ (g, \mathfrak{g})$ has $h = fg$ and $\mathfrak{h}_n = \mathfrak{f}_{g(n)}\mathfrak{g}_n$ with a fixed arrow $\mathfrak{h}_i = \mathfrak{f}_{g(i)}\mathfrak{g}_i : A_i \rightarrow A_i$, and any composition $(g, \mathfrak{g}) \circ (f, \mathfrak{f})$ has $h' = gf$ and $\mathfrak{h}'_m = \mathfrak{g}_{f(m)}\mathfrak{f}_m$ with a fixed arrow $\mathfrak{h}'_i = \mathfrak{g}_{f(i)}\mathfrak{f}_i : A_i \rightarrow A_i$.

In general, we do not have $\mathfrak{h}_i = f_i \circ \mathfrak{g}_i$ being the same arrow as $\mathfrak{h}_{i'} = \mathfrak{g}_i \circ \mathfrak{f}_i$.

However, we notice here that any arrows $\mathfrak{g}_p \circ \mathfrak{f}_p$ and $\mathfrak{f}_k \circ \mathfrak{g}_k$ always exist in both collections $\{\mathfrak{h}\}$ and $\{\mathfrak{h}'\}$. Thence, arrows $\mathfrak{h}_i = \mathfrak{f}_i \circ \mathfrak{g}_i$ and $\mathfrak{h}_{i'} = \mathfrak{g}_i \circ \mathfrak{f}_i$ are members of both collections $\{\mathfrak{h}\}$ and $\{\mathfrak{h}'\}$, which are identical as collections. The aforementioned equality thence follows.

For any other hom-collection $[(A_I, A_i), (B_J, B_j)]$, we need to establish now the distributivity of the composition of arrows, whenever it is defined, over the operation of addition:

$$(f, \mathfrak{f}) \circ [(g, \mathfrak{g}) + (h, \mathfrak{h})] = (f, \mathfrak{f}) \circ (g, \mathfrak{g}) + (f, \mathfrak{f}) \circ (h, \mathfrak{h})$$

$$[(g, \mathfrak{g}) + (h, \mathfrak{h})] \circ (k, \mathfrak{k})] = (g, \mathfrak{g}) \circ (k, \mathfrak{k}) + (h, \mathfrak{h}) \circ (k, \mathfrak{k})$$

Here $(g, \mathfrak{g})$ and $(h, \mathfrak{h})$ are any arrows of the hom-collection $[(A_I, A_i), (B_J, B_j)]$, and $(f, \mathfrak{f})$ and $(k, \mathfrak{k})$ are arrows whose compositions with them are defined. These now follow from the law of composition of the involved arrows.

Every hom-collection $[(A_I, A_i), (B_J, B_j)]$ of category $p\mathbb{F}\text{amily}(\mathbb{C})$ has an additive structure, at least, of a commutative monoid[1] and is a locally compact topological space as well, for every monoid of addition (commutative monoid) can be viewed in this manner.

Collection $Cp\mathbb{F}\text{amily}(\mathbb{C}))$ of all the arrows of category $p\mathbb{F}\text{amily}(\mathbb{C})$ has the structure of the product of locally compact topological spaces (or of monoids of addition) and always has the associated product topology. It is useful to remember this topology of $Cp\mathbb{F}\text{amily}(\mathbb{C}))$ when analyzing (functors) functions from it to collection $C(\mathbb{A})$ of another category $\mathbb{A}$.

[1]A 1-1 correspondence of a commutative monoid of any hom-collection of $p\mathbb{F}\text{amily}(\mathbb{C})$ to that of the addition of cardinal numbers is, in some situations, also possible.

An inclusion functor $\mathcal{J}_{\mathbb{C}}v\colon \mathbb{C} \to p\mathbb{F}\text{amily}(\mathbb{C})$ is now a 1-1 association of any $\mathbb{C}$-arrow $A \xrightarrow{g} B$ with an arrow $(\{A\}, A) \xrightarrow{(f, \{g\})} (\{B\}, B)$, having $(f, \{g\})\ A = B$, of category $p\mathbb{F}\text{amily}(\mathbb{C})$. This inclusion is, however, not always a full embedding.

Then for any functor[2] $\mathfrak{M} : p\mathbb{F}\text{amily}(\mathbb{C}) \to \mathbb{A}$, the composition of functors $\mathfrak{M} \circ \mathcal{J}_{\mathbb{C}} : \mathbb{C} \to \mathbb{A}$ associates any $\mathbb{C}$-arrow with a unique $\mathbb{A}$-arrow. For additive category $\mathbb{A}$, quantities additive over families of $\mathbb{C}$-objects thus get represented by functors $\mathfrak{M} \circ \mathcal{J}_{\mathbb{C}} : \mathbb{C} \to \mathbb{A}$, called the $\mathbb{A}$-valued categorical measures over category $\mathbb{C}$. Categorical measures : $\mathfrak{M} \circ \mathcal{J}_{\mathbb{C}} : \mathbb{C} \to \mathbb{A}$ are then partial binary algebra, identity, and compositions preserving *continuous* functions from the collection of arrows $\mathcal{C}(\mathbb{C})$ to the collection of arrows $\mathcal{C}(\mathbb{A})$.

Category $p\mathbb{F}\text{amily}(\mathbb{C})$ is only a via-media to define the additivity of such measures. Being functorial, such measures are independent of the mathematical structure of the $\mathbb{C}$-objects.

All additive conceptions associated with physical bodies, for example, inertia and energy, are then of a similar mathematical nature, are measures, independent of the mathematical representation of bodies by categorical objects.

10.2.2 Additive Categories of Real Numbers

For historical reasons, physical notions are based on the system of real numbers. We will therefore be interested in two additive categories, namely, that of the commutative group and that of the commutative monoid of addition of real numbers. Knowing the way in which these categories arise helps with the broader[3] view.

To see how these categories arise, consider collections $\mathcal{R} = (-\infty, \infty)$ of all real numbers and $\mathcal{R}_+ = (0, \infty)$ of all non-negative real numbers.

Adding a non-negative real number r to collection $\mathcal{R}_+$ then yields a collection $r + (0, \infty) = (r, \infty)$. Then denote the collection of all such

[2]Functors are considered modulo their natural isomorphism.

[3]An additive category is not required to be that of real numbers. Quantities like distance (defined later) need not be real-valued. Physics need not be based on the system of real numbers, then. But we can always associate a real number with a physical quantity and can always base physics on the system of real numbers. There are equivalent ways of doing the same physics, as our present considerations show.

collections (r, ∞) by R_+ and imagine the operation of the addition of a non-negative real number r as an arrow $0 : R_+ \rightarrow R_+$.

There exists an identity arrow $0 : R_+ \rightarrow R_+$, any two arrows $R_+ \xrightarrow{r_1} R_+$ and $R_+ \xrightarrow{r_2} R_+$ can also be composed to obtain $R_+ \xrightarrow{r_2 \circ r_1} R_+ = R_+ \xrightarrow{r_2 + r_1} R_+$, and the composition of any arrows $R_+ \xrightarrow{r_1} R_+ \xrightarrow{r_2} R_+ \xrightarrow{r_3} R_+$ is associative, that is,

$$R_+ \xrightarrow{(r_3 \circ r_2) \circ r_1} R_+ = R_+ \xrightarrow{r_3 \circ (r_2 \circ r_1)} R_+ = R_+ \xrightarrow{r_3 + r_2 + r_1} R_+ .$$

The commutative monoid $(R_+, +)$ of addition of real numbers is an additive category with object R_+. We denote this category by $\mathbb{R}_+^+$.

By extending the operation of addition to negative real numbers and imagining it as an arrow $r : \mathcal{R} \rightarrow \mathcal{R}$, we then obtain a commutative group of addition of real numbers as an additive category with a single object R. We denote this category by $\mathbb{R}^+$.

10.3 Quantifiers of Motion

Separation, thence, scalar distance, of a body is never with respect to itself, but always relative to another body, and is their symmetric binary relation vanishing for any pair of identical bodies. A change in the scalar distance is also a similar symmetric binary relation of objects. Scalars of the quantifiers of motion like speed and acceleration are also similar symmetric binary relations of objects. Motion is conceivable only as relative motion.

We can capture these symmetric binary relations of objects using a functorial measure over a generic category as follows.

10.3.1 Distance and Changes to Distance

Any functor maps an identity arrow in one category to an identity arrow in another category. With $A_I^{SF} = \{A, A, \ldots\}$ as a self-family of any $\mathbb{C}$-object A, consider a functor $p\mathbb{F}\text{amily}(\mathbb{C})\ \mathfrak{D} : p\mathbb{F}\text{amily}(\mathbb{C}) \rightarrow \mathbb{R}_+^+$ making the associations of arrows of $p\mathbb{F}\text{amily}(\mathbb{C})$ to those of $\mathbb{R}_+^+$ as:

- $[(A_I^{SF}, A_i), (B_J^{SF}, B_j)] \mapsto a \in \mathbb{R}_+^+$
- $[(B_I^{SF}, B_i), (A_J^{SF}, A_j)] \mapsto a \in \mathbb{R}_+^+$
- $[(A_I^{SF}, A_i), (A_J^{SF}, A_j)] \mapsto 0 \in \mathbb{R}_+^+$

The composition of functors $\mathfrak{D} \circ \mathfrak{I}_{\mathbb{C}} : \mathbb{C} \to \mathbb{R}_{+}^{+}$, with $\mathcal{J}_{\mathbb{C}}$ being the inclusion functor of category $\mathbb{C}$ to $p\mathbb{F}$amily($\mathbb{C}$), then associates arrows of (not necessarily additive) category $\mathbb{C}$ to those of (additive) category $\mathbb{R}_{+}^{+}$ as:

(i) $A \xrightarrow{h} B \mapsto R_{+} \xrightarrow{a} R_{+}$

(ii) $B \xrightarrow{f} A \mapsto R_{+} \xrightarrow{a} R_{+}$

(iii) $A \xrightarrow{g} A \mapsto R_{+} \xrightarrow{0} R_{+}$

(iv) $A \xrightarrow{h} B \xrightarrow{k} \bar{C} \mapsto R_{+} \xrightarrow{a} R_{+} \xrightarrow{b} R_{+} = R_{+} \xrightarrow{a+b} R_{+}$ for $A \xrightarrow{h} B \mapsto R_{+} \xrightarrow{a} R_{+}$ and $B \xrightarrow{k} C \mapsto R_{+} \xrightarrow{b} R_{+}$

This provides a binary function, $\ell : \mathcal{C}(\mathbb{C}) \times \mathcal{C}(\mathbb{C}) \to \mathcal{R}_{+}$, which is then object-symmetric, $\ell(A, B) = \ell(B, A)$, by associations (i) and (ii) above; vanishing, $\ell(A, A) = 0$, for every pair of identical $\mathbb{C}$-objects by association (iii) above; and for which Pythagorean law or the triangle equality $\ell(A, B) + \ell(B, C) = \ell(A, C)$ holds by association (iv) above. The binary function $\ell(\#, \#)$ is then a distance function over the collection of objects of (generic or arbitrary) category $\mathbb{C}$.

The same associations of functor $\mathfrak{D}$ also define the scalars of speed and acceleration. Clearly, *mutual* distances, speeds, and accelerations of *all* $\mathbb{C}$-objects are provided by these associations of functor $\mathfrak{D}$.

10.4 Physical Space Is 3D, Always!

Now, how is Pythagorean law or triangle inequality of the 3D (Euclidean) space holding in an arbitrary situation?

Consider a binary relation $\mathbf{d}(\#, \#)$ of $\mathbb{C}$-objects such that it assigns a non-negative real number to a pair of objects in such a way that $\mathbf{d}(A, A) = 0$ for all $\mathbb{C}$-objects A and $\mathbf{d}(A, B) > 0$ for all $\mathbb{C}$-objects A and B. Then, given $\mathbf{d}(A, B) = d_1$ and $\mathbf{d}(B, C) = d_2$, we can always assign, as a binary relation $\mathbf{d}$, the value $d_1 + d_2$ to $\mathbf{d}(A, C)$.

Now, when more than two objects are involved, we obtain an additional condition over the associations of the binary relation $\mathbf{d}$, which does not have any knowledge about additional conditions that can be imposed on it. Therefore, if the binary relation $\mathbf{d}$ is additive over an additive category like $\mathbb{R}_{+}^{+}$, then it assigns the value $d_1 + d_2$ to $\mathbf{d}(A, C)$.

Moreover, when a binary relation $\mathbf{d}$ assigns value $\mathbf{d}(A, B) = d_1$, it has no knowledge that in making the assignment $\mathbf{d}(B, C) = d_2$, we will be or will not be using the same ruler to measure the distance $\mathbf{d}(B, C)$, as we used to measure the distance $\mathbf{d}(A, B)$, to assign it a non-negative real value d_1. Thence, units can always be chosen to satisfy the triangle equality, as illustrated below.

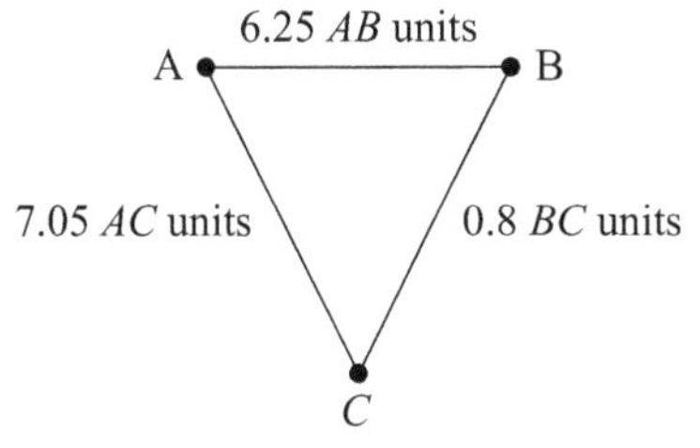

Let non-negative real numbers $\mathbf{d}(A, B) = d_1$, $\mathbf{d}(B, C) = d_2$, and $\mathbf{d}(A, C) = d_3$ be such that $d_1 + d_2 = d_3$. The condition of the sameness of units for distances $\mathbf{d}(A, B)$, $\mathbf{d}(B, C)$, and $\mathbf{d}(A, C)$ now means the action of an appropriate endo-functor $\mathfrak{Scale}_{\mathfrak{D}} : \mathbb{R}_+^+ \to \mathbb{R}_+^+$ because by choosing the same unit for all the distances, we are only reassigning new real numbers to them.

10.4.1 Geometry Is a Free Choice

Now, having the same units for *all* the distances separating $\mathbb{C}$-objects by way of the scale functor $\mathfrak{Scale}_{\mathfrak{D}}$, we can also impose the condition of whether the distance function satisfies the triangle inequality of the Euclidean or that of any non-Euclidean nature by way of the action of another endo-functor $\mathfrak{E} : \mathbb{R}_+^+ \to \mathbb{R}_+^+$. Once again, we are only reassigning, even in this case, new real numbers to involved distances. In particular, we will denote by $\mathfrak{E}u$ the functor leading to Euclidean distances separating $\mathbb{C}$-objects.

Thus, we consider functorial composition $\mathfrak{E} \circ \mathfrak{Scale}_{\mathfrak{D}} \circ \mathfrak{D} \circ \mathfrak{J}_{\mathbb{C}} : \mathbb{C} \to \mathbb{R}_+^+$ for various geometrical considerations. For fixed $\mathfrak{Scale}_{\mathfrak{D}} \circ \mathfrak{D} \circ \mathfrak{J}_{\mathbb{C}}$, the entirety of geometrical structures over $\mathbb{C}$ then rests with all the possible endo-functors $\mathfrak{E} : \mathbb{R}_+^+ \to \mathbb{R}_+^+$ of the monoid of addition of real numbers.

For the same category, we can then obtain[4] Euclidean as well as non-Euclidean geometries using different endo-functors. The geometry of a category is[5] completely our free choice.

In general, using an appropriate scale functor $\mathfrak{Scale}_{\mathfrak{M}}$, we can select the same units for measure $\mathfrak{M}$. Henceforth, we will assume, without explicit mention, the sameness of units for any measure $\mathfrak{M}$, that is, $\mathfrak{M} \equiv \mathfrak{Scale}_{\mathfrak{M}} \circ \mathfrak{M}$.

10.4.2 Vectors and Changes to Vectors

Consider an object antiasymmetric functor $\vec{\mathfrak{D}}: p\mathbb{F}\text{amily}(\mathbb{C}) \to \mathbb{R}^+$ making the following associations:

(v1) $[(A_I^{\text{SF}}, A_i), (B_J^{\text{SF}}, B_j)] \mapsto a \in \mathbb{R}^+$.

(v2) $[(B_I^{\text{SF}}, B_i), (A_J^{\text{SF}}, A_j)] \mapsto -a \in \mathbb{R}^+$.

(v3) $[(A_I^{\text{SF}}, A_i), (A_J^{\text{SF}}, A_j)] \mapsto 0 \in \mathbb{R}^+$.

This differs from functor $\mathfrak{D}$ only with (v2) in place of its association (2).

Then composition $\vec{\mathfrak{D}} \circ \mathfrak{I}_{\mathbb{C}} : \mathbb{C} \to \mathbb{R}^+$ makes these associations:

(va) $A \xrightarrow{h} B \mapsto R \xrightarrow{a} R$.

(vb) $B \xrightarrow{f} A \mapsto R \xrightarrow{-a} R$.

(vc) $A \xrightarrow{g} A \mapsto R \xrightarrow{0} R$.

(vd) $A \xrightarrow{h} B \xrightarrow{k} C \mapsto R \xrightarrow{a} R \xrightarrow{b} R = R \xrightarrow{a+b} R$ with assignments $A \xrightarrow{h} B \mapsto R \xrightarrow{a} R$ and $B \xrightarrow{k} C \mapsto R \xrightarrow{b} R$.

This differs from functor $\mathfrak{D} \circ \mathfrak{I}_{\mathbb{C}}$ only with (vb) in place of its association (b).

We then obtain a function $\vec{\mathfrak{d}} : \mathcal{C}(\mathbb{C}) \times \mathcal{C}(\mathbb{C}) \to \vec{\mathcal{R}}$ such that, for all the objects A, B, $C \in \mathbb{C}$, we have:

- $\vec{\mathfrak{d}}(A, A) = 0$.
- $\vec{\mathfrak{d}}(A, B) = -\vec{\mathfrak{d}}(B, A)$.
- $\vec{\mathfrak{d}}(A, B) + \vec{\mathfrak{d}}(B, C) = \vec{\mathfrak{d}}(A, C)$.

[4]The notion of an extensionless point exists in both the Euclidean and in any non-Euclidean geometries. This is the same as our being able to define different geometries for objects of the same category.

[5]As we represent physical bodies by objects of a category, there does not arise any issue of the geometry of the universe of physical bodies. The geometry of physical space is completely our free choice as well.

This above function $\vec{\mathfrak{d}}: \mathcal{C}(\mathbb{C}) \times \mathcal{C}(\mathbb{C}) \to \vec{\mathcal{R}}$, as can be easily seen, defines vectors of a vector space of linear algebra over the field of real numbers, with $\vec{\mathcal{R}}$ being the vector space of real numbers. This is the point of view of a vector being a directed line segment between two objects.

As noted before, we can choose a scale functor suitable for the measure over the category. With the choice of the scale functor, aforementioned vectors then acquire the corresponding property for their magnitudes.

In particular, we obtain the notion of a distance vector $\vec{d}(A,B)$, from any $\mathbb{C}$-object A to any other $\mathbb{C}$-object B. Because of the antisymmetry of functor $\vec{\mathfrak{D}}$, vector $\vec{d}(B,A)$ from B to A will then be oppositely directed than the vector $\vec{d}(A,B)$. However, both vectors $\vec{d}(A,B)$ and $\vec{d}(B,A)$ have the same amplitude $\ell(A, B)$ for them.

Functorial associations of $\vec{\mathfrak{D}}$ are also those of the vectors of velocity and acceleration. Once again, this is so because all such quantities are conceptually similar as object antisymmetric binary relations over $\mathcal{C}(\mathbb{C})$ having a vanishing self-value for every $\mathbb{C}$-object.

All these quantities are defined independently of the mathematical structure of categorical objects; and, of course, for *all* of the $\mathbb{C}$-objects, we may emphasize these aspects once again.

10.4.3 Motion and Time

Now, with an endo-functor $\mathfrak{F}$: $\mathbb{C} \to \mathbb{C}$, 1-1 and onto on $\mathcal{C}(\mathbb{C})$, associations of measures are, by composition, $\mathfrak{M} \circ \mathfrak{I}_{\mathbb{C}} \circ \mathfrak{F}$.

Then let functor $\mathfrak{D}$ be such that it produces associations for $\mathfrak{D} \circ \mathfrak{I}_{\mathbb{C}}$, and also for $\mathfrak{D} \circ \mathfrak{I}_{\mathbb{C}} \circ \mathrm{id}_{\mathbb{C}}$, where $\mathrm{id}_{\mathbb{C}}$ is an identity endo-functor of category $\mathbb{C}$, resulting in distances as $\ell(A, B) = a$, $\ell(B, C) = b$, and $\ell(C, A) = c$, for $\mathbb{C}$-objects A, B, and C. (Of course, the mutual distances involving all the other $\mathbb{C}$-objects are also specified by functor $\mathfrak{D}$.)

Now, let $\mathfrak{F} : \mathbb{C} \to \mathbb{C}$ be such that $\mathfrak{F}A = B$, $\mathfrak{F}B = C$, and $\mathfrak{F}C = A$, while mapping all the other $\mathbb{C}$-objects to themselves. Then under the action of $\mathfrak{F}$, that is, from composition $\mathfrak{D} \circ \mathfrak{I}_{\mathbb{C}} \circ \mathfrak{F}$, we have $\ell(\mathfrak{F}A, \mathfrak{F}B) = b$, $\ell(\mathfrak{F}B, \mathfrak{F}C) = c$, and $\ell(\mathfrak{F}C, \mathfrak{F}A) = a$. The distance between A and B changes from being a to being b, etc. The distances of A, B, and C relative to all the other objects also change, for example, the distance $\ell(A, D)$ changes for a $\mathbb{C}$-object D. However, any changes to mutual distances of these remaining objects, for example, $\ell(D, E)$ between $\mathbb{C}$-objects D and E, do not result from functor $\mathfrak{F}$ as above.

Under iterates of $\mathfrak{F}$, producing sequences of $\mathbb{C}$-objects, or equivalently as sequences of $\mathbb{C}$-objects of its functorial associations, periodic motions of objects A, B, and C result, evidently, only as relative motions.

Changes to a functorial measure over category $\mathbb{C}$ leading to a binary relation of objects, like the distance relation, then arise due to its endo-functors, with iterates of its identity endo-functor $\mathrm{id}_{\mathbb{C}}$, which produce only the self-families of *all* the $\mathbb{C}$-objects, not causing such changes.

Changes to binary relations, like the distance relation, alone provide us a sense of *time*. This is the only sense of time available to us here, within the present categorical framework.

Iterates of $\mathbb{C}$-endo-functor $\mathfrak{G} : \mathbb{C} \to \mathbb{C}$ produce sequences of $\mathbb{C}$-objects. A universal label for *all* the sequences of $\mathbb{C}$-objects is then the *universal time*. (Notice that this universal time is defined by characteristics that are logically just enough for the definition of time.)

Clearly, we need more labels to assign to more of $\mathbb{C}$-objects of their sequences. Universal time then always increases, larger and larger sequences of $\mathbb{C}$-objects resulting with its passage. This increment defines *an arrow of universal time*. The concordance of this arrow of universal time for *all* changes to *all* measures over any category $\mathbb{C}$ is then evident. Thus, we get the notion of a universal arrow for that of universal time.

Periodicity in one sequence of $\mathbb{C}$-objects may now be used to label $\mathbb{C}$-objects of another sequence, beginning with any $\mathbb{C}$-object of either sequences. Then periodic motions of $\mathbb{C}$-objects can count the passage of universal time, albeit without affecting it, because time is merely a label of $\mathbb{C}$-objects. *Universal time is then absolute but defined using $\mathbb{C}$-objects.*

That this universal uime is absolute here does not conflict with the premise of the principle of relativity, however. Because we can use $\mathbb{C}$-objects to universally label all their sequences, universal time too is definable using a categorical measure $\mathfrak{T} : p\mathbb{F}\text{amily}(\mathbb{C}) \to \mathbb{R}^+$, making the following associations of arrows of category $p\mathbb{F}\text{amily}(\mathbb{C})$ with those of category $\mathbb{R}^+$:

(t1) $[(A_I^{\mathrm{SF}}, A_i), (A_J^{\mathrm{SF}}, A_j)] \mapsto 0 \in \mathbb{R}^+$.

(t2) $[(A_I, A_i), (A_I, A_i)] \mapsto a \in \mathbb{R}^+$.

(t3) whenever $[(A_I, A_i), (A_I, A_i)] \mapsto a \in \mathbb{R}^+$ and $[(B_J, B_j), (B_J, B_j)] \mapsto b \in \mathbb{R}^+$, we always have $a \leq b$ for $I \subseteq J$.

(For collections, the relation $I \subseteq J$ is a well-defined notion.)

By property (t1), functor $\mathfrak{T} \circ \mathfrak{J}_{\mathbb{C}}$ then associates every $\mathbb{C}$-arrow $A \xrightarrow{g} A$ with the arrow 0 of $\mathbb{R}^+$. (Thus, we can consider A's motion beginning from any moment of time, it being a *local* origin of time.) Property (t2) associates real value a as the length of family A_I of $\mathbb{C}$-objects, and property (t3) compares lengths of any two families. All characteristics of universal time thence get represented by this functor.

In the absence of any periodic motions whatsoever, time does not get counted by any changes to $\mathbb{C}$-objects. Nevertheless, distances of $\mathbb{C}$-objects can change as entirely nonperiodic motions and correspond to sequences $\{A_i\}_{i \in I}$ of objects with $A_i \neq A_j$ for $i \neq j$. As considerations are functorial, *all* these motions correspond only to such sequences.

In case there are no changes whatsoever to properties of $\mathbb{C}$-objects due to iterates of $\mathrm{id}_{\mathbb{C}}$, no sense of time exists, that is, it cannot be said as to how long such a state exists, for no $\mathbb{C}$-object counts labels of iterates of identity functor. We call this the absolute zero of time.

The origin of time is when $\mathbb{C}$-objects begin their motions. As the object-symmetric scalar distance $\ell(\#, \#)$ or even the object antisymmetric vectorial distance $\vec{\mathfrak{d}}(\#, \#)$, is functorial defined, the motion of even a single $\mathbb{C}$-object sets the entire universe of them in motion. This is evidently because the motion of a body is a relative (binary) notion.

Clearly, an absolute zero of universal time can be reached more than once, that is, when some motion occurs, then all motions of *all* objects stop and then some motions occur again and again.

Within the present framework, we cannot consider just a single sequence of the objects of a category $\mathbb{C}$ for studying the motion of any object from that sequence without considerations of any other sequences of $\mathbb{C}$-objects.[6] Any sequence of objects of category $\mathbb{C}$ is an integral part of all sequences of a functorial 1-1 map of category $\mathbb{C}$ onto itself.

The property of a $\mathbb{C}$-object is then determined by the totality of $\mathbb{C}$-objects, by way of the functorial associations of measures over category $\mathbb{C}$. Thus, observable bodies, represented by objects of any category whatsoever, are the only ones that are responsible for their behavior, with no spurious references other than themselves for the definitions of physical quantities ascribable to them. This framework is truly Machian, therefore.

10.5 Physical Coordinate System

Now, when any $\mathbb{C}$-object A moves to collide with $\mathbb{C}$-object B, associations of distance functor $\mathfrak{D}$ must change for distance $\ell(A, B)$ to change. This must then happen under the action of some $\mathbb{C}$-endo-functor $\mathfrak{C}$. To merge, however, $\ell(A, B)$ must eventually vanish. Then both A and B must, under the action of functor $\mathfrak{C}$, that is, as $\mathfrak{C}$-associations, become the same $\mathbb{C}$-object, say D, for distance vanishes only as the self-distance of any $\mathbb{C}$-object. It is thus a change to either one of them, or to both the $\mathbb{C}$-objects A and B, as the distance between them becomes smaller and smaller with time.

We may let object A be any physical body and let object B be a coordinate axis. The coordinate axis then has no special status here.

When motion is considered, motion relative to the system of coordinates is usually implied. For Newtonian mechanics, for Einstein's general relativity using curved space-time, and for quantum theory, a coordinate system is implied to be a collection of practically rigid physical bodies. It is further assumed also that such rigid bodies can be positioned to be at rest relative to each other and in common with the imaginary bodies of the assumed underlying (Euclidean or non-Euclidean) geometry.

A rigid body is approximately achievable in nature, but not to any desired accuracy. The concept of a rigid body does not then possess meaning, specifiable strictly without ambiguity. Thus, to first base

[6]Recall that the Newtonian time is entirely independent of physical bodies. Within the Newtonian framework, we could also consider the motion of only a single physical body with complete disregard for the motions of all the other bodies.

the laws of physics on such an ambiguous concept as is that of a rigid body and then to reconstruct that body out of different elementary laws, which in turn are based on the notion of another elemental rigid body, is logically unjustifiable.

Here, we overcome this problem in the following manner. Objects A and B will have values for measures changing with the passage of time, for the (sequence of) endo-functors maps them to different $\mathbb{C}$-objects for their distance to change. Changes to measures then incorporate changes to physical bodies, which are represented by the objects of the category. Functorial relations, although object independent, incorporate these changes.

Now, at any time t, distances of (all) bodies are provided by distance functor $\mathfrak{D}$. The endo-functor of category $\mathbb{C}$ causes changes to these distances. For any fixed object A, these changes may be such as to keep the distances of some of the other bodies unchanging from object A. (If distances of object A relative to all the bodies were unchanging, the endo-functor of category $\mathbb{C}$ would be its identity endo-functor $\mathrm{id}_{\mathbb{C}}$.)

For any object A, and for general motions relative to it, we then only have a certain collection of $\mathbb{C}$-objects that are at rest relative to A and also relative to each other. That is to say, the distance $\ell(\#, \#)$ separating any two of them does not change with the passage of time $\mathfrak{T}$. (We note that such a collection may as well be an empty one.)

We therefore do not have a coordinate system of an observer, coordinate axes (objects at rest) and clock bodies (objects undergoing periodic motion), that fills the entire space represented by the distance measure $\mathfrak{D}$.

Nevertheless, an observer (any $\mathbb{C}$-object) can use some bodies (objects at rest) to be coordinate axes and some other bodies (objects undergoing periodic motion) to be clocks at different locations. Which body serves as a coordinate axis and which as a clock is represented by the corresponding $\mathbb{C}$-endo-functor. We could say that the setting up of the coordinate system by an observer can be described by the $\mathbb{C}$-endo-functor. This description of the coordinate system holds with the underlying geometry being any and the associated universal time being absolute and universal.

10.6 Characteristics of "Individuality" of Objects

Now, let us consider some measures that do not lead to binary relations of the objects of category $\mathbb{C}$ but can possibly have a nonvanishing $\mathbb{A}$-value for any $\mathbb{C}$-object A, that is, for the functorial composition $\mathfrak{M} \circ \mathfrak{J}_{\mathbb{C}}(A)$.

Then when we form a self-family of n copies of, say, object A, that family has an $\mathbb{A}$-value that is n times the $\mathbb{A}$-value for object A. Such measures define additive characteristics intrinsic to $\mathbb{C}$-objects.

Such measures define then the fundamental identifiers or characteristics of the individuality of a $\mathbb{C}$-object.

Consider then category $\mathbb{R}_+^+$ of the commutative monoid of the addition of real numbers and a functor $\mathfrak{m}$ from $p\mathbb{F}$amily($\mathbb{C}$) to $\mathbb{R}_+^+$.

Any functor $\mathfrak{m}$ maps identity arrows $(A_I, A_i) \xrightarrow{\mathrm{id}_{(A_I, A_i)}} (A_I, A_i)$ of category $p\mathbb{F}$amily($\mathbb{C}$) to the (only) identity arrow $R_+ \xrightarrow{0} R_+$ of $\mathbb{R}_+^+$.

Consider now the hom-collection $[(\{A\}, A), (\{A\}, A)]$. Let some arrow $(f, \mathfrak{f})$ from this hom-collection be mapped by functor $\mathfrak{m}$ to arrow r in the (only) hom-collection $[R_+, R_+]$ of category $\mathbb{R}_+^+$. Let another arrow $(g, \mathfrak{g})$ of $[(\{A\}, A), (\{A\}, A)]$ be mapped by functor $\mathfrak{m}$ to arrow r' in $[R_+, R_+]$. Since a functor preserves the composition of arrows, we have:

$$
\begin{array}{ccccc}
(\{A\}, A) & \xrightarrow{(f,\, \mathfrak{f})} & (\{A\}, A) & \xrightarrow{(g,\, \mathfrak{g})} & (\{A\}, A) \\
{\scriptstyle \mathfrak{m}}\big\downarrow & & \big\downarrow & & \big\downarrow \\
R_+ & \xrightarrow[r]{} & R_+ & \xrightarrow[r']{} & R_+
\end{array}
$$

$$
\equiv \quad
\begin{array}{ccc}
(\{A\}, A) & \xrightarrow{(g,\, \mathfrak{g}) \circ (f,\, \mathfrak{f})} & (\{A\}, A) \\
\big\downarrow & & \big\downarrow \\
R_+ & \xrightarrow[r' \circ r]{} & R_+
\end{array}
$$

By the addition functions defined over these involved hom-collections, namely $[(\{A\}, A), (\{A\}, A)]$, and $[R_+, R_+]$, we then have

$$((g, \mathfrak{g}), (f, \mathfrak{f})) \mapsto (g, \mathfrak{g}) + (f, \mathfrak{f}) \equiv (h, \mathfrak{h})$$

$$(r', r) \mapsto r' + r \equiv r''$$

as the corresponding categories are additive.

The zero arrow of any hom-collection $[(A_I, A_i), (B_J, B_j)]$ gets mapped by functor $\mathfrak{m}$ to the zero arrow of the hom-collection $[R_+, R_+]$.

Consider next the action of composition $\mathfrak{m} \circ \mathfrak{I}_{\mathbb{C}} : \mathbb{C} \to \mathbb{R}_+^+$. Then we have the following as a diagram:

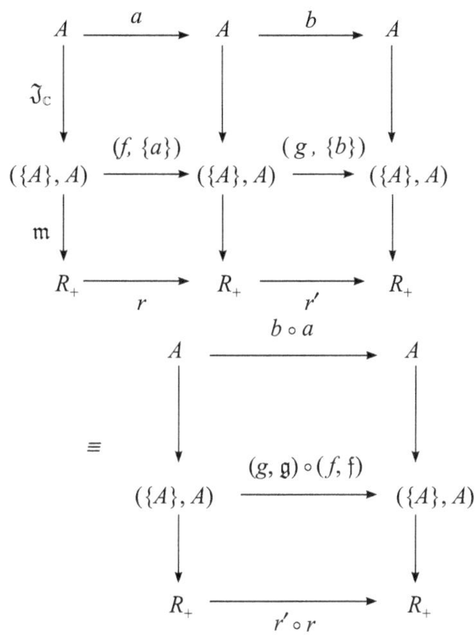

The identity arrow id_A of object A is mapped by any functor $\mathfrak{m}$ to the identity arrow $\mathrm{id}_{R_+} = 0$, and we have other real values that can be assigned to the remaining arrows of the form $A \to A$. Let these be $(0, r_1, r_2, \ldots)$. We call these basic values of measure $\mathfrak{m} \circ \mathfrak{I}_{\mathbb{C}}$.

Notice that we can normalize these basic values of the measure $\mathfrak{m} \circ \mathfrak{I}_{\mathbb{C}}$ so that the largest of them is 1. This can be achieved by using an appropriate scale functor $\mathfrak{S}\mathrm{cale}_{\mathfrak{m}} : \mathbb{R}_+^+ \to \mathbb{R}_+^+$.

Now, when we concatenate N copies of object A, the value of the measure $\mathfrak{m} \circ \mathfrak{I}_{\mathbb{C}}$ increases in integral multiples, by N, of the basic values.

We have therefore illustrated above a real-valued, non-negative measure $\mathfrak{m} \circ \mathfrak{J}_{\mathbb{C}}$ for object A.

We can now substitute the additive category $\mathbb{R}_+^+$ of the above considerations by any another additive category $\mathbb{A}$ to obtain, of course, an $\mathbb{A}$-valued functorial measure for object A of category $\mathbb{C}$.

In particular, we may substitute $\mathbb{R}_+^+$ by the additive category $\mathbb{R}^+$ of the group of addition of real numbers to obtain a measure, say $\bar{\mathfrak{m}} \circ \mathfrak{J}_{\mathbb{C}}$, admitting basic values, positive as well as negative, including zero. These basic values can also be normalized with respect to the modulus of the largest of them to obtain basic values within interval (−1, +1), with 1 always being zero.

Thus, noting that both categories, $p\mathbb{F}$amily($\mathbb{C}$) and $\mathbb{R}_+^+$, are additive categories, consider a measure $\mathfrak{m} : p\mathbb{F}\text{amily}(\mathbb{C}) \to \mathbb{R}_+^+$ that assigns a possibly nonzero element $a \in \mathbb{R}_+^+$ to an object A of $\mathbb{C}$ with $[(\{A\},A),(\{A\},A)] \mapsto a \in \mathbb{R}_+^+$, except for the identity arrow of this hom-collection. With this association of $\mathfrak{m}$, functor $\mathfrak{m} \circ \mathfrak{J}_{\mathbb{C}}$ then assigns an arrow $R_+ \xrightarrow{a} R_+$ in $\mathbb{R}_+^+$ to the hom-collection $[A, A]$ in $\mathbb{C}$, except id_A.

If $[(\{A\},A),(\{A\},A)] \mapsto a \in \mathbb{R}_+^+$, then $[(A_I^{SF}, A_i),(A_I^{SF}, A_i)] \mapsto a\alpha \in \mathbb{R}_+^+$ where α is the cardinality of collection I. If functor $\mathfrak{m}$ makes assignments $[(A_I^{\mathrm{SF}}, A_i),(A_I^{\mathrm{SF}}, A_i)] \mapsto a\alpha \in \mathbb{R}_+^+$ and $[(B_J^{\mathrm{SF}}, B_j),(B_J^{\mathrm{SF}}, B_j)] \mapsto b\beta \in \mathbb{R}_+^+$, then we have[7] $a\alpha \le b\beta$ whenever $a \le b$ and $I \subseteq J$.

This above then defines functor $\mathfrak{m}$ as:

Definition 154 A functor $\mathfrak{m} : p\mathbb{F}\text{amily}(\mathbb{C}) \to \mathbb{R}_+^+$ makes assignments, excepting the identity arrow mapped to 0:

- $[(\{A\},A),(\{A\},A)] \mapsto a \in \mathbb{R}_+^+$,
- If $[(A_I^{\mathrm{SF}}, A_i),(A_I^{\mathrm{SF}}, A_i)] \mapsto a\alpha \in \mathbb{R}_+^+$ and $[(B_J^{\mathrm{SF}}, B_j),(B_J^{\mathrm{SF}}, B_j)] \mapsto b\beta \in \mathbb{R}_+^+$, then $a\alpha \le b\beta$ whenever $a \le b$ and $I \subseteq J$, with α and β being cardinalities of collections I and J, respectively.

Functor $\mathfrak{m} \circ \mathfrak{J}_{\mathbb{C}}$ then has, for its physical relevance, all the properties of the inertia or mass of a physical body. To name one, all

[7]By the well ordering principle and as the operation of $\subseteq$ is meaningful for arbitrary collections, this property of functor $\mathfrak{m}$ is well defined for the cardinalities α of collection I and β of collection J.

the transformations of a physical body represented by object A do not make any difference to its inertia. That is why all the endo-arrows of A, except the identity, are mapped to the same real value a.

As was noted before, the scalars of speed for *all* the $\mathbb{C}$-objects arise from associations similar to those of functor $\mathfrak{D}$. To this end, we need to consider a functor $\mathfrak{V}$ making associations similar to those of functor $\mathfrak{V}$, except that whenever $\mathfrak{V}$ associates an $\mathbb{R}_+^+$-value a with the distance of an object, this functor $\mathfrak{V}$ associates an $\mathbb{R}_+^+$-value $0 \leq v = d\ell/dt \leq 1$ with it, where $d\ell/dt$ is the Lebesgue–Radon–Nikodym derivative from measure theory. (Then we can Taylor expand the total change $\Delta\ell$ in the distance $\ell(\#, \#)$.) By choosing the units for distance and time, we have ensured that the largest speed is unity. We now use only these units.

Functor $\mathfrak{e}_K$ of kinetic energy is then definable with the corresponding associations, except that whenever $\mathfrak{V}$ associates an $\mathbb{R}_+^+$-value v to any object, functor $\mathfrak{e}_K$ associates an $\mathbb{R}_+^+$-value $mv^2/2$ to that object, whose inertia is m as assigned to it by the inertia functor considered earlier.

Next, we have the rest energy functor $\mathfrak{e}_R$, which assigns a value m to an object whenever $\circ\, \mathbb{C}\, \mathfrak{m} \circ \mathfrak{J}_{\mathbb{C}}$ assigns to that object inertia m.

Now, the concept of potential energy has origins in the Newtonian concept of force, which equates the temporal derivative of momentum with a function of coordinates (Newtonian force) to obtain displacement along a trajectory of a physical body from this equation.

Within the present categorical framework, the total change $\Delta\ell$ in the distance is a functor, naturally defined in the sense that any nonidentity endo-functor of category $\mathbb{C}$ produces it. Any Lebesgue–Radon–Nikodym derivative is also naturally definable as a functor. Then derivatives of the distance to all orders are naturally definable functors. Thus, the temporal derivative of momentum is a naturally definable functor.

However, the Newtonian force is not a functor naturally definable in the sense described, for the action of any nonidentity endo-functor of category $\mathbb{C}$ does not indicate any function to serve as the definition of force. Moreover, the total change $\Delta\ell$ is not always obtainable from speed and acceleration appearing in Newton's three postulates of motion. The notion of potential energy is not a naturally definable functor quantity either.

However, we can define an intrinsic energy functor $\mathfrak{e}_I$ that assigns an $\mathbb{R}_+^+$-value E^i to a $\mathbb{C}$-object. Justification for the intrinsic energy arises[8] from the zero-inertia objects, which can possess nonzero energy even when the kinetic energy and the rest energy vanish for such objects.

We now define the total energy functor $\mathfrak{e}_T$ as:

Definition 155 The total energy functor $\mathfrak{e}_T : p\mathbb{F}\text{amily}(\mathbb{C}) \to \mathbb{R}_+^+$ has

$$\mathfrak{e}_T(A) = \mathfrak{e}_K(A) + \mathfrak{e}_R(A) + \mathfrak{e}_I(A) = E^T = \frac{1}{2}mv^2 + m + E^i,$$

where, for brevity, we have denoted it as the sum of functors, when each of its assignments is actually a sum of the (component) assignments.

The total energy $\mathfrak{e}_T \circ \mathfrak{J}_{\mathbb{C}}(A)$ of any $\mathbb{C}$-object A can be decomposed into appropriate components that are functorial values.

Thus, for a zero-inertia object, we have $m = 0$ and $E^T = E^i$. The energy of a zero-inertia object is entirely of intrinsic type. Any object with nonzero inertia then also has an extrinsic component of energy, the kinetic energy, the rest energy being speed independent.

Now, all the transformations of any mathematical structure (representing a physical body) to itself also do not make any difference to its electric charge. Then a measure $\mathfrak{Q}$ corresponding to an electric charge of a physical body is, similarly, defined by assignments:

Definition 156 The (electric) charge measure is a functor $\mathfrak{Q}$: $p\mathbb{F}\text{amily}(\mathbb{C}) \to \mathbb{R}^+$ with assignments (except identity arrows):

- $[(\{A\}, A), \{A\}, A] \mapsto a \in \mathbb{R}^+$.
- If $[(\{A\},A),\{A\},A] \mapsto a \in \mathbb{R}^+$, then $[(A_I^{\mathrm{SF}}, A_i),(A_I^{\mathrm{SF}}, A_i)] \mapsto a\alpha \in \mathbb{R}^+$ with α being the cardinality of collection I.

An electric charge of an object can then be positive, negative, or zero. The usual properties of the additivity of the electric charge clearly also hold for the above (electric) charge measure.

The above are defined as real-valued measures. However, if $\mathbb{M}$ is a category equivalent to category $\mathbb{R}_+^+$ and $\mathbb{G}$ is a category equivalent

[8]Intrinsic energy, in a sense, incorporates the notion of potential energy because potential energy is obtainable, although in a specific way, from a function (force) related to a functor of the temporal rate of the momentum.

to $\mathbb{R}^+$, then the assignments equivalent to those of measures $\mathfrak{m}$ and $\mathfrak{Q}$ are made by some functors $\bar{\mathfrak{m}}: p\mathbb{F}\text{amily}(\mathbb{C}) \to \mathbb{M}$ and $\bar{\mathfrak{Q}}: p\mathbb{F}\text{amily}(\mathbb{C}) \to \mathbb{G}$, respectively. We need not distinguish $\mathfrak{m}$ from $\bar{\mathfrak{m}}$ and so on.

Clearly, every $\mathbb{C}$-object has *all* these characteristics of individuality definable for it. Only the $\mathbb{A}$-values of the characteristics of individuality can now be distinguishing $\mathbb{C}$-objects from each other.

In particular, an $\mathbb{A}$-value for an object of category $\mathbb{C}$ may correspond to the zero arrow of category $\mathbb{A}$ or, equivalently, the additive identity of its additive structure.

Now, if any particular object is mapped by the functorial measure, say $\mathfrak{Q} \circ \mathfrak{J}_\mathbb{C}$, to the additive identity of category $\mathbb{A}$, then concatenating it as a singleton family to any other object of its category does not change the $\mathbb{A}$-value from that of the other object. (An electrically neutral body does not change the electric charge when combined with another body.)

This above is exactly the "measure zero" situation of the standard theory of measures, when we consider $\mathcal{C}(p\mathbb{F}\text{amily}(\mathbb{C}))$ as a topological product of the monoids of addition.

Action of the endo-functor $\mathfrak{C}: \mathbb{C} \to \mathbb{C}$ changes the measure $\mathfrak{M} \circ \mathfrak{J}_\mathbb{C}$ over any category $\mathbb{C}$ by way of composition $\mathfrak{M} \circ \mathfrak{J}_\mathbb{C} \circ \mathfrak{C}$. Such changes are what we call the categorical dynamics.

The dynamics of $\mathbb{C}$-objects have changes to binary (relative) quantities and to characteristics of individuality of $\mathbb{C}$-objects both. Evidently, relative dynamics of $\mathbb{C}$-objects (dealing only with their binary relations) cannot then provide laws about changes to characteristics of individuality of $\mathbb{C}$-objects.

10.7 Categorical Dynamics

We are therefore led to studies of changes, due to $\mathbb{C}$-endo-functors, to $\mathbb{A}$-values given by measures to $\mathbb{C}$-objects.

To visualize this aspect in its generality, consider an endo-functor $\mathfrak{C}: \mathbb{C} \to \mathbb{C}$. Then, in general, given a measure $\mathfrak{M} : p\mathbb{F}\text{amily}(\mathbb{C}) \to \mathbb{A}$ over any category $\mathbb{C}$, the composition of functors $\mathfrak{M} \circ \mathfrak{J}_\mathbb{C}$ associates a $\mathbb{C}$-arrow $A \xrightarrow{g} B$ with an $\mathbb{A}$-arrow

$$\mathfrak{M}\left(\mathfrak{J}_\mathbb{C}(A)\right) \xrightarrow{\mathfrak{M}\left(\mathfrak{J}_\mathbb{C}(g)\right)} \mathfrak{M}\left(\mathfrak{J}_\mathbb{C}(B)\right).$$

Now, on composition with any $\mathbb{C}$-endo-functor $\mathfrak{E}$, that is, $\mathfrak{M} \circ \mathfrak{J}_{\mathbb{C}} \circ \mathfrak{E} : \mathbb{C} \to \mathbb{A}$, this association is with the following $\mathbb{A}$-arrow:

$$\mathfrak{M}\left(\mathfrak{J}_{\mathbb{C}}\left(\mathfrak{E}(A)\right)\right) \xrightarrow{\mathfrak{M}\left(\mathfrak{J}_{\mathbb{C}}\left(\mathfrak{E}(g)\right)\right)} \mathfrak{M}\left(\mathfrak{J}_{\mathbb{C}}\left(\mathfrak{E}(B)\right)\right)$$

In general, this last is a different $\mathbb{A}$-arrow than the earlier one, except when $\mathfrak{E}$ is $\mathrm{id}_{\mathbb{C}}$, for $\mathfrak{E}(\mathrm{A}) \neq A$, $\mathfrak{E}\,(g) \neq g$, etc.

Thus, all $\mathbb{C}$-endo-functors $\mathfrak{E} : \mathbb{C} \to \mathbb{C}$ except $\mathrm{id}_{\mathbb{C}}$ change $\mathbb{A}$-values given to some $\mathbb{C}$-arrows by a categorical measure $\mathfrak{M}$, and such changes are given by composition $\mathfrak{M} \circ \mathfrak{J}_{\mathbb{C}} \circ \mathfrak{E}$.

Given additive categories $\mathbb{A}_1$ and $\mathbb{A}_2$, consider then two measures over the same category $\mathbb{C}$, say $\mathfrak{M}_1 \circ \mathfrak{J}_{\mathbb{C}} : \mathbb{C} \to \mathbb{A}_1$ and $\mathfrak{M}_2 \circ \mathfrak{J}_{\mathbb{C}} : \mathbb{C} \to \mathbb{A}_2$. However, there always exists a functor $\mathfrak{R} : \mathbb{A}1 \to \mathbb{A}_2$ relating the concerned additive categories. We may always make each triangle of the following diagram commute:

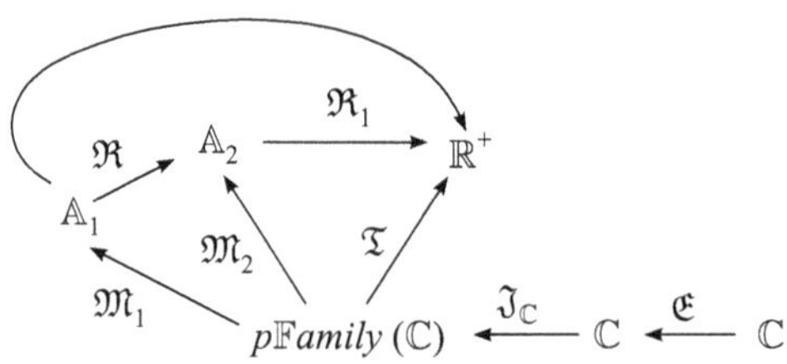

Because we can simply make the associations involved in this triangle commutation and thereby define functor $\mathfrak{R}$, there always exists a functor $\mathfrak{R}$ such that we have $\mathfrak{R}\mathfrak{M}_1\mathfrak{J}_{\mathbb{C}}\mathfrak{E} = \mathfrak{R}\mathfrak{M}_2\mathfrak{J}_{\mathbb{C}}\mathfrak{E}$. (For the convenience of writing, we omit the symbol of composition as well as separating brackets.)

We have also indicated the time measure $\mathfrak{T} : p\mathbb{F}\mathrm{amily}(\mathbb{C}) \to \mathbb{R}^+$. Then the above diagram means that any change in $\mathbb{A}_1$-measure $\mathfrak{M}_1$ may itself be used as (local) time when such changes occur. Notice also that the time measure $\mathfrak{T}$ allows a one-parameter treatment of the dynamics of $\mathbb{C}$-objects.

The universal nature of time is clearly manifest then in the fact that the $\mathbb{A}_1$-measure $\mathfrak{M}_1$ can be any additive physical quantity and the $\mathbb{A}_2$-measure $\mathfrak{M}_2$ can be any another physical quantity. The single parameter t of time functor $\mathfrak{T}$ always labels (continuously) all the sequences of $\mathbb{C}$-objects for them.

We then study changes due to $\mathbb{C}$-endo-functors to $\mathbb{A}$-values given by measures to $\mathbb{C}$-objects. This is categorical dynamics. We also obtain the following equivalent descriptions for categorical dynamics.

As collection $\mathcal{C}(p\mathbb{F}\text{amily}(\mathbb{C}))$ of the arrows of category $p\mathbb{F}\text{amily}(\mathbb{C})$ is the product of locally compact topological spaces (or of monoids of addition) with associated product topology, functors $\mathfrak{M}$: $p\mathbb{F}\text{amily}(\mathbb{C}) \to \mathbb{R}^+$ are the partial binary algebra-preserving, identity-preserving, and compositions-preserving continuous functions from $\mathcal{C}(p\mathbb{F}\text{amily}(\mathbb{C}))$ to $\mathcal{C}(\mathbb{R}^+)$.

Because we can now use the metric topology induced by $\ell(\#, \#)$ for collection $\mathcal{C}(\mathbb{C})$, the defined real-valued categorical measures are then the partial binary algebra-preserving, identity-preserving, and compositions-preserving continuous functions from collection $\mathcal{C}(\mathbb{C})$ to collection $\mathcal{C}(\mathbb{R}_+^+)$, possessing always the decomposition $\mathfrak{M} \circ \mathfrak{J}_{\mathbb{C}}$.

Thus, for changes in measures over category $\mathbb{C}$, we can, alternatively, consider Borel automorphisms of the (Borel) space $\mathcal{C}(p\mathbb{F}\text{amily}\ (\mathbb{C}))$or $\mathcal{C}(\mathbb{C})$.

Using the metric $\ell(\#, \#)$ for $\mathcal{C}(\mathbb{C})$, we can also consider it as a manifold and use its differential structure to consider its diffeomorphisms.

The following four situations for the mathematical analysis of changes to collection $\mathcal{C}(\mathbb{C})$, with T_t: $\mathcal{C}(\mathbb{C}) \to \mathcal{C}\ (\mathbb{C})$ as a 1-1 and onto map labeled by a single parameter t, then arise:

(i) $\mathcal{C}\ (\mathbb{C})$ is a topological space, and T_t is a homeomorphism.
(ii) $\mathcal{C}\ (\mathbb{C})$ is a measure space, and T_t is a measure-preserving transformation.
(iii) $\mathcal{C}\ (\mathbb{C})$ is a differentiable manifold, and T_t is a diffeomorphism.
(iv) $\mathcal{C}\ (\mathbb{C})$ is a category and T_t is an endo-functor.

Case (i) above is the topological dynamics, (ii) is the measurable dynamics, and (iii) is the differentiable dynamics. The fourth (iv) is the new situation that we have called the categorical dynamics.

Of course, the above cases overlap, and we can switch from one to another situation, as per convenience, and can view a situation from any of the above perspectives as per convenience.

Now, for categorical dynamics, consider a sequence $\mathfrak{C}_t$ of compositions of endo-functors of category $\mathbb{C}$ labeled by a parameter $t \in \mathcal{R}$, where $\mathcal{R}$ denotes the real line. From the law of composition of functors, we then have $\mathfrak{C}_{s+t} = \mathfrak{C}_s \circ \mathfrak{C}_t$. Then in tune with the terminology of dynamical systems, we will call $\mathfrak{C}_t$ a flow or a group action of $(\mathcal{R}, +)$ on $\mathcal{C}(\mathbb{C})$, that is, action of the additive group $(\mathcal{R}, +)$ of real numbers on $\mathcal{C}(\mathbb{C})$.

Evidently, categorical dynamics then constitute studies of changes to all the physical conceptions additive over physical bodies, that also independently of their representation by any mathematical structure whatsoever. This is achievable by way of the one-parameter dynamical group.

Within the methods of the aforementioned categorical dynamics, there are no preferred states of motion of $\mathbb{C}$-objects. In other words, any reference $\mathbb{C}$-object is equivalent to any another $\mathbb{C}$-object, even that moving with the speed of light, for formulating the laws of such changes. This aspect is Einstein's general principle of relativity.

Therefore, this mathematical formalism of categorical dynamics is in conformity with the general principle of relativity and is independent also of the mathematical representation of bodies of nature.

Singularity-Free Nature of the Categorical Dynamics

Now, some bodies can "collect" together, with their mutual distances vanishing at some t. The dynamical system $(\mathcal{C}(\mathbb{C}), T_t)$ allows for this possibility, when the corresponding assignments of the distance functor are to the element 0 of $\mathbb{R}_+^+$. Such situations are singular for the usual considerations.

All the distances between *all* the $\mathbb{C}$-objects can also be vanishing. This too is a singular situation for the usual considerations. The distance measure, in this case, maps all the arrows of $p\mathbb{F}$amily$(\mathbb{C})$ to the element 0 of $\mathbb{R}_+^+$ and is still a well-defined concept.

Assignments of time, speed, and acceleration (measures) also remain well defined for this situation above. Not only these, but assignments of all other measures also remain well defined for this situation. Nothing within the present categorical formalism "breaks down" what can be called as singular behavior of the categorical dynamical system.

Nevertheless, when all the distances between all or some of the $\mathbb{C}$-objects vanish, involved objects need to be mapped to the same object by some endo-functor of category $\mathbb{C}$ for this to happen dynamically. This is, however, an impossibility as any endo-functor is a 1-1 map of category $\mathbb{C}$ onto itself and does not map two identity arrows to the same identity arrow. Singular situations cannot therefore arise dynamically.

However, objects can be arbitrarily close to each other. The behavior of the categorical dynamical system $(\mathcal{C}(\mathbb{C}), T_t)$ can also be analyzed arbitrarily close to the (dynamically impossible) situation of zero distance of objects of $\mathbb{C}$. The behavior of the dynamical system $(\mathbb{C})(\mathcal{C}(\mathbb{C}), T_t)$ is entirely nonsingular at all stages. This is the resolution of the singularities, then.

10.8 Inertialess Objects

As seen before, we associate with any (categorical) object the notion of total energy, which has three (functorial definable) components, namely, kinetic energy, rest energy, and intrinsic energy. We also noted that any object having zero inertia can possibly possess only the nonvanishing intrinsic energy E^i, kinetic and rest energy components of its total energy being vanishing. Such inertialess objects get characterized also by the vanishing of their momentum relative to all the other objects, even when possessing nonvanishing total energy. Such momentumless energy objects are of interest to us now.

In the context of such inertialess objects, we need to firstly note that any such object possesses a definite direction of motion, just like any another object. (Clearly, the associations of the distance functor $\mathfrak{D}$ are independent of those of the inertia functor $\mathfrak{m}$.)

Thus, processes involving momentumless objects will all be directional, just exactly to the extent as the processes involving any other objects can also be considered to be directional.

Secondly, when we concatenate an inertialess object to a family of other general objects, the energy of the new family increases by E^i, but not its momentum and not its inertia, from values of that of the original family.

Now, consider the merger of objects A and B, or the absorption of object A by object B, to form a (family) object of category $p\mathbb{F}\text{amily}(\mathbb{C})$, whose characteristics can be attributed to some object C. Similarly, we can consider the process of splitting up of the (family) object of category $p\mathbb{F}\text{amily}(\mathbb{C})$, whose characteristics can be attributed to some object C, into objects A and B or the transmutation of object C into objects A and B.

(Because the objects themselves can be considered as categories, and we are requiring the existence of some object C possessing the characteristics of individuality of any family object, we will be dealing here with the category of all categories.)

In these merging and splitting processes, characteristics of individuality of object C are exactly those obtained by the addition of the characteristics of objects A and B. We can thus think of object C as being composed of objects A and B, that is, as a compound object.

Relevant here is not just a particular family of objects constituting object C but all permissible families of objects constituting it. A compound object may, in general, transmute into component objects, permissible depending on the values of the characteristics of individuality, which ought to be obtainable as sums of those of the constituent objects.

With these general observations, let us consider the processes of merging and splitting involving inertialess or momentumless objects.

In this context, consider an object C of inertia m and let it possess the total energy $E^T(C) = \frac{1}{2}mv_b^2 + m + E^i(C)$ relative to a reference object O.

Let object C split up into two objects, A and B, with A being a momentumless object with $E^T(A) = E_\gamma$. Because of the additivity of energy as a measure, we must then have $E^T(C) = E^T(A) + E^T(B) \equiv E_\gamma + E^T(B)$.

The question now arises as to what the values are of the components of energy $E^T(B)$ of object B.

From the additivity of rest energy as a measure, the rest energy of object B is the same as that of object C. From the additivity of internal energy as a (separate) measure, object B should now possess the internal energy $E^i(\mathrm{B}) = E^i(C) - E^i(A) = E^i(C) - E_\gamma$.

However, the dynamics of object B allows for three possibilities here. When $E^i(B) = E^i(C) - E_\gamma$, the kinetic energy of object B is the same as that of object C. On the other hand, when $E^i(B) = E^i(C)$, the kinetic energy of B is different than that of C. The third possibility is that of the kinetic energy of B being different than that of C and the internal energy of B also being different than that of C. Clearly, these possibilities also crop up when we consider the merger of objects A and B to form object C.

In the case that the intrinsic energy of object B is the same as that of object C, we have $E^T(B) = \frac{1}{2}mv_a^2 + m + E^i$, relative to object O. In other words, we have that $E^T(C) - E^T(B) = E_\gamma = \frac{1}{2}m(v_b^2 + v_a^2)$ is entirely the change in the kinetic energy relative to reference object O.

This change in kinetic energy is now the change in the amplitude of the momentum $P(C) - P(B) \equiv m(v_b - v_a) = 2E_\gamma/(v_b + v_a)$.

But from the additivity of the momentum as a measure, we clearly have $P(C) = P(A) + P(B) \equiv P(B)$, for object A is momentumless.

It therefore follows that the principle of conservation of momentum does not hold for every process of merger and splitting up involving momentumless bodies. (The law of momentum is also empty of contents for inertialess bodies within the Newtonian framework.)

In the same context, consider now that object C is composed of three objects A, B, and D, with object A being inertialess.

Then by additivity of the mass measure, we have $m_C = m_B + m_D$; by that of the internal energy measure, we have $E_C^i = E_\gamma + E_B^2 + E_D^i$; and by that of the kinetic energy measure, we have $\frac{1}{2}m_C v_C^2 = \frac{1}{2}m_B v_B^2 + \frac{1}{2}m_D v_D^2$.

And by additivity of the measure of the momentum, we then also have $m_C\vec{v}_C = m_B\vec{v}_B + m_D\vec{v}_D$. For this case under consideration, the law of momentum would then hold in situations.

We may now force the applicability of the law of momentum onto all the processes by ascribing momentum $2E_\gamma/(v_b + v_a)$ to object A. In this case, as the momentum of object A cannot depend on the velocities v_a and v_b, we may set the quantity $2/(v_a + v_b)$ to be unity, the speed of light in vacuum, and get the momentum of the inertialess body A to be E_γ.

However, in this case, the possibility of the kinetic energy of object B being the same as that of object C is completely lost, in spite of it being allowed by general considerations of the dynamics of object B.

Providing grace to the law of momentum by associating, certainly fictitious, momentum[9] to inertialess bodies cannot then be justified, as doing so is not in conformity with the general dynamical considerations.

We thus need to analyze such cases to arrive at the understanding of changes to the measure-values of objects. This analysis would deal with the preservation of values of measures under the actions of endo-functors.

Nevertheless, statistical considerations of momentumless objects also lead to Planck's formula for energy density and to Einstein's formula for the momentum fluctuations of an atom interacting with such objects under equilibrium. These formulas owe themselves to the assumptions of statistical independency of the motions of such objects and of their interactions with the atom and to the existence of the nonzero lowest-energy state of the atom, independently of whether they have momentum or not.

In 1927, Dirac had stated that

> the light quantum has the peculiarity that it apparently ceases to exist when it is in . . . the zero state in which its momentum, and therefore its energy, is zero. When a light quantum is absorbed it can be considered to jump into the zero state, and when one is emitted it can be considered to jump from the zero state to one in which it is in physical evidence, so that it appears to have been created.

10.9 Emission Origin for the Wave of Quanta

After explaining photoelectric effect and thereby bringing physics back to Newton's corpuscular picture of light in 1905, Einstein had envisaged in 1909 that

---

[9]A Light quantum is an inertialess body. Einstein could have associated momentum E/c with his light quantum when he associated energy $E = h\nu$ with it in 1905. Interestingly, Einstein did not explicitly do so for many years. In 1916, when Einstein studied momentum fluctuations of an atom immersed in a radiation bath, he stated, still with a very careful selection of words, that "if a bundle of radiation causes a molecule to emit or absorb an energy amount $h\nu$, then a momentum $h\nu/c$ is transferred to the molecule, directed along the bundle for absorption and opposite the bundle for [induced] emission."

> the next phase in the development of theoretical physics will bring us a theory of light that can be interpreted as a kind of fusion of the wave and the emission[10] theory. . . . [The] wave structure and [the] quantum structure are not to be considered as mutually incompatible. . . .

During 1916–1917, Einstein addressed the issue of the equilibrium of a gas of Bohr-type atoms with Planck-type radiation. As a consistency condition for the Maxwell distribution of the atomic gas and the Planck distribution of radiation, he was led to postulate the process of spontaneous emission of radiation quanta that he, at that time, could treat only statistically. He also compared his statistical treatment of spontaneous emission with that of Rutherford's for radioactivity.

However, no one could show, neither then nor at any time later, how this statistical treatment could be supplemented with deterministic treatment for the process of spontaneous emission of quanta.

Einstein remained unsatisfied with the statistical nature of his theory of spontaneous emission of quanta. Perhaps the lack of satisfactory understanding of Planck's well-known relation $\varepsilon = h\nu$, relating corpuscular energy ε to wave frequency ν, remained his concern.

Moreover, if the quanta of light are inertialess indeed, then their momentum, defined according to Galileo and Newton, must also be vanishing. But Compton-like effects can be explained assuming momentum $p = h\nu/c$ for inertialess quanta. Such mutually contradictory results too contributed to his dilemma about light quanta. Quantum theory does not explain $\varepsilon = h\nu$, but assumes this.

Here, we show that the emission aspects of the quanta, whether inertialess or not, determine the wave characteristics of their flux at any location; and this explains $\varepsilon = h\nu$, in a definite sense. The (Newtonian) momentum of the inertialess quanta is then allowed to be vanishing.

The wavy characteristics of quanta arise here quite analogously to that of a hose pipe putting out a straight jet of water through a nozzle, where changing the location of the nozzle changes the flux of water passing a location. As can be easily verified experimentally, the oscillatory motion of the nozzle also produces an oscillatory motion of the water jet.

---

[10]It is of certain interest that Einstein refers to the emission aspects of quanta here.

10.9.1 Statistical Considerations

Noteworthy first is S. N. Bose's statistical derivation of Planck's formula for the spectral energy density of black-body radiation, for it assumes bodies of light as inertialess particles of (nonkinetic) energy $\varepsilon = h\nu$ to distribute them in energy boxes,[11] under equilibrium.

Bose offered no explanation for how a corpuscular property ε can be so related to a wave property ν as in the relation $\varepsilon = h\nu$. Still, consider assumptions of his derivation of Planck's formula.

Bose considered *indistinguishable* objects having[12] nevertheless *individuality* of their own. A number N of these objects are to be distributed in a number m of boxes, with the boxes being labeled by energy $E_i = (i - 1)\varepsilon$, with ε being a constant and the index i taking values $i = 1, 2, \ldots, m$. Then a state of this system has n_i objects in the ith box, with $\sum_{i=1}^{m} n_i = N$, where N is not necessarily a constant. Both N and m are assumed to be very large.

Under additional assumptions leading to the statistical independence of objects while getting distributed in energy boxes, objects not preferring any particular box, and allowing a single box to be populated by more than one object, the most probable state of the system has, as can be easily shown under canonical distribution, the mean occupation number of the ith box

$$\bar{n}_i = 1/(e^{\varepsilon/\Theta} - 1)$$

and the mean energy of the ith box

$$\bar{E}_i = \varepsilon/(e^{\varepsilon/\Theta} - 1),$$

where $\Theta \equiv kT$ is the modulus of the distribution, with T as the temperature of the system and k as Boltzmann's constant.

As can also be easily shown, the system, when in the most probable state, displays the mean square fluctuations in number and in energy as

$$\overline{n_i^2} = \bar{n}_i/[1 + \bar{n}_i] \text{ and}$$

$$\overline{E_i^2} = \bar{E}_i/[\varepsilon + \bar{E}_i],$$

---

[11]The "boxes" will, in general, refer to the momentum part of the phase space of the system and do not represent its spatial dimensions.

[12]For an interesting discussion of the two concepts involved herein, see Cheng, Chi-Ho in https://arxiv.org/abs/0903.4748.

respectively. The above are statistical results and are to be interpreted in terms of motions of objects (whether inertia is vanishing or not).

The most probable state of the system is now taken to be its equilibrium state. For the physical problem of radiation enclosed in a cavity, the most probable state is then the state of equilibrium of radiation, which is also interacting with the walls of the cavity containing it.

The first term of these fluctuations is the particle term, which can be interpreted to be due to fluctuations resulting from the corpuscular collisions of objects undergoing otherwise rectilinear motions.

On the other hand, for radiation in a cavity, the second term of these fluctuations is interpreted as being due to their wave character as per Maxwell's electrodynamics, and it is called the wave term.

Now, objects can themselves possess wavy motions. But no physical mechanism exists here for the wavy motions of objects, for it would require (Hooke's law) force, which does not exist within this situation.

Noticeably, in the past, interaction of objects (of radiation) with walls of the cavity was not considered to account for the second fluctuation term. This interaction is precisely the subject of the emission theory that Einstein referred to in 1909. But emission aspects were ignored as no forces causing wavy motions of quanta all over the space within the cavity can arise even from such interactions.

Objects have been assumed here to possess an individuality of their own and are particulate or corpuscular. It is thence a complete mystery now as to how the second term of their purely statistical fluctuations can be interpreted as being due to the wavy character or wavy motion of the particles or corpuscles.

Nevertheless, the thesis here is that emission aspects provide an explanation of the second term of the fluctuations, without invoking forces causing wavy motions of quanta. As will be seen, wavy changes in the number of objects at any location inside the cavity are due to the emission aspects. In fact, all of the wave aspects should emerge out of emission aspects, even when the emitted quanta continue unabated their rectilinear motion till, for example, their collisions take place.

10.9.2 Emission and Formation of Wave of Quanta

Of significance now is the fact that the emission of light quanta does not occur unless the kinetic energy of their emitter changes. The kinetic energy of the emitter is emitted as the (nonkinetic) energy of the light quanta.

Furthermore, the emitter of light quanta also needs to emit them spherically symmetrically, else Maxwell and Lorentz's scenario of wave emission in all directions away from its source is hard to maintain vis-à-vis experiments, that is, the failure of Maxwell's equations should have been evident in many experiments of the past.

The above two peculiarities of the emission of quanta are crucial and suffice for our purpose here of showing that their emission aspects are responsible indeed for all of their wave aspects.

Let an emitter of radiation emit n light quanta that move rectilinearly with the same speed, forming a spherical front centered at the location of their emitter. For the change ΔE in the kinetic energy of the emitter, each of the emitted quantum then has energy $\Delta E/n$ if distributed equally among the emitted n quanta.

How does the kinetic energy of the emitter get distributed among quanta? How many quanta get emitted at an instant? What are the laws underlying emission? These issues are related to aspects of Maxwell's theory dealing with the power radiated by an accelerated charge.

We, however, do not need, at this stage, these details as well as those of how the oscillatory motion of the source arises, and so on. We only need to know that emitters of quanta undergo oscillatory motions at the cavity walls. This is so for the following reasons:

As emission of quanta takes place only when the kinetic energy of the emitter changes, the next such spherical front, after one is emitted, is centered at a different location of their emitter. This, however, means that the flux of quanta passing any (distant or not) location would change according to changes in the location of their emitter. Thus, for oscillatory changes in the location of the emitter, oscillatory changes would be seen in the flux of quanta at any location as well.

This is, precisely, the nozzle-and-jet-of-water situation mentioned before. Therefore, the above peculiarities of the emission of light quanta cause the wavy behavior of their flux passing any location.

No radiation object undergoes wavy motion. So, no need exists for force causing the wavy motion of quanta. The wavy behavior of the flux of quanta results only from the oscillatory motion of their emitter. Wave aspects of quanta originate in their emission aspects.

Now, clearly, the frequency of oscillations of the number of quanta at a location is the frequency of oscillations of their emitter. An emitter of quanta could be assumed to be a linear oscillator. Then the maximum energy in quanta is emitted when the acceleration of the emitter is maximum.

We thus have Planck's oscillators at the walls of the cavity containing radiation and in equilibrium with it. This clearly means that Planck's law for cavity or black-body radiation is obtained here.

Let us then note that wave modes of radiation, when existing within a cavity in equilibrium, would be standing modes. Under equilibrium, the number $f(\nu)\delta\nu$ of standing wave modes within a cubical cavity of each side of length ℓ and within the frequency range ν to $n + \delta\nu$ is[13]

$$f(\nu)\delta\nu = \frac{4\ell^3\nu^2}{c^3}\delta\nu\,,$$

allowing for two polarization states[14] for wave modes, with c being the wave speed. Then the number δn_ν of objects or quanta *perceived to be undergoing (an apparent) wavy motion* within this frequency range is, simply, given by

$$\delta n_\nu = \bar{n}_i f(\nu)\delta\nu\,,$$

with the energy of radiation within the same frequency range then being

$$\delta E_\nu = \varepsilon\delta n_\nu.$$

The spectral energy density of the cavity radiation is therefore obtained as

$$\frac{\delta E_\nu}{\ell^3} = \frac{(4\varepsilon\nu^2 / c^3)\delta\nu}{e^{\varepsilon/kT} - 1}\,.$$

---

[13]Here, the wavelength λ of the wave mode within the cavity needs to be much smaller than the cavity length scale ℓ.

[14]We offer no explanation here for this assumption. But it should relate to results established in Wigner, E. P. (1939). On unitary representations of the inhomogeneous Lorentz group. *Ann. Math.*, **40**, 149.

This yields Planck's law, assuming relation $\varepsilon = h\nu$, with consequently correct results for fluctuations also.

We addressed only the issue of the origin of wavy fluctuations in the number of objects within the cavity, particularly without invoking force for them. Aspects of emission of quanta yield Planck's formula.

Then our considerations show that this origin for the wavy behavior of the number of quanta at a cavity location is consistent with black-body radiation laws. The second term of the statistical fluctuations of the most probable state of cavity radiation arises indeed due to the wavy nature of the number of quanta at a cavity location, but this wavy nature of their flux owes itself to peculiarities of their emission at the walls of the cavity they are in equilibrium with.

10.9.3 Remarks

We assumed that inertialess light quanta move rectilinearly once emitted. The emitter of these quanta must undergo a change in its kinetic energy for their emission. Quantal emission is to be spherically symmetric, with change ΔE in kinetic energy of the emitter to be distributed among the n emitted quanta, with the energy of a light quantum being, say, ε.

We then pointed out that the flux of light quanta at a location must change according to the motion of the emitter. If the motion of the emitter is any oscillatory with a frequency ν, then the flux of the light quanta at any location is also oscillatory, with the same frequency ν.

We demonstrated that this origin of the wavy behavior of the flux of light quanta explains the laws of black-body radiation, following Bose's statistical method for the most probable state of the system of quanta that are to be distributed in energy boxes $E_i = (i - 1)\varepsilon$.

However, energy ε of a quantum is *not related* to the frequency ν of the Planck oscillator. Planck's relation $\varepsilon = h\nu$ is, therefore, an ad hoc assumption relating a corpuscular property ε to a wave property ν.

Nevertheless, the quantum energy ε is vanishing when the frequency ν of the Planck oscillator is vanishing, because no quanta are emitted unless the kinetic energy of the emitter changes. It is for this reason that Planck's postulate $\varepsilon = h\nu$ is able to explain the observed laws of black-body radiation. In a definite sense, we therefore "explain" this relation by Planck. Hence, Planck's constant h arises due to this postulate and also has an ad hoc status.

We note here that the above is consistent with Einstein's views regarding the role of fundamental constants of nature in a theory:

> In a sensible theory there are no [dimensionless] numbers whose values are determinable only empirically. I can, of course, not prove that . . . dimensionless constants in the laws of nature, which from a purely logical point of view can just as well have other values, should not exist.

Now, if we focus only on the wavy nature of the number of quanta passing a location and ignore their emission origin, then we would be dividing the problem of the spectrum of cavity radiation into two parts—the wave part and the emission part—to conquer it.

Under this divide-and-conquer strategy of problem solving, the wave part has a description by way of wave equations. Maxwell's theory has one wave equation for inertialess objects, while we postulate varieties of Schrödinger-type equations for objects of nonvanishing inertia.

Quantum theory now provides the probabilistic description for quanta within the wave description. In a sense, we then also understand here as to how the description of quantum theory arises: it is a result of the divide-and-conquer strategy of problem solving and considers only a partial aspect of this problem.

Evidently now, quantum theory neglects certain crucial aspects of the emission of quanta. Needless to say then, but important nevertheless to point out here, that the description of the physical world provided by quantum theory is[15] an incomplete one in the aforementioned context.

We emphasize that we have not used the concept of force, or even that of field, to explain here the wave of quanta.

---

[15]Einstein did not consider quantum theory to be a complete description of the physical world. Of pivotal importance to his thinking was Boltzmann and Gibbs's statistical theory of many particle systems, which is incomplete in relation to Newtonian mechanics. For him, quantum theory was similarly an incomplete description of physical systems, a statistical theory that is incapable of describing the individual constituents of a many-body system.

In 1954, Einstein considered "it quite possible that physics cannot be based on the field concept, i.e., on continuous structures." Then with remarkable openness for ideas, he had said[16] further that "[in that case], nothing remains of my entire castle in the air, gravitation theory included, [and of] the rest of modern physics." "Continuous structures" referred to by Einstein above are to represent the objects themselves and not the probability of their being at some space location at some instant of time. This is the field concept of Faraday and Maxwell or Lorentz and Einstein.

The present results then imply that Einstein's aforementioned extreme opinion may have partial merit, for we have not used, in particular, such a concept of field to represent quanta and to also explain their wave. *We have not even specified here which mathematical structure represents objects of our statistical considerations.*

Spontaneous emission of a quantum (of radiation) by atom may have to be dealt with statistically, but the computation of its probability can then have basis in deterministic ideas not based on the concepts of force and field by both Faraday–Maxwell and Lorentz–Einstein.

We also conclude here that a deterministic basis underlies probabilistic methods of quantum theory, and it is rooted neither in the Newtonian notion of force nor in the field concept of Faraday–Maxwell and Lorentz–Einstein. The possibility of such a deterministic theory was explored[17] by Hertz with an intention to free Newtonian mechanics from the notion of the potential energy, which he considered to be too unsatisfactory a concept.

10.10 Concluding Remarks

To conclude, let us note that categorical dynamics refer to the study of changes to measures and we can expect them to provide relations of changes in one measure to those in other measures. Primarily, physical laws are the relations of changes in one quantity with those of other quantities.

---

[16]See Pais, A. (1982). *Subtle Is the Lord*. Oxford University Press: Oxford. p. 467.
[17]See Einstein's *Autobiographical Notes* (edited by P. A. Schlipp and printed in 1979 by Open Court, Chicago).

Laws of physics within the present categorical framework will thus be the laws of the dynamical system $(\mathcal{C}(\mathbb{C}), T_t)$, which consider changes to all the measures definable over the category of all the categories, because this is the structure of all mathematical structures. Clearly, such laws are completely independent of any specific mathematical structure.

We represent physical bodies with some mathematical structure and study changes in it to *model* changes in them. Then any model explaining *all* the observations, the physical entirety, is a theory of everything, and we associate the notion of the physical reality with it.

We have demanded the physical reality to be independent of mathematical representation of bodies of nature. Then by the epistemological principle that concepts and distinctions are only admissible to the extent that observable facts can be assigned to them without ambiguity, concepts of universal relativity, for example, Euclidean distance, having relational existence, should unambiguously be representing all of our physical conceptions. We may expect them to explain all the phenomena observable in nature then.

The categorical dynamical system provides then mathematical foundations for us to conceptually grasp the physical entirety. Following Einstein, as quoted toward the end in Section 10.1, we have begun with the "whole of the laws" and have applied the stipulation of meaning to this whole first. Advancing with it now involves analyzing the categorical dynamical system and putting it, step-by-step, to the world of experience.

We may also note that the laws of the dynamical system $(\mathcal{C}(\mathbb{C}), T_t)$ will also find applications in other fields of human knowledge, where mathematical structures are used to represent our concepts.

> Physicists who have studied attentively the laws of nature will feel that such simplicity and such close relations between the different elements of the phenomenon are conclusive in favour of the hypothesis on which they are based.
>
> —**Augustine Jean Fresnel (1788–1827)**

Chapter 11

Epilogue

Now, we have, in Part II of this book, seen how the mathematics of category theory is useful for formulating most general physical laws.

The question is that of any permissible generalization of category theory itself. In this context, we note the following:

Set $\mathcal{N}$ of natural numbers is a category under the usual total order $\leq$ since it is a totally partially ordered set and all of its members are comparable under this order.

Now, consider set $\mathcal{N}$ under the (partial) order of divisibility that for any $A, B \in \mathcal{N}$, $A \sqsubseteq B$ if and only if (iff) there exists a nonzero $c \in \mathcal{N}$ such that $B = A \times c$. We then have an arrow $c : A \to B$ or $A \xrightarrow{c} B$. Firstly, we note that number 1 is a factor of every other member of set $\mathcal{N}$ and, therefore, we have the identity arrow to correspond to every object $A \in \mathcal{N}$ as

$$A = A \times 1 \Rightarrow A \xrightarrow{id_A} A.$$

Now, if A and B in set $\mathcal{N}$ are comparable under the order of divisibility such that $A \xrightarrow{c} B$, then there exists an empty arrow from B to A. We shall refer to this as emptiness of the first kind (EFK) and shall denote an empty arrow of this kind by

$$A \xrightarrow{c} B \quad \text{and} \quad EFK\,(B, A) : B \to A.$$

In this case, we have compositions

$$A \xrightarrow{c} B \xrightarrow{EFK(B,A)} A = A \xrightarrow{id_A} A$$

and

$$B \xrightarrow{EFK(B,A)} A \xrightarrow{c} B = A \xrightarrow{id_A} B.$$

Such compositions of arrows exist also under the usual total order $\leq$ on set $\mathcal{N}$.

Furthermore, the composition of two EFK arrows is also an EFK arrow. That is to say, we have

$$C \xrightarrow{EFK(C,B)} B \xrightarrow{EFK(B,A)} A = C \xrightarrow{EFK(C,A)} A$$

because, we have

$$A \sqsubseteq B \text{ and } B \sqsubseteq C \Rightarrow A \sqsubseteq C.$$

Next, if P and Q are incomparable objects of $\mathcal{N}$ under the order of divisibility, then there exist two empty arrows, one from P to Q and the second from Q to P. We shall refer to this situation as emptiness of the second kind (ESK) and shall denote an empty arrow of this kind by

$$ESK(P, Q) : P \to Q \text{ and } ESK(Q, P) : Q \to P.$$

Then we have

$$P \xrightarrow{ESK(P,Q)} Q \xrightarrow{ESK(Q,P)} P = P \xrightarrow{id_P} P$$

and

$$Q \xrightarrow{ESK(Q,P)} P \xrightarrow{ESK(P,Q)} Q = Q \xrightarrow{id_Q} Q.$$

Notice that such a composition does not exist with the total order $\leq$ on set $\mathcal{N}$ as all members of $\mathcal{N}$ are comparable under it. Under the total partial order, only one of these two arrows is an empty arrow.

Now, the composition of two ESK arrows of incomparable objects of set $\mathcal{N}$ is not[1] necessarily empty. As an example, we have $2 \xrightarrow{ESK(2,3)} 3 \xrightarrow{ESK(3,8)} 8 = 2 \xrightarrow{4} 8$. We also have the possible composition $2 \xrightarrow{ESK(2,3)} 3 \xrightarrow{ESK(3,5)} 5 = 2 \xrightarrow{ESK(2,5)} 5$.

Thence, set $\mathcal{N}$ is not a category under the (partial) order of divisibility because a collection of its arrows is not a partial binary algebra under the law of their composition.

Now, this situation is suggestive of possible generalization of a category, which we shall refer to as a *super category* or simply as a *SuperCat*. For defining a SuperCat, we may relax the condition of the collection of arrows as forming a partial binary algebra.

We may, for example, posit the following definition for a SuperCat:

Definition 157 A SuperCat is a quadruple

$$\mathbb{A} = (\mathrm{Ob}(\mathbb{A}), [-,-]_{\mathbb{A}}, \mathrm{id}, \circ)$$

[1]I am grateful to Professor George Janelidze for pointing out my error of assuming the contrary during TACT-2016, September 2016, held at UNISA, Florida Campus, Johannesburg, South Africa. Thanks, George.

consisting of

- A collection Ob($\mathbb{A}$), whose members are called objects.
- For each $\mathbb{A}$-object A, an arrow $\mathrm{id}_A : A \to A$, called the $\mathbb{A}$-identity of A.
- For each pair (A, B) of $\mathbb{A}$-objects, a collection $[A, B]_{\mathbb{A}}$, whose members are called the $\mathbb{A}$-arrows from A to B.

  (The collection of *all* $\mathbb{A}$-arrows is defined to be the union of all the collections $[A, B]_{\mathbb{A}}$ in $\mathbb{A}$ and is denoted by Mor($\mathbb{A}$).)
- A composition law associating each $\mathbb{A}$-arrow $A \xrightarrow{f} B$ and each $\mathbb{A}$-arrow $B \xrightarrow{g} C$ with an $\mathbb{A}$-arrow $A \xrightarrow{g \circ f} C$, called the composite of f and g.

  These data are then subject to the following conditions:
  - $\mathbb{A}$-identity arrows act as identities with respect to composition, that is, for $\mathbb{A}$-arrows $A \xrightarrow{f} B$, we have $\mathrm{id}_B \circ f = f$ and $f \circ \mathrm{id}_A = f$.
  - The collections $[A, B]_{\mathbb{A}}$ are pairwise disjoint.

Clearly, any category is always a super category of the above definition. However, conditions of the above definition for a SuperCat are insufficient to ensure that the collection Mor($\mathbb{A}$) forms a partial binary algebra under the law of composition of its members. In other words, a SuperCat is not necessarily a category. Set $\mathcal{N}$ of natural numbers under the order of divisibility is such an example here.

But we notice that for members P and Q of set $\mathcal{N}$ that are incomparable under the partial order of divisibility, the total order distinguishes one of the two ESK arrows as being nonempty and identifies its complementary ESK arrow with the empty EFK arrow.

The aforementioned is then suggestive that we distinguish emptiness whenever we encounter the situation of the existence of two complementary ESK arrows, in general. We then refer to one of such ESK arrows as an ESK arrow and identify its complementary arrow as the usual empty arrow, to be referred to as an EFK arrow. The ESK arrow is then treated as nonempty with respect to the empty EFK arrow. This amounts to differentiating the two types of emptiness as referred to above.

It then follows that set $\mathcal{N}$ of natural numbers becomes a category under the emptiness distinguishing strategy, for it amounts to adopting total order on this set and a totally partially ordered set, poset, is a category under that total order.

Now, it is important to notice that a SuperCat of the above definition fails to be a proper category only because of the situations of emptiness of the second kind arising within its considerations. Then the strategy of distinguishing types of emptiness arising in category theory shows that such a SuperCat can always be rendered to be a category. In other words, the aforementioned definition of *a SuperCat does not constitute a generalization of that of a category.*

Consequently, the notion of categorical measures will be obtainable for a SuperCat, as well.

Furthermore, it is well known that concatenation of categories does not result in a category. However, the strategy of distinguishing emptiness now implies the following:

In general, consider two (or more) categories $\mathbb{A} = \{A_1, A_2, A_3, A_4, \ldots\}$ and $\mathbb{B} = \{B_1, B_2, B_3, B_4, \ldots\}$, with their collections of arrows being Mor($\mathbb{A}$) and Mor($\mathbb{B}$), respectively. Let the law of composition in category $\mathbb{A}$ be $\circ$ and that in category $\mathbb{B}$ be $*$.

For it to be a category, the law of composition $\boxtimes$ of arrows for the concatenation collection Mor($\mathbb{A}$)$\uplus$Mor($\mathbb{B}$) is then taken to be $\circ$ when the arrows being composed are $\mathbb{A}$-arrows and $*$ when the arrows being composed are $\mathbb{B}$-arrows. Every other situation is that of ESK for the composition law $\boxtimes$ now. That is to say, for $A_i \in \mathbb{A}$ and $B_j \in \mathbb{B}$, we have

$$A_i \boxtimes B_j : A_i \xrightarrow{ESK(A_i, B_j)} B_j \quad \text{and} \quad B_j \boxtimes A_i : B_j \xrightarrow{ESK(B_j, A_i)} A_i .$$

Then under the emptiness distinguishing strategy, we make assignments

$$A_i \boxtimes B_j : A_i \xrightarrow{ESK(A_i, B_j)} B_j \quad \text{and} \quad B_j \boxtimes A_i : B_j \xrightarrow{EFK(B_j, A_i)} A_i$$

with arrow ESK(A_i, B_j) being nonempty with respect to the empty arrow (of the first kind) EFK(B_j, A_i).

It can now be verified in a straightforward manner that with the above assignments prescribed by the emptiness distinguishing strategy, the concatenated collection Mor($\mathbb{A}$)$\uplus$Mor($\mathbb{B}$) is a category indeed, under the law $\boxtimes$ of the composition of its members.

Distinguishing emptiness has provided total order[2] for the concatenated collection to make it a category, to be denoted by $\mathbb{A}\uplus\mathbb{B}$.

[2]A collection, in general, and a set, in particular, can be well ordered. The strategy of distinguishing emptiness is then rooted in the principle of well ordering.

Furthermore, $\mathbb{A}$ and $\mathbb{B}$ are obvious full subcategories of their concatenation category $\mathbb{A}\uplus\mathbb{B}$. A category dual to $\mathbb{A}\uplus\mathbb{B}$ will be obtained on changing above assignments or the roles of ESK and EFK arrows of the emptiness distinguishing strategy.

In conclusion, the concepts of EFK and ESK arise quite naturally in category theory. Distinguishing types of emptiness arising in category theory leads us to a strategy that is closely related to the well ordering principle, which always renders any SuperCat of the above definition to be a category.

Appendix A

Set Theoretical Foundations

A.1 Collection: Meaning

The word "collection" will refer to "gathering" or "accumulating" of "basic" members or elements. We denote by the statement $x \in X$ that x is a member of collection X.

A.2 Essential Preliminaries

Consider a collection X and another collection Y. Consider an assignment to each $x \in X$ of a unique element $f(x) \in Y$. A **rule** can then be defined to be a triple (X, f, Y), where f, as a subcollection of $\{(x, y) | x \in X \text{ and } y \in Y\}$, is an assignment such that for each $x \in X$ there exists a unique $y \in Y$ with $(x, y) \in f$. We write $y = f(x)$ or $x \mapsto f(x)$. The notation is $f : X \to Y$ or XfY. (Such a rule is called a **function** or a **map.)** For XfY and YgZ, the **composite rule** $Xg \circ fZ$ is defined by $x \mapsto g(f(x))$.

If A is a subcollection of X and $f : X \to Y$ is a rule, then we may define a new rule $g : A \to Y$ by defining $g(x) = f(x)$ for $x \in A$. This new rule is called the **restriction of** f **to** A and is written as $f|A$.

We express the above also in the following manner: Given $A \subset X$ and its inclusion function $i_{AX} : A \to X$, $B \subset Y$ and its inclusion function $i_{BY} : B \to Y$, and $f : X \to Y$ with $f[X] \subset B$, the unique function $h : A \to B$ for which we have $i_{BY} \circ g = f \circ i_{AX}$ is called as the restriction of f to A and is denoted, in this case, by $f \uparrow_A^B$.

The collection G of pairs $(x, f(x))$ is called the **graph** of f. A graph has, for each $x \in X$, a unique pair in G whose first element is x.

We call X the **source** or the **domain** of f and call Y as the **target** or the **codomain** of f. The collection of values taken by f, ie, $\{y \in Y | (\exists x) [y = f(x)]\}$, is called the **range** of f. In general, the range of f can be smaller than Y. When the range of f is equal to the collection Y, we call f an **onto** or a **surjective** rule.

An **ordered pair** is then a rule whose domain is the collection $\{1,2\}$. An n-**tuple** or a **finite sequence** is a rule whose domain is the first n natural numbers, that is, $\{i \in \mathbb{N} | i < n\}$. An **infinite sequence** is then a rule whose domain is the collection $\mathbb{N}$ of natural numbers.

A collection is called **finite** if it is either empty or the range of a finite sequence. It is called **countable** or **denumerable** if it is either empty or the range of a (finite or infinite) sequence.

If A denotes the subcollection of X, the **image** under f of A is defined to be the collection of elements in Y such that $y = f(x)$ for some $x \in X$. We denote this image by $f[A]$. Then, any rule $f : X \to Y$ is onto iff $Y = f[X]$.

If B is a subcollection of Y, an **inverse image** $f^{-1}[B]$ of B is the (sub-)collection of elements of X for which $f(x) \in B$. Then, f is onto or surjective only if the inverse image of each nonempty subcollection of Y is nonempty.

We call the rule $f : X \to Y$ as **one-to-one (1-1)** or **injective** or **univalent** if $f(x_x) = f(x_2)$ only when $x_1 = x_2$.

We call the rule $f : X \to Y$ that is injective (1-1) as well as surjective (onto) a **1-1 correspondence** or a **bijection** or **bijective correspondence.** In this situation, there exists a rule $g : Y \to X$ such that for all $x \in X$ and for all $y \in Y$, we have $g(f(x)) = x$ and $f(g(y)) = y$. The rule g is called the **inverse of** f and is also denoted by f^{-1}.

Two given entities x and y may be related in many ways. In general, R denotes a relation between them if x stands in **Relation** R to y or if x does not stand in Relation R to y. Then, Relation R is said to be a relation on a collection X if xRy implies that $x \in X$ and $y \in X$. The graph of the relation R is then a collection $\{(x, y) | xRy\}$. Two relation R and S such that if $(xRy) \Leftrightarrow (xSy)$ are considered to be equivalent. Consequently, each subcollection of $\{(x, y) | x \in X$ and $y \in X$ defines a relation on the collection X.

A relation R is said to be **transitive** on the collection X, provided that xRy and yRz imply xRz for all $x, y, z \in X$. A relation R is said to be **symmetric** on X, provided that xRy implies yRx for all $x, y \in X$. A relation R is said to be **reflexive** on X, provided that xRx for all $x \in X$.

A relation R is said to be **antisymmetric** on X, provided that xRy and yRx imply $x = y$ for all $x, y \in X$.

A transitive, reflexive and symmetric relation is called an **equivalence relation,** or simply an **equivalence on** X. A collection of all elements of X that are equivalent to a particular $x \in X$ is called an **equivalence class of** x and is denoted by E_x. Either $E_x = E_y$ (identical) when $y \in E_x$ or $E_x \cap E_y = \emptyset$ (disjoint) when $y \notin E_x$. The collection of equivalence classes of X under an equivalence = is called the **quotient of** X **relative to** = and is denoted by $X/=$. The assignment $x \rightarrow E_x$ is called the **natural map of** X **onto** $X/=$.

If R is any equivalence relation on X and * is any other relation on X, then the equivalence relation is called a **congruence relation** R **on** X **or congruence modulo** *, provided that, given aRa' and bRb' for any $a, a', b, b' \in X$, we always have $(a * a')R(b * b')$.

A **binary operation on** X is an assignment from $\{(x, y) | x \in X$ and $y \in X]$ to X. An equivalence = is said to be compatible with a binary operation + on X if $x = x'$ and $y = y'$ imply that $(x + y) = (x' + y')$ and, in this case, + defines an operation on the quotient $X/=$.

A relation < is called a **partial order** on X if it is transitive and antisymmetric. A partial order on X for which either $x < y$ or $y < x$ for any elements $x, y \in X$ is called the **linear order.** For a subcollection E of X, an element $a \in E$ is called the **first (last)** or **smallest (largest) element in** E if, whenever $x \in E$ and $x \neq a$, we have $a < x$ $(x < a)$. A member $a \in E$ is called a **minimal (maximal) element of** E if there is no $x \in E$ with $x \neq a$ and $x < a$ $(a < x)$. In general, minimal (maximal) member of E need not be the smallest (largest) member of E.

A transitive and reflexive relation is called a **preorder.** If for a partial order <, $x < x$ for all $x \in X$, we call < a **reflexive partial order.** If we have $x \nless x$ for all $x \in X$, we call < a **strict partial order.** For example, < is a strict partial order, while $\leq$ is a reflexive partial order on real numbers. To any partial order there is associated a unique strict partial order and a unique reflexive partial order that agree with < for all (x, y) with $x \neq y$.

A strict linear order < is said to **well order** X if every nonempty subcollection of X contains a first element. For example, natural numbers are well ordered by <, but real numbers are not well-ordered by the same relation "less than".

Well-ordering principle: Every collection X can be well ordered, that is, there is a relation < on X which well orders X.

Proposition: There is an uncountable collection X that is Well-ordered by a relation $<$ in such a way that

(1) there is a last element Ω in X
(2) if $x \in X$ and $x \neq \Omega$, then the collection $\{y \in X | y < x\}$ is countable.

Proof: If Y is any uncountable collection, there exists a well order $<$ on Y by the well-ordering principle. If Y does not have a last element, consider element $\alpha \notin Y$, replace Y by $Y \bigcup \{\alpha\}$ and extend the well order by setting $y < \alpha$ for all $y \in Y$. This new Y has a last element and is well-ordered by $<$. Now, a collection of $y \in Y$ for which the collection $\{x \in Y | x < y\}$ is uncountable is nonempty, as it has the last element of Y. Let Ω be the smallest element in this collection, and let $X = \{x \in Y : x < \Omega \text{ or } x = \Omega\}$. This X is the required collection. ■

It can be shown that X of the above proposition is unique up to 1-1 order-preserving correspondence. The last element $\Omega \in X$ is called the **first uncountable ordinal** and X is called the **collection of ordinals less than or equal to the first uncountable ordinal.** Elements $x < \Omega$ are called **countable ordinals.** If $\{y | y < x\}$ is finite, we call x a **finite ordinal.** If ω is the first **nonfinite ordinal,** then $\{x | x < \omega)\}$ is the collection of finite ordinals and is equivalent, as an ordered collection, to the set $\mathbb{N}^+$ of positive integers.

A.3 Definition of a Set

A collection with which the following constructions can be performed will be called a **set:**

(1) For each set X and each **property** P, we can form the set $\{x \in X | P(x)\}$ of all members/elements of X having property P.
(2) For each set X, we can form the set $\mathcal{P}(X)$ of all **subsets** of X—called the **power set of** X.
For any sets X and Y, we can form the following sets:
   (a) the **two-member set** $\{X, Y\}$ whose members are exactly X and Y
   (b) the **ordered pair** (X, Y) (likewise n tuples of sets, for any natural number $n > 2$) with the first coordinate X and the second coordinate Y
   (c) the **union** $X \bigcup Y = \{x | x \in X \text{ or } x \in Y\}$
   (d) the **intersection** $X \bigcap Y = \{x | x \in X \text{ and } x \in Y\}$

(e) the **Cartesian product** $X \times Y = \{(x, y) | x \in X \text{ and } y \in Y\}$
(f) the **relative complement** $X/Y = \{x | x \in X \text{ and } x \notin Y\}$
(g) the set Y^X of all **functions** $f: X \to Y$ from X to Y

A **family** $(A_i)_{i \in I}$ **of sets** is a function $A : I \to \mathcal{U}$, where $\mathcal{U}$ is the collection of all sets, called the **universe,** is **not** a set, sending $i \in I$ to $A(i) = A_i$, where I is a collection of sets. If I is a set, then $(A_i)_{i \in \mathrm{I}}$ is called the **set-indexed** family of sets.

(4) For any set I and any set-indexed family $(X_i)_{i \in I}$ of sets, we can form the following sets
 (a) the **image** $\{X_i \,| i \in I\}$ of the indexing function $A : I \to \mathcal{U}$, Note that a family and its image are different entities and that a family is **not** determined by its image. We denote the family also by $(A_i)_I$.
 (b) the **union** $\bigcup_{i \in I} X_i = \{x | x \in X_i \text{ for some } i\}$
 (c) the **intersection** $\bigcap_{i \in I} X_i = \{x | x \in X_i \text{ for all } i\}$, provided that $I \neq \emptyset$
 (d) the **Cartesian product** $\Pi_{i \in I} X_i = \{f : I \to \bigcup_{i \in I} X_i | f(i) \in X_i$ for each $i \in I\}$
 (e) the **disjoint union** $\biguplus_{i \in I} X_i = \bigcup_{i \in I} X_i \times \{i\})$

(5) We can form the following sets:
 $\mathbb{N}$ of all **natural numbers**
 $\mathbb{Z}$ of all **integer numbers**
 $\mathbb{Q}$ of all **rational numbers**
 $\mathbb{R}$ of all **real numbers**
 $\mathbb{C}$ of all **complex numbers**

Note: By means of these above constructions, we **cannot** form the **set of all sets.**

A.4 Definition of a Class

A **class** is a collection such that

(1) the members of each class are sets,
(2) for every property P, we can form the class of sets with property P

 A **set,** as defined above, is a property of the collection. Hence, there exists a **largest class** $\mathbb{U}$—the universe. The classes are precisely the subcollections of $\mathbb{U}$. Given classes A and B, we can form classes $A \cup B$, $A \cap B$, $A \times B$, and define functions, equivalence relations etc. on classes.

A **family** $(A_i)_{i \in I}$ **of sets** is a function then $A : I \rightarrow \mathcal{U}$ sending $i \in I$ to $A(i) = A_i$, where I is a collection of sets.

(3) if X_1, X_2, X_n are classes, then so is the n tuple (X_1, X_2, X_n) a class

(4) every set is a class, or, equivalently, every member of a set is a set.

Sets are special classes. Classes that are not sets are called **proper classes** and these cannot be members of any class. Russell's famous paradox of the set theory now translates into a harmless statement that the class of all sets that are not members of themselves is a proper class. Therefore, $\mathcal{U}$ is a proper class.

(5) there is **no surjection** (an onto function) from a set to a proper class. This is called the **axiom of replacement.**

The axiom of replacement is a consequence of property 4a of the definition of a set. Consequently, each set must then have "fewer" elements than any proper class.

Sets are called **small classes** and proper classes are called **large classes.**

Note: Since members of a class must be sets, as defined, we cannot form a class $\{\mathcal{U}\}$ as $\mathcal{U}$ is **not** a set, or a class whose members are all the subclasses of or all functions from to . To be able to deal with these mathematically "valid" entities, we need the next level of conception given later.

A.5 Definition of a Conglomerate

A **conglomerate** is a collection of classes such that

(1) Every class is a conglomerate.

(2) For every property P, we can form a conglomerate of all classes having property P.

(3) Conglomerates are **closed** under all of the constructions (operations on the conglomerate-indexed families) that have been outlined under the definition of a set.

Hence, the conglomerates of all classes, functions between conglomerates, and families of conglomerates are well-defined concepts.

(4) For each surjection (onto map) between conglomerates $f: \rightarrow Y$, there is an injection (1-1 map) $g : Y \rightarrow X$ with $f \circ g = \mathrm{id}_Y$. This is referred to as the **axiom of choice for conglomerates.**

A conglomerate X is called **codable** by a conglomerate Y if there exists a surjection $Y \rightarrow X$, or, equivalently, an injection $X \rightarrow Y$.

Conglomerates that are codable by a class (set) are called **legitimate (small)** and will be treated like classes (sets).

Example: $\{\mathcal{U}\}$ is a small conglomerate and $\mathcal{U} \cup \{\mathcal{U}\}$ is a legitimate conglomerate.

Conglomerates that are not legitimate are called **illegitimate conglomerates.** Example: The power set $\mathcal{P}(\mathcal{U})$ of the universe is an illegitimate conglomerate.

Appendix B

Propositional Calculus

This appendix summarizes certain important issues in logic. In particular, we discuss propositional calculus away from the usual set theoretical Boolean algebra structure resulting from the use of the concept of a subset of a set. But the concept of a subset is not naturally available in category theory, except for the categories *Set* and alike. As noted in Chapter 1, a given object of a category may not possess any subobjects. This "peculiar" categorical situation therefore forces on us explorations of alternative logical structures for them. The Heyting algebra structure is such an alternative logical structure that will be discussed here.

B.1 Propositional Calculus

Propositional calculus deals with propositions, to be denoted by lowercase letters such as $p,q,r,\cdots$, that can be combined under the operations or the propositional connectives of "and," "or," "implies," and "not". The notation that is often used for these operations of the propositional calculus is the following:

p "and" q	$p \wedge q$
p "or" q	$p \vee q$
p "implies" q	$p \Rightarrow q$
"not" q	$\neg q$

Now, we need an *interpretation* for each of these propositional operations in order to obtain their *representation.* (This is similar to obtaining the representation for an abstract notion of a group.) This representation is usually called a *model* for propositional calculus.

Consider, therefore, the usual set theory to obtain a model of propositional calculus. Any given *fixed* set S then has various subsets to be denoted by $P, Q, R, \ldots$. Consider then an element $x \in S$. Each proposition p can then be replaced by an equivalent proposition about x being related to a given subset of S, so each proposition has the aforementioned operations that become operations on the subsets of S.

As is well known, we have "and" ($\wedge$) as the *intersection* of subsets of S, "or" ($\vee$) as the *union* of subsets of S, and *implication* $(P \Rightarrow Q)$ is $\neg P \wedge Q$, where "not" ($\neg P$) is the *complement* of a subset P relative to the fixed set S. Under these "usual set theoretical realizations" of the operations of propositional calculus, these operations satisfy further identities, and we then obtain a Boolean algebra structure for the subsets of a fixed set S and, hence, for propositional calculus. That is precisely how Boolean algebra is the *algebraiccorrelate* of classical propositional calculus.

However, if were to follow intuitionistic logic, we would end up with an algebraic system different than the Boolean system. This formulation of an appropriate algebraic system for intuitionistic logic or intuitionistic propositional calculus was what Heyting achieved as a model using *sameusual* set theory. In doing so, he only considered, instead of all the subsets of a fixed set S, all the open subsets of a fixed set S, which is taken to be a topological space. Then, the interpretations of operations of "and" as well as "or" remain the same as before—those of intersection and union of open subsets of the fixed topological space S. But the remaining propositional operations of implication $\Rightarrow$ and *negation* $\neg$ need to be *re-interpreted* in order to obtain open subsets of S on using them.

Heyting considered that $P \Rightarrow Q$ is the largest open subset R of S such that $R \wedge P \subset Q$, while $\neg P$ is the *interior* of the complement of P, that is, the largest open subset of S disjoint from that of P.

But then the algebraic structure underlying propositional calculus is not that of Boolean algebra, as can be easily checked. The *new* algebraic structure that is obtained in this aforementioned manner is called, obviously, Heyting algebra.

As the set of all subsets of a fixed set S, the power set of S, is a lattice (of subsets), and the set of all open subsets of a fixed set S is also a lattice (of open subsets), it is instructive, for being general in consideration, to formulate the Heyting algebra structure using a generic lattice. Then, the *realdifference* between these algebraic structures is also revealed.

To fix ideas, let us, therefore, note that a lattice L is a partially ordered set (poset) that considered as a category $\mathcal{L}$ has all the binary products and all the binary coproducts. Notice that for elements (objects) $x, y \in L, x \leq y$ if and only if there exists a unique arrow $x \rightarrow y$. Then, the coproduct of objects $x, y \in \mathcal{L}$ is the least upper-bound (lub or sup) $x \vee y$ and their product is the greatest Lower-bound (glb or inf) $x \wedge y$. The category $\mathcal{L}$ then has an initial object 0 and a terminal object 1 iff the lattice L has elements 0 and 1 such that $0 \leq x \leq 1$ for all $x \in L$.

The lattice with 0 and 1 can also be defined equationally as a set with distinguished elements 0 and 1 such that there exist two associative and commutative binary operations V and A on L such that

$$x \wedge x = x, \qquad x \vee x = x$$

$$1 \wedge x = x, \qquad 0 \vee x = x,$$

$$x \wedge (y \vee x) = x = (x \wedge y) \vee x$$

Notice then that these equations on the operations:

$$\wedge : L \times L \rightarrow L, \qquad V : L \times L \rightarrow L, \qquad 0,1 : 1 \rightarrow L$$

help us define a *latticeobject* in any category having finite products and with an initial object *0* and a terminal object 1.

A distributive lattice L is then a lattice for which the identity

$$x \wedge (y \vee z) = (x \wedge y) \vee (x \wedge z)$$

holds for all the elements *x,y,z E L*. One would also like to notice here that the dual of the above identity, ie,

$$x \vee (y \wedge z) = (x \vee y) \wedge (x \vee z)$$

follows from the first identity, the associativity of the composition of arrows in L, and the basic equations on the involved operations. It is therefore not a statement that is independent of the first identity.

We now define a complement of an element $x \in L$ with 0 and 1 as that element $a \in L$ for which we have

$$x \wedge a = 0, \qquad x \vee a = 1$$

(This is analogous to the situation with the subsets of a set S: $P \cap P^c = \emptyset$ and $P \cup P^C = 5$, where P^c denotes the complement of the subset

P relative to S.) It is then easy to see that in a distributive lattice the complement a of an element x, as equationally defined above, is unique in L, whenever it exists: If a' is some another complement of the element x, then we have

$$\begin{aligned} a' = a' \wedge 1 = a' \wedge (x \vee a) &= (a' \wedge x) \vee (a' \wedge a) \\ &= (x \wedge a) \vee (a' \wedge a) \\ &= (x \wedge a') \wedge a \\ &= a. \end{aligned}$$

Let us denote the unique complement of any element x in a distributive lattice L by the symbol $\neg x$. Then, a Boolean algebra B is a distributive lattice B with distinguished elements 0 and 1, for which every element x has a complement $\neg x$ in B with

$$x \wedge \neg x = 0 \text{ and } x \vee \neg x = 1$$

Then, with these additional equations, the following De Morgan's laws also hold in B:

$$\neg(x \vee y) = (\neg x) \wedge (\neg y), \neg(x \vee y) = (\neg x) \vee (\neg y), \neg(\neg x) = x$$

A power set of a given set is then partially ordered by inclusion and is always a Boolean algebra. A fundamental theorem by M. H. Stone then asserts that every Boolean algebra is isomorphic to the algebra, by way of inclusion, of the power set of some set.

Now, a Brouwerian lattice is a distributive lattice in which the complement, as equationally defined earlier, is not available for every of its elements. In general, the equation $x \vee \neg x = 1$ fails to hold for every of its elements; hence, $\neg(\neg x) = x$ also fails to hold. Thus, $\neg x$ is not the complement of x in a Brouwerian lattice. An algebraic structure of a Brouwerian lattice is, in general, non-Boolean, and this is the subject of the following considerations of Heyting algebra.

B.2 Heyting Algebra

Let us now define an operation called an *exponentiation* for the elements of a poset as follows: Given elements x and y of a lattice L, let the least upper bound of all those elements be z in L such that $z \wedge x \leq y$ be denoted by y^x and be called the *exponential*. Clearly, $y \leq y^x$. Furthermore, $z \leq y^x$ if and only if $z \wedge x \leq y$. Diagrammatically, we then have

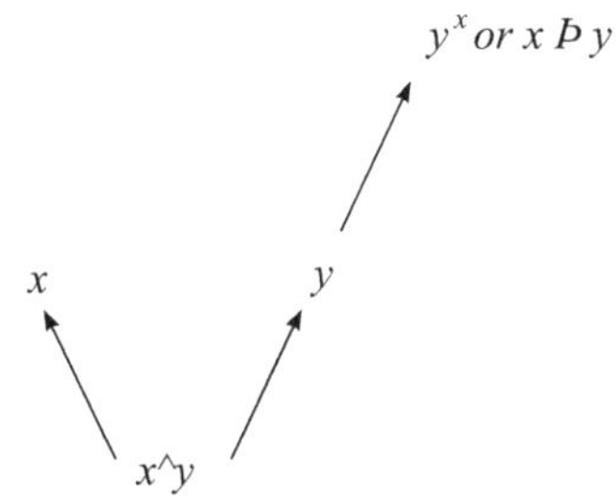

(There exists one and only one arrow from an element p to an element q of a poset if and only if $p \leq q$.) We often write y^x as $x \Rightarrow y$.

Now, we can define Heyting algebra as follows:

(Heyting algebra): A Heyting algebra H (or a Brouwerian lattice) is a poset with distinguished elements 0 and 1 such that each pair of its elements x and y has an exponential y^x or $x \Rightarrow y$.

Clearly, the underlying lattice of a Heyting algebra then has all the small products and all the small coproducts. We call a set with distinguished elements 0 and 1 a *lattice* and with all the small products and all the small coproducts as a complete lattice. A Heyting algebra is a then complete and distributive lattice. Conversely, a complete and (infinitely) distributive lattice can be shown to be a Heyting algebra.

As an example of a Heyting algebra, consider the set *Open(S)* of all the open subsets of a set S, including, of course, Ø and S. Binary unions and binary intersections of any open subsets of S are also open. Then, this set *Open(S)* is a lattice under the operation of inclusion. For any open subsets P and Q of S, we have the union $\bigcup R_i$ of all the open subsets R_i of S for which $R_i \cap P \subset Q$ as being open.

Furthermore, we have

$$(\bigcup R_i) \cap P = \bigcup (R_i \cap P) \subset Q$$

Hence, we define

$$Q^P \text{ or } (P \Rightarrow Q) = (\bigcup R_i)$$

Then, all the exponentials Q^P exist in Open(S), which is a Heyting algebra. But Open(S) is not a Boolean algebra, because the complement of P relative to S is not necessarily open.

We now notice the following: In any Boolean algebra B, we have, for all of its elements x, y, z:

$$z \leq (\neg x \vee y) \text{ if and only if } z \wedge x \leq y$$

Proof: The first if part of the proof is immediate:

$$z = z \wedge 1 = z \wedge (\neg x \vee x) = (z \wedge \neg x) \vee (z \wedge x) \leq (\neg x \vee y)$$

For the only if part, we have

$$z \wedge x \leq (\neg x \vee y) \wedge x \leq y \wedge x \leq y.$$

Therefore, $z \leq (\neg x \vee y)$ if and only if $z \wedge x \leq y$. □

Therefore, we have the following theorem: Every Boolean algebra is a Heyting algebra. □ The converse of this above theorem is, of course, false, as seen earlier from the example of the set of all open subsets of a fixed set (topological space). A Heyting algebra is therefore a more general algebraic structure than is a Boolean algebra.

B.3 Some Results in Heyting Algebra

The following results and identities then hold in any Heyting algebra: For any elements $x, y, z \in H$, we have

(1) $1^x = 1$ or $(x \Rightarrow 1) = 1$
(2) $x^1 = x$ or $(1 \Rightarrow x) = x$
(3) $x \leq (x \wedge y)^y$ or $x < (y \Rightarrow (x \wedge y))$
(4) $y \wedge x^y \leq x$ or $y \wedge (y \Rightarrow x) \leq x$
(5) $(y \wedge z)^x = (y^x) \wedge (z^x)$ or $(x \Rightarrow (y \wedge z) = (x \Rightarrow y) \wedge (x = z)$
(6) $x^{y \wedge z} = (x^y)^z$ or $((y \wedge z) \Rightarrow x) = (z \Rightarrow (y \Rightarrow x))$
(7) $((x \vee z) \wedge y) = (x \wedge y) \vee (z \wedge y)$
(8) $z \leq y^x$ iff $z \wedge x < y$ iff $x \wedge z \leq y$ iff $x \leq y^z$, in other words, $z \leq (x \Rightarrow y)$ iff $z \wedge x \leq y$ iff $x \leq z \leq y$ iff $x \leq (z \Rightarrow y)$
(9) $y^{x \vee z} = (y^x) \wedge (y^z)$ or $((x \vee z) \Rightarrow y) = (x = y) \wedge (z \Rightarrow y)$ □

Now, interpreting x, y, z as propositions, A as "and," and V as "or," the above are the familiar properties of the connector implication $\Rightarrow$ for the propositions of propositional calculus.

We now define the negation of proposition x in any Heyting algebra as

$$\neg x = 0^x = (x \Rightarrow 0)$$

That is to say, "not x" now means "x implies falsity B or "x implies Absurdity. " Then, we have that

$$y \leq \neg x \text{ iff } y \wedge x = 0$$

That is to say, "not x" is now, in a complete lattice, the join of all those y that meet x in 0.

Consequent to the above, we see that in the case of Open(S), the negation of the open $P \subset S$ means the union of all open subsets of S that do not meet P. It is then the largest open subset disjoint with P and, hence, is the interior of the set theoretical complement of P. But then, $\neg(\neg P)$ can be larger than P. Therefore, $\neg\neg x$ need not be equal to x in a Heyting algebra, although it is still true that $x \wedge \neg x = 0$ in it. It then also follows that $x \vee \neg x = 1$ need not be true in a Heyting algebra.

The following usual properties of the negation connector still hold in a Heyting algebra: In a Heyting algebra, the following hold:

(1) $x \wedge \neg x = 0$
(2) $x \leq \neg\neg x$
(3) $x \leq y$ implies $\neg y < \neg x$
(4) $\neg x = \neg\neg\neg x$
(5) $\neg\neg(x \wedge y) = (\neg\neg x) \wedge (\neg\neg y)$ □

In a Heyting algebra, the complement of an element x, if it exists, must be equal to $\neg x$. □

In a Heyting algebra H, the following hold for all elements $x, y, z \in H$:

(1) $x^x = (x \Rightarrow x) = 1$
(2) $x \wedge (y^x) = x \wedge (x \Rightarrow y) = x \wedge y$
(3) $y \wedge (y^x) = y \wedge (x \Rightarrow y) = y$
(4) $(y \wedge z)^x = (x \Rightarrow (y \wedge z)) = (y^x) \wedge (z^x) = (x \Rightarrow y) \wedge (x \Rightarrow z)$

Conversely, if in a lattice L with distinguished elements 0 and 1, there exists a binary operation * satisfying the above equalities, then that binary operation * must be the implication connector, "$\Rightarrow$" of a Heyting algebra structure on that lattice L. □

A Heyting algebra H is a Boolean algebra if and only if $\neg\neg x = x$ for all $x \in H$ or if and only if $x \vee \neg x = 1$ for all $x \in H$. □

B.4 Heyting Algebra and a Category

To end this brief discussion about Heyting algebra, we note that the collection of all the subobjects of a given object in a category forms a poset under the preorder $\leq$ and is always a Heyting algebra that is a Boolean algebra only in some special situations. That is the importance of a Heyting algebra structure vis-à-vis categories.

Bibliography

General Physics, Atomic Theory, Astronomy, and Astrophysics

1. Berry A (1961). *A Short History of Astronomy*, Dover, New York.
2. Einstein A (1933). *On the Method of Theoretical Physics*, Oxford University Press, Oxford.
3. Einstein A and Infeld L (1956). *The Evolution of Physics*, Simon and Schuster, London.
4. Feynman RP (1984). *The Character of Physical Law*, MIT Press, Cambridge, MA.
5. Galileo G (1955). *Dialogue on the Great World Systems*, in the translation by T. Salsbury. Edited, corrected, annotated, and provided with a historical introduction by Giorgio de Santillana, University of Chicago Press, Chicago.
6. Herzberg G (1944). *Atomic Spectra and Atomic Structure*, Dover, New York.
7. Heisenberg W (1945). *Physics, and Beyond*, Harper Torch Books, New York.
8. Jenkins FA and White HE (1957). *Fundamentals of Optics*, McGraw-Hill, New York.
9. Kibble TWB (1981). *Classical Mechanics*, John Wiley & Sons, London (ELBS).
10. Lindsay RB (1941). *Introduction to Physical Statistics*, Dover, New York.
11. Mach E (1872). *Die Geschichte und die Wurzel des Satzes von der Erhaltung der Arbeit* (Prag: J.G. Calvelsche K. u. K. Universitäts Buchhandlung). English translation by Philip B. Jourdain (1911): *History and Root of the Principle of the Conservation of Energy*, Open Court,Chicago, IL.

12. Mach E (1933). *Die Mechanik in ihrer Entwicklung* (Leipzig: F.A. Brockhaus, 9te Auage). English translation by Thomas J. McCormack (1960): *The Science of Mechanics*,6th Edition with revisions through the 9th German Edition, Open Court, La Salle, IL.
13. Maxwell JC (1954). *A Treatise on Electricity and Magnetism*, Vols. 1 & 2, Dover, New York.
14. Nagel E (1984). *The Structure of Science: Problems in the Logic of Scientific Explanation*, Routledge & Kegan Paul, England (Indian edition by Macmillan India).
15. Narlikar JV (1983). *Introduction to Cosmology*, Jones, and Bartlett, Boston.
16. Narlikar JV (2000). *A Cosmic Adventure*, Rajhansa, Pune, India.
17. Pais A (1982). *Subtle Is the Lord ... The Science and the Life of Albert Einstein*, Clarendon Press, Oxford.
18. Peacock JA, Heavens AF, and Davies AT (Eds.) (1989). *Physics of the Early Universe*, SUSSP.
19. Rosenthal-Schneider I (1980). *Reality and Scientific Truth*, p. 74, Wayne State University Press.
20. Saha MN and Srivastava BN (1965). *Treatise on Heat*, Indian Press, Allahabad.
21. Schlipp PA (Ed.) (1970). *Albert Einstein: Philosopher Scientist*, Open Court, Library of Living Philosophers, Vol. VII, La Salle, IL.
22. Wagh SM and Deshpande DA (2013). *Essentials of Physics*, Vol. I, PHI Learning, New Delhi.
23. Wagh SM and Deshpande DA (2013). *Essentials of Physics*, Vol. II, PHI Learning, New Delhi.
24. Whittaker ET (1910). *History of the Theories of Ether and Electricity*, Longman Green, London.

Quantum Theory

25. Bjorken SD and Drell SD (1964). *Relativistic Quantum Mechanics*, McGraw-Hill, New York.
26. Bjorken SD and Drell SD (1964). *Relativistic Quantum Field Theory*, McGraw-Hill, New York.
27. Bohr N (1913). *Phil. Mag.*, **26**, 1.
28. Bose SN (1924). *Z. Physik*, **26**, 178–181.
29. Compton AH (1923). *Phys. Rev.*, **21**, 483.

30. Dirac PAM (1927). *Proc. R. Soc. A*, **114**, 243–265.
31. Einstein A (1905). *Ann. Physik*, **17**, 132.
32. Einstein A (1916). *Verh. Deutsch. Phys. Ges.*, **18**, 318.
33. Einstein A (1916). *Mitt. Phys. Ges. Zürich*, **16**, 47.
34. Einstein A (1916). *Phys. Zeitschr.*, **17**, 101.
35. Einstein A (1917). *Phys. Zeitschr.*, **18**, 121.
36. Feynman RP (1985). *QED: The Strange Story of Light and Matter*, Princeton University Press, Princeton.
37. Heisenberg W (1949). *The Physical Principles of the Quantum Theory*, Dover, New York.
38. Itzykson C and Zuber J (1985).*Quantum Field Theory*, McGraw-Hill, New York.
39. Planck M (1900). *Verh. Deutsch. Phys. Ges.*, **2**, 202.
40. Rutherford LM (1911). *Phil. Mag.*, **21**, 669.
41. Schweber SS (1961). *An Introduction to Relativistic Quantum Field Theory*, Harper and Row, New York.
42. Stuewer RH (1975). *The Compton Effect*, Science History, New York.

Doppler Effect, Lorentz Transformations, and Applications

43. Ashby N (2003). *Living Rev. Relativity*, **6**, 1, http//www.livingreviews.org/lrr-2003-1.
44. Gaisser TK (1990). *Cosmic Rays and Particle Physics*, Cambridge University Press, Cambridge.
45. French AP (1966). *Special Relativity*, W. W. Norton, New York.
46. Ives HE and Stilwell GH (1938). *J. Opt. Soc. Am.*, **28**(7), 215.
47. Ives HE and Stilwell GH (1940). *J. Opt. Soc. Am.*, **31**, 369.
48. Martin BR and Shaw G (1997). *Particle Physics*, Wiley, New York.

Einstein and Relativity

49. Einstein A (1923). *Fundamental Ideas and Problems of the Theory of Relativity*, Lecture at the Nordic Assembly of Naturalists at Gothenburg, July 11, 1923.
50. Einstein A (1938). *Yale University Library Gazette*, **6**, 3.
51. Einstein A (1968). *Relativity: The Special, and the General Theory*, Methuen, London.

52. Lorentz HA, Einstein A, Minkowski H, and Weyl H (1952). *The Principle of Relativity: A Collection of Original Papers on the Special and General Theory of Relativity*. Notes by A. Sommerfeld, Dover, New York.

Category Theory

53. Adámek J, Herrlich H, and Strecker GE (2004). *Abstract and Concrete Categories: Joy of Cats*, http://katmat.math.uni-bremen.de/acc.
54. Barr M and Wells C (2002). *Toposes, Triples, and Theories*, Version 1.1, http://www.cwru.edu/artsci/math/wells/pub/ttt.html.
55. Johnstone PT (1977). *Topos Theory*, Academic Press, New York.
56. Lawvere FW (1966). *The Category of Categories as a Foundation for Mathematics*, pp. 1–20, La Jolla Conference on the Categorical Algebra, Springer-Verlag.
57. MacLane S (1998). *Categories for the Working Mathematician*, 2nd Edition, Springer-Verlag, New York.
58. MacLane S and Moerdijk I (1992). *Sheaves in Geometry, and Logic: A First Introduction to Topos Theory*, Springer-Verlag, New York.

Measure Theory and Dynamical Systems

59. Cornfeld IP, Fomin SV, and Sinai Ya G (1982).*Ergodic Theory*, Springer-Verlag, New York.
60. Halmos PR (1978). *Measure Theory* (Springer International Student Edition), Narosa, New Delhi, and references therein
61. Nadkarni MG (1995). *Basic Ergodic Theory, Texts, and Readings in Mathematics No. 6*, Hindustan Book, New Delhi.
62. Oxtoby JC (1971). *Measure and Category*, Springer-Verlag, New York.
63. Rohlin VA (1962). *Am. Math. Soc. Trans.*, **10**(1), 1.
64. Rohlin VA (1966). *Am. Math. Soc. Trans.*, **49**(2), 171.

Universal Relativity

65. Wagh SM (2005). *Foundations of a Universal Theory of Relativity*, and references therein, http://arxiv.org/physics/0505063.
66. Wagh SM (2005). *Progress with a Universal Theory of Relativity*, Talk delivered at the South African Relativity Society's Einstein Centennial Meeting, Durban, September 25–26, 2005, http://arxiv.org/physics/0602032.

67. Wagh SM (2006). *Universal Theory of Relativity and the Unification of Fundamental Physical Interactions*, http://arxiv.org/physics/0602042.
68. Wagh SM (2006). *Beyond Einstein: Are We All Afraid of the Truth?*, http://arxiv.org/physics/0603149.
69. Wagh SM and Wagh AH (2006). *The Gravity and The Quantum: A Bohr-Inspired Synthesis*, http://arxiv.org/physics/0608027.
70. Wagh SM (2006). *Measures in the Categorical Context*, preprint: smw-cat-m-02-06, preprint circulation date: September 2, 2006.
71. Wagh SM (2007). *Category Theory in Relativity*, http://arxiv:physics/0709.3631.
72. Wagh SM (2007). *Categorical Foundations of Physics: Program at a Glance*, http://arxiv:physics/0708.2468.
73. Wagh SM (2009). *Emission Origin for the Wave of Quanta*, unpublished preprint.
74. Wagh SM and Wagh AH (2009). *Explaining Michelson-Morley without Special Relativity*, http://arxiv:physics/1009.1883.
75. Wagh SM (2011). *Universal Relativity and Categorical Dynamics*, Talk at the Tercentenary Conference of the Laplace-Runge-Lenz Vector, Salt Rock, Durban, November 2011.
76. Wagh SM (2013). *Pramana*, **81**(3), 439–448, doi:10.1007/s12043-013-0584-6.
77. Wagh SM (2013). *J. Mod. Phys.*, **4**(8A), 102–104, doi:10.4236/jmp.2013.48A008.
78. Wagh SM (2013). *Subtlety in Relativity*, Talk at the SPIE Conference on the Nature of Light. *What Are Photons? V: Part of Optics, and Photonics Event*, Paper no. 8832-25, San Diego, California, August 25–29,2013.
79. Wagh SM (2014). *Questioning the Relativity of Time*, Talk during the International Conference on Matters of Gravity and the Universe, Center for Theoretical Physics, Jamia Millia Islamia, New Delhi, October 27–30, 2014.
80. Wagh SM (2014). *J.Astrophys. Astr.*, **35**(4), 595–605, doi:10.1007/s12036-014-9304-y.
81. Wagh SM (2015). *Abandoning the Relativity of Time*, Proceedings of International Meeting: Physical Interpretations of Relativity Theory, Bauman Moscow State Technical University, Moscow, 29 June–2 July 2015, Russia, p. 537–545.
82. Wagh SM (2015). *J. Adv. Phys.*, **10**(1), 2534–2537.
83. Wagh SM and Baboolal D (2017). *On Types of Emptiness in Category Theory*, to be submitted.

Index

For Product Safety Concerns and Information please contact our EU representative GPSR@taylorandfrancis.com
Taylor & Francis Verlag GmbH, Kaufingerstraße 24, 80331 München, Germany

www.ingramcontent.com/pod-product-compliance
Ingram Content Group UK Ltd.
Pitfield, Milton Keynes, MK11 3LW, UK
UKHW022159190926
475702UK00003B/117
9789814774444